Springer Series in

ADVANCED MICROELECTRONICS 21

Springer Series in

ADVANCED MICROELECTRONICS

The Springer Series in Advanced Microelectronics provides systematic information on all the topics relevant for the design, processing, and manufacturing of microelectronic devices. The books, each prepared by leading researchers or engineers in their fields, cover the basic and advanced aspects of topics such as wafer processing, materials, device design, device technologies, circuit design, VLSI implementation, and subsystem technology. The series forms a bridge between physics and engineering and the volumes will appeal to practicing engineers as well as research scientists.

18 **Microcontrollers in Practice**
By I. Susnea and M. Mitescu

19 **Gettering Defects in Semiconductors**
By V.A. Perevoschikov and V.D. Skoupov

20 **Low Power VCO Design in CMOS**
By M. Tiebout

21 **Continuous-Time Sigma-Delta A/D Conversion**
Fundamentals, Performance Limits and Robust Implementations
By M. Ortmanns and F. Gerfers

Volumes 1–17 are listed at the end of the book.

M. Ortmanns F. Gerfers

Continuous-Time Sigma-Delta A/D Conversion

Fundamentals, Performance Limits and Robust Implementations

With 122 Figures

Springer

Dr.-Ing. Maurits Ortmanns
ASIC Engineering Center
sci-worx GmbH
Garbsener Landstraße 10
30419 Hannover, Germany
E-mail: Maurits.Ortmanns@sci-worx.com

Dr. Friedel Gerfers
Philips Semiconductors
Von-der-Tann-Straße 10
82319 Starnberg, Germany
E-mail: Friedel.Gerfers@philips.com

Series Editors:

Dr. Kiyoo Itoh
Hitachi Ltd., Central Research Laboratory, 1-280 Higashi-Koigakubo
Kokubunji-shi, Tokyo 185-8601, Japan

Professor Thomas Lee
Stanford University, Department of Electrical Engineering, 420 Via Palou Mall, CIS-205
Stanford, CA 94305-4070, USA

Professor Takayasu Sakurai
Center for Collaborative Research, University of Tokyo, 7-22-1 Roppongi
Minato-ku, Tokyo 106-8558, Japan

Professor Willy M. C. Sansen
Katholieke Universiteit Leuven, ESAT-MICAS, Kasteelpark Arenberg 10
3001 Leuven, Belgium

Professor Doris Schmitt-Landsiedel
Technische Universität München, Lehrstuhl für Technische Elektronik
Theresienstrasse 90, Gebäude N3, 80290 München, Germany

ISSN 1437-0387

ISBN-10 3-540-28406-0 Springer Berlin Heidelberg New York

ISBN-13 978-3-540-28406-2 Springer Berlin Heidelberg New York

Library of Congress Control Number: 2005934298

Springer is a part of Springer Science+Business Media.

springer.com

Printed in The Netherlands

Typesetting: Data prepared by the Author and by SPI Publisher Services using a Springer TEX macro package
Cover concept by eStudio Calmar Steinen using a background picture from Photo Studio "SONO".
Courtesy of Mr. Yukio Sono, 3-18-4 Uchi-Kanda, Chiyoda-ku, Tokyo
Cover design: *design & production* GmbH, Heidelberg

Printed on acid-free paper SPIN: 11012009 62/3100/SPI Publisher Services - 5 4 3 2 1 0

For my brother Markus

For Nicole and Lea

Preface and Acknowledgments

This book reflects our work as research assistants at the Institute of Microelectronics at the Saarland University, Saarbrücken, the Albert-Ludwigs-University, Freiburg and thereafter in our current positions in the field of continuous-time sigma–delta modulator design. During this time, the interest in this promising field of A/D converter architectures drastically increased. This is obvious from the steadily increasing number of publications seen at conferences and in journals, from today's number of industry contributions in this field and also from the increasing number of special sessions at a number of conferences; all of them discuss the ongoing research of continuous-time sigma–delta modulators.

In contrast to this common interest, no standard book exists but only very specific thesis reports are available. Therefore, it has been our intention to put together a book which presents an overview and a summary of some of the most important basics and design issues when considering continuous-time $\Sigma\Delta$ modulators. This collection is based on our own work, and also includes many aspects of various contributors to the field.

At this point, we would like to thank the people, who directly contributed to or otherwise supported this book to come into being. For reviewing the manuscript and for contributing valuable comments for its improvement, many thanks to Prof. Philippe Benabes, Ecole Sup. Elect., Gif Sur Yvette, France, to Dr. Lucien Breems, Philips Research, Eindhoven, The Netherlands, and to Prof. Fernando Medeiro, Institute of Microelectronics, IMSE-CNM, Sevilla, Spain. In particular, we also express our thankyou to Prof. Yiannos Manoli, Albert-Ludwigs-University, Freiburg, Germany, for giving us the possibility to work in the field of continuous-time $\Sigma\Delta$ modulators at his departments at the Universities of Saarbrücken and Freiburg, for supporting us and for his assistance over many years.

Finally, our most grateful thanks are reserved for our families and our parents, for their encouragement, patience and support, and for always believing in us and having made all of this possible. Also many thanks also to our friends and colleagues, for the time when we worked and for the ones when we did not!

Hannover, Starnberg
June 2005

Maurits Ortmanns
Friedel Gerfers

Contents

List of Symbols and Acronyms

Physical Symbols

K	Boltzmann constant
T	Absolute temperature

Electrical Symbols

a_i	Discrete-time feedback scaling coefficient within ith loop
a_{f}	Flicker noise exponent
α	Relative time of the rising edge of DAC pulse
α_{ideal}	Ideal relative time of the rising edge of DAC pulse
α_{h}	Relative time of the rising edge of HRZ DAC pulse
A_{dc}	Amplifier dc gain
$A(s)$	Frequency dependent amplifier gain
A_{NRZ}	Activity factor using a NRZ DAC pulse shape
A_{RZ}	Activity factor using a RZ DAC pulse shape
A_{QUAD}	Activity factor using a quadratic DAC pulse shape
A_{SCR}	Activity factor using a SCR DAC pulse shape
A_{TRI}	Activity factor using a triangular DAC pulse shape
$A_{V_{\mathrm{T}}}$	Threshold voltage mismatch parameter
$A_{V_{\mathrm{T_n}}}$	Threshold voltage mismatch parameter of an NMOS
$A_{V_{\mathrm{T_p}}}$	Threshold voltage mismatch parameter of a PMOS
A_{β}	Transistor gain mismatch parameter
B	Resolution in bit
B_{int}	Resolution of the internal quantizer in bit
β	Relative time of the falling edge of a DAC pulse
β_{ideal}	Ideal relative time of the falling edge of DAC pulse

β_{h}	Relative time of the falling edge of HRZ DAC pulse
β_{p}	Transistor transconductance parameter of a PMOS device
c	Ratio between GBW and the f_{S}
c_i	DT interstage scaling coefficient between the ith stage
clk	Clock
C	Capacitance, integrator capacitor
C_{DAC}	DAC capacitance
C_{GS}	Gate–Source capacitance
C_{L}	Load capacitance
C_{ox}	Gate oxide capacitance per unit area
C_{p}	Parasitic capacitance
C_{R}	SCR-DAC capacitor
Corr	Correction factor
Corr_i	Correction factor in the ith stage
d_{I}	Duty cycle of integration phase
δ_{CE}	Mismatch of the correction factor
δ_{RC}	Mismatch of the RC time constant
$\varDelta$	Quantizer step size
$\varDelta_i$	Quantizer step size within the ith stage
$\varDelta_M$	Quantizer step size of the Mth (last) stage
ΔV_{T}	Threshold voltage mismatch
DF_i	Digital recombination filter within the ith loop
DNL	Differential nonlinearity
DR	Dynamic range
$e(n)$	Quantization error
$e_i(n)$	Quantization error introduced in the ith stage
$E_{\mathrm{d}}(s)$	Continuous-time DAC error signal
$E_i(s)$	Continuous-time ith integrator error signal, $i = 1, \ldots, \infty$
$E_{\mathrm{j}}(s)$	Clock jitter induced error signal
$E_{\mathrm{q}}(s)$	Continuous-time quantizer error signal
$\mathfrak{E}_{Q_{\mathrm{DAC}}}$	DAC feedback error charge
f	Frequency
f_{B}	Signal bandwidth
$f_{\mathrm{m}i}$	ith intermodulation frequency
f_{N}	Nyquist frequency
f_{S}	Sampling frequency
f_{sig}	(Input) signal frequency
f_u	Lower bound of the flicker noise integration interval
$\mathrm{FF}(s)$	Continuous-time feed-forward filter
$\mathrm{FF}(z)$	Discrete-time feed-forward filter
$F_{\mathrm{AAF}}(s)$	CT frequency response of the antialiasing filter
FOM	Figure of Merit
FS	Full scale signal range (peak-to-peak)
g_i	DT interstage scaling coefficient
GBW	Gain bandwidth product

gm	Transconductance
gnd	Negative power supply rail
γ_i	Scaling coefficient within a local feedback loop
GE_i	Gain error in the ith integrator
GE_{ik}	Gain error in the kth integrator within the ith loop
$\mathrm{GE}_{\mathrm{GBW}}$	Gain error caused by finite GBW in the integrators
GE_{RC}	Gain error in presence of filter time constant mismatch
h	Hysteresis of a comparator
$h(t)$	Impulse response of a loop filter function
$H(s)$	Continuous-time (CT) loop filter
$H(z)$	Discrete-time (DT) loop filter
$H_{\mathrm{LP}}(s)$	Continuous-time, low-pass filter transfer function
HD	Harmonic distortion
HD_2	Second harmonic distortion
HD_3	Third harmonic distortion
i, k, l	integer variables
$\bar{i}^2_{\mathrm{n,OTA}}$	Input refereed noise current of the amplifier
$\bar{I}$	Mean current of a unit current DAC cell
$I(s)$	Continuous-time integrator transfer function
$I(z)$	Discrete-time integrator transfer function
I_{B}	Bias current flowing in a amplifier branch
I_{DS}	Drain to source current
I_{DAC}	DAC current
$\hat{I}_{\mathrm{DAC}}$	DAC current amplitude
I_{unit}	DAC unit current amplitude (LSB)
IBN	Integrated in-band noise
$\mathrm{IBN}_{\mathrm{OSR}}$	IBN of an oversampled ADC
$\mathrm{IBN}_{\mathrm{Q}}$	IBN considering only quantization noise
$\mathrm{IBN}_{\mathrm{Q}^*}$	IBN considering quantization noise and loop filter scaling
$\mathrm{IBN}_{\mathrm{Q}^{**}}$	IBN considering quantization noise of a multibit ADC
IBN_1	IBN of a first order $\Sigma\Delta$ modulator
IBN_N	IBN of an Nth order $\Sigma\Delta$ modulator
$\mathrm{IBN}_{\mathrm{casc}}$	IBN of a cascaded $\Sigma\Delta$ modulator
IBN_{σ_t}	Integrated clock jitter induced in-band noise
INL	Integral nonlinearity
$\mathrm{IS}_{\mathrm{diff,max}}$	Maximal differential input swing
$\mathrm{ITF}(s)$	CT integrator transfer function
$\mathrm{ITF}_{\mathrm{GBW}}(s)$	CT integrator transfer function with finite GBW
$\mathrm{ITF}_{RC}(s)$	CT integrator transfer function with gain mismatch
j	Complex variable, $\sqrt{-1}$
k_{Adder}	Number of digital adders in the cancellation logic
$k_{\mathrm{C_{CT}}}$	Continuous-time capacitive excess factor
$k_{\mathrm{C_{DT}}}$	Discrete-time capacitive excess factor
$k_{\mathrm{q,crit}}$	Critical still stable quantizer gain

k_{Delay}	Number of flip-flops in the DT filter of a multistage ΣΔ
k_{eff}	Effective quantizer gain
k_{GBW}	Gain bandwidth excess factor
k_i	CT scaling coefficient within ith loop
$k_{i,n}$	CT scaling coefficient between the loops of a cascaded ΣΔ
k_i^*	CT scaling coefficient modified for error compensation
$k_{i_{\mathrm{QUAD}}}$	CT scaling coefficient within ith loop for a quadratic DAC
$k_{i_{\mathrm{SCR}}}$	CT scaling coefficient within ith loop for a SCR-DAC
$k_{i_{\mathrm{TRI}}}$	CT scaling coefficient within ith loop for a triangular DAC
k_{q}	Quantizer gain
$k_{\mathrm{q}M}$	Quantizer gain in the Mth stage
k_{noise}	Noise excess factor
k_{OTA}	Number of current branches in the OTA
$k_{\Sigma\Delta}$	Sum of scaling coefficients
k_{sig}	CT scaling coefficient in front of the entire modulator
k_{V}	Forward continuous-time scaling coefficient
K_{f}	Process dependent flicker noise parameter
L	Transistor length
LC	Inductor–capacitor combination
$\mathrm{LF}(s)$	Continuous-time loop filter
$\mathrm{LF}(z)$	Discrete-time loop filter
m	Delay factor of the modified $\mathcal{Z}$-transform
M	Total number of cascaded stages
Mi	ith transistor
n	Index variable, $n = 1, \ldots, \infty$
n_{A}	Number of input paths of an integrator
n_{Adder}	Size of a digital adder converted into equivalent inverters
n_{Delay}	Size of a register converted into equivalent inverters
$n_{\mathrm{e,f}}$	Flicker noise excess factor
$n_{\mathrm{e,th}}$	Thermal noise excess factor
n_{I}	Number of current branches in an amplifier
n_{p}	Number of points (for FFT calculation)
n_{R}	Number of resistors
n_{Q}	Equivalent quantizer hardware converted in CMOS gates
N	ΣΔ modulator order
N_{casc}	Overall order of a multiloop modulator
N_i	Filter order of the ith stage
NRZ	Non return-to-zero DAC pulse shape
NTF	Noise-transfer function
NTF_i	NTF of the ith stage in a cascaded modulator
$\mathrm{NTF}_{\mathrm{GE}}$	NTF under the influence of gain error
NTF_{∞}	NTF gain at $f_{\mathrm{S}}/2$
$\mathrm{NTF}_{\mathrm{casc}}$	NTF of a cascaded ΣΔ modulator
OL	ΣΔ modulator overload level
$\mathrm{OS}_{\mathrm{diff,max}}$	Maximal differential output swing

OSR	Oversampling ratio
P_{dynamic}	Dynamic power drain
$P_{\text{dynamic,input}}$	Dynamic power drain cause at the input of the integrator
P_{inv}	Power drain of a CMOS inverter
P_{static}	Static power drain
$P_{\text{static,DAC}}$	Static power drain cause by the DAC
$P_{\text{static,OpAmp}}$	Static power drain cause by the OpAmp
P_{digital}	Power drain of the digital part
P_{digital^*}	Power drain of the digital part (multibit)
$P_{\text{dynamic}^{**}}$	Dynamic power drain of a multistage modulator
P_{DC}	dc signal power
$P_{\text{FS/2}}$	Full scale input signal power
P_{sig}	Input signal power
$P_{\text{min},A_{VT}}$	Minimal power drain in matching dominated circuits
$P_{\text{min},\Sigma\Delta_{\text{CT}}}$	Minimal power drain in CT $\Sigma\Delta$ ADCs
$P_{\text{min},\Sigma\Delta_{\text{DT}}}$	Minimal power drain in DT $\Sigma\Delta$ ADCs
$P_{\text{min,SNR}}$	Minimal power drain in noise-dominated circuits
$P_{\text{min,SNDR}}$	Minimal power drain in noise-dominated and distortion-dominated circuits
psd	Power spectral density
φ_{m}	Phase margin
φ_{max}	Maximal allowed voltage feed through
$q(n)$	Discrete-time quantizer input signal
$q(t)$	Continuous-time quantizer input signal
Q	Quality factor
Q_{DAC}	DAC feedback charge
R	Resistance, integrator resistor
R_z	Excess phase cancellation resistor
r_{dsi}	Output resistance of the ith CMOS device
r_{out}	Output resistance
$r_{\text{DAC}}(t)$	Impulse responses of a general DAC pulse
$R_{\text{DAC}}(s)$	Laplace transform of a general DAC pulse
$r_{\text{COS}}(t)$	Impulse responses of a cosine-shape DAC pulse
$r_{\text{LS}}(t)$	Impulse responses of a linear-slope DAC pulse
$R_{\text{LS}}(s)$	Laplace transform of $r_{\text{SL}}(t)$
$r_{\text{NRZ}}(t)$	Impulse responses of a NRZ-DAC pulse
$R_{\text{NRZ}}(s)$	Laplace transform of $r_{\text{NRZ}}(t)$
$r_{\text{NLS}}(t)$	Impulse responses of a nonlinear-slope DAC pulse
$R_{\text{NLS}}(s)$	Laplace transform of $r_{\text{NLS}}(t)$
R_{R}	SCR-DAC resistor
$r_{\text{RZ}}(t)$	Impulse responses of a RZ-DAC pulse
$R_{\text{RZ}}(s)$	Laplace transform of $r_{\text{RZ}}(t)$
$r_{\text{SCR}}(t)$	Impulse responses of a SCR-DAC pulse
$R_{\text{SCR}}(s)$	Laplace transform of $r_{\text{SCR}}(t)$
$r_{\text{QUAD}}(t)$	Impulse responses of a quadratic DAC pulse

$R_{\mathrm{QUAD}}(s)$	Laplace transform of $r_{\mathrm{QUAD}}(t)$
$r_{\mathrm{TRI}}(t)$	Impulse responses of a triangular-slope DAC pulse
$R_{\mathrm{TRI}}(s)$	Laplace transform of $r_{\mathrm{TRI}}(t)$
$\mathcal{R}(z)$	Real part of the $\mathcal{Z}$-domain transfer function
RZ	Return-to-zero DAC pulse shape
s	Laplace-domain variable
S_e	Power spectral density of the quantization noise
S_u	Power spectral density of the input signal
S_{Q}	Power spectral density of the shaped quantization noise
SFDR	Spurious free dynamic range
SNDR	Signal-to-noise + distortion ratio
$\mathrm{SNDR_p}$	Peak signal-to-noise + distortion ratio
SNR	Signal-to-noise ratio
$\mathrm{SNR_p}$	Peak signal-to-noise ratio
$\mathrm{SNR_p}\|_{\mathrm{Nyquist}}$	Peak signal-to-noise ratio for a Nyquist rate converter
SR	Slew rate
SR_{ik}	Slew rate in the kth integrator within the ith stage
STF	Signal transfer function
STF_i	Signal transfer function of the ith stage
$\mathrm{STF_{GE}}$	Signal transfer function under the influence of gain error
THD	Total harmonic distortion
σ_e	Standard deviation of the quantization noise
σ_I	Standard deviation of the current
σ_t	Standard deviation of the sampling instant, clock jitter
t	Time
t_d	DAC pulse delay time
$t_{\mathrm{D}i}$	Modeled (GBW) delay for an ith order feedback path
t_i	ith time instant, $i = 1, \ldots, N$
t_n	nth time instant
t_{p}	DAC pulse duration
$t_{\mathrm{sr}i}$	ith rise/fall time
t_{j}	Timing jitter error
T_{S}	Sampling period
τ_{d}	Normalized DAC pulse delay time
τ_{D1st}	Modelled relative delay for a first-order feedback path
τ_{D2nd}	Modelled relative delay for a second-order feedback path
τ_{DAC}	DAC time constant for exponentially sloping DAC waveform
τ_{DAC_i}	Normalized DAC time constant within the ith feedback loop
τ_{SR}	ith normalized DAC pulse rise/fall time
$\hat{u}$	Input signal amplitude
$u(t)$	Continuous-time $\Sigma\Delta$ modulator input voltage
$u_a(t)$	Filtered continuous-time $\Sigma\Delta$ modulator input voltage
$u(n)$	Discrete-time $\Sigma\Delta$ modulator input voltage
$U(z)$	Discrete-time $\Sigma\Delta$ modulator input voltage
$\overline{v}^2_{\mathrm{n,in}}$	Input refereed noise power

$\overline{v}_{\mathrm{n}}$	Noise voltage
$\overline{v}_{\mathrm{n}^2}$	Noise power
$\overline{v}^2_{\mathrm{n,OTA}}$	Input refereed noise power introduced by the amplifier
$\overline{v}^2_{\mathrm{n},R}$	Noise power introduced by the resistor R
$\overline{v}^2_{\mathrm{n},R_{\mathrm{DAC}}}$	Noise power introduced by the resistor R_{DAC}
$\overline{v}^2_{\mathrm{n},R_Z}$	Noise power introduced by the resistor R_Z
$\overline{v}^2_{\mathrm{in,1/f}}$	Input refereed noise power of the flicker noise part
$\overline{v}^2_{\mathrm{in,thermal}}$	Input refereed noise power of the thermal noise part
$V_{\mathrm{B}i}$	ith bias reference voltage
V_{CM}	Common-mode voltage
V_{dd}	Positive power supply rail
V_{DS}	Drain to source voltage
$V_{\mathrm{DS,lin}}$	Drain to source voltage in triode region
$V_{\mathrm{DS,sat}}$	Drain to source saturation voltage
V_{GS}	Gate to source voltage
V_{in}	Input voltage
V_{out}	Output voltage
V_{ref}	Reference voltage
V_{T}	Transistor threshold voltage
W	Transistor width
ω_{A}	Dominate pole of the OpAmp used in the integrator
ω_{C}	Bandpass $\Sigma\Delta$ modulator center frequency
ω_{I}	Continuous-time integrator corner frequency
ω_{Is}	Actual continuous-time integrator corner frequency
ω_{p}	Pole frequency
$\omega_{\mathrm{p}i}$	ith integrator pole frequency
$x(n)$	Discrete-time integrator output states
$x(t)$	Continuous-time integrator output states
$x_{\mathrm{d}}(n)$	Discrete-time difference modulator input signal
$y(n)$	Discrete-time $\Sigma\Delta$ modulator output signal
$y_{\mathrm{d}}(n)$	Filtered and downsampled DT $\Sigma\Delta$ ADC output signal
$Y(z)$	Discrete-time $\Sigma\Delta$ modulator output signal
y_i	Output signal of the ith stage
$Y_i(z)$	Discrete-time output of the ith stage
z	Discrete-time frequency variable, $z = \mathrm{e}^{s\,T_{\mathrm{s}}}$
Z_{F}	Integrator feedback impedance

Acronyms

1p	One pole
2p1z	Two poles one zero
AAF	Antialiasing filter
A/D	Analog-to-digital
ADC	Analog-to-digital converter

AHDL	Analog hardware description language
CM	Common mode
CMFB	Common mode feedback
CMOS	Complementary metal oxide semiconductor
CS	Current source
CT	Continuous-time
DAC	Digital-to-analog converter
DP	Double polylayer process
DPDM	Double polylayer double metal layer process
DPTM	Double polylayer triple metal layer process
DPFM	Double polylayer five metal layer process
DPQM	Double polylayer six metal layer process
DPSM	Double polylayer single metal layer process
DT	Discrete-time
FFT	Fast Fourier transformation
I/O	System input/output
IF	Intermediate frequency
INT	Integrator
ISI	Inter symbol interference
LSB	Least significant bit
MASH	Multistage noise-shaping topology
MOSFET	Metal oxide semiconductor field effect transistor
MT	Multithreshold process
mb	Multibit
OTA	Operational transconductance amplifier
OpAmp	Operational amplifier
sb	Single-bit
S/H	Sample and hold
SC	Switched capacitor
SI	Switched current
SOFO	Second order, first order (cascaded modulator)
SPTM	Single poly layer triple metal layer process
TM	Triple metal layer process
VerilogA	Analog extension of Verilog
VLSI	Very large scale integration

1 Introduction

1.1 Motivation and History

The common trend of moving functionality to the digital domain in very large scale integration (VLSI) systems, the rapid growth of the market for portable, battery operated systems for communications, computer and consumer electronics, and finally the shift of the analog-to-digital interface to the system input/output (I/O) resulted in an enormously increasing interest in analog to digital converter (ADC) design over the last decade.

This is because digital signal processing is simple in design and test and produces robust, flexible and programmable structures. At the same time the requirements on the A/D interface circuit increase rapidly making it a bottleneck for the overall system, concerning speed and resolution. This becomes even more complicated, if the possibility of scaling in the digital circuit design is fully adopted, i.e., for low power the supply voltage is heavily reduced while for high speed design the most current complementary metal oxide semiconductor (CMOS) technologies are employed. For the analog circuitry this results in serious constraints, because these modern processes with their reduced supply voltages usually provide the worst analog features and are a great challenge for achieving an acceptable dynamic range. At the same time, in modern ADC applications the demand for high performance concerning speed and resolution steadily increases.

There exist plenty of possibilities to implement an A/D conversion, while for different ranges of resolution and speed also different conversion methods are adopted. One favorable option, especially in VLSI systems, is the sigma–delta ($\Sigma\Delta$) ADC. Since its invention by Cutler in 1960 and its first description in the published literature by Inose and Yasuda in 1962 [1], this analog signal encoder, which is based on redundant temporal data through oversampling, has received an exponentially increasing attention.

The reason for this is manifold: while conventional A/D conversion has been highly sensitive to circuit imperfections or at least needs correction mechanisms, the $\Sigma\Delta$ modulator shows a very low sensitivity to the nonidealities of most building blocks. This is achieved through extensive use of digital signal postprocessing, which actually is a very favorable feature in modern VLSI technology, where the implementation of dense and fast digital circuits can be better realized than accurate analog functions.

Nonetheless, from their invention in the 1960s it took almost 20 years for the breakthrough of the ΣΔ technology, when the implementation of integrated switched capacitor circuits and filters became a common and well understood technique, which made an easy mapping of the modulator mathematics onto the circuit level possible. Since then, especially after Candy's widely cited paper in 1985 on the implementation of the double integration modulator [2], the number of applications increased tremendously: from audio through measurement to almost wide bandwidth communication A/D conversion, while an enormous number of papers and books have been published concerning purely theoretical as well as architectural and implementation issues.

But especially the demand of increased conversion bandwidths in the communications industry and the desired further application of the ΣΔ modulator resulted in the need of architectural alternatives. This is due to settling-time constraints in typical discrete-time (DT) implementations and the resulting limited sampling frequency. Because of this and the required oversampling the signal bandwidths have stayed rather narrow.

Therefore in the mid 1990s the implementation of continuous-time (CT) ΣΔ modulators became popular again, while already the early implementation of the ΣΔ ADCs were implemented in the CT domain: here, the loop filter is built out of continuous-time circuits such as transconductances or integrators and implementations of up to gigahertz-sampling modulators confirmed the approach. Principally, the advantages of the CT ΣΔ ADC are all based on the displacement of the sampler inside the modulator loop, thus:

- Allowing the filters to be implemented as continuous-time circuits
- Consequently reducing the speed requirements of the filters
- Providing an implicit antialiasing filter
- Lacking a precise input sample and hold circuit

Consequently, the resulting CT architectures principally show a better power efficiency and the ability to realize very high sampling rate ΣΔ modulators. But in spite of these advantages and some promising realizations and advances, CT modulators have not become commonly implemented technique for A/D conversion, but rather remained an academic field of research – for which the reasons are twofold.

First, the advances in DT modulator implementation have made high-resolution, high-bandwidth converters possible using low oversampling ratios. This has been achieved by extensive use of either cascaded or multibit modulator implementations, whose performance is intrinsically much higher compared to standard designs. Additional architectural and theoretical research have reduced the required speed constraints. Therefore, the ongoing and successful research on and design of DT, switched capacitor ΣΔ modulators imposes a very substantial competition to a real take-off of the CT ΣΔ implementation.

Second, and even more important, the continuous-time modulators suffer from circuit nonidealities which make their practical advantage not finally conclusive:

- Most notably the sensitivity to system clock jitter, which is orders of magnitudes higher than in the switched capacitor implementations
- Process dependent integrator gain variations
- Further timing nonidealities, e.g., the feedback loop delay

1.2 Intention of this Work

While continuous-time sigma–delta ADCs received increasing intention over the last decade, up to now only very specific thesis reports are available covering this topic. Therefore, it is the intention of this book to present an overview and a summary of some of the most important basics and design issues when considering continuous-time ΣΔ modulators. Thus, the following chapters cover the previous work of the authors, while at many points also aspects of various contributors to the field of CT ΣΔ modulation have been included.

In order to keep the extent of this book in a feasible limit, the topics considered in detail had to be chosen carefully, while other aspects are only briefly discussed or referenced for the interested reader. After a short introduction into the theory of ΣΔ conversion, which is given to allow beginners to read into the topic, the differentiation between DT and CT modulator implementations, their advantages and drawbacks are discussed. Thereafter, the main focus of this work considers the loop filter design and implementation as well as the susceptibility of CT modulators to their most critical nonidealities. In addition, various possibilities for the reduction or correction of the nonideal behavior are given, as well as successful design examples presented.

The analysis and considerations mainly cover modulators with low-pass loop filter characteristic, and without loss of generality most simulation results were performed for the distributed feedback architecture. When multibit modulators are concerned, the necessary linearization techniques are not considered in detail, because special literature exists on this topic, which is referenced in detail.

1.3 Further Recommended Literature

It is a favorable part of papers or books which survey a particular area to include many of the important references in that area. But, due to the immense number of publications on the topic of ΣΔ A/D conversion, and the ever increasing number of work dealing with the continuous-time implementation, it is difficult to give a complete survey of papers in this area or to completely reference them. Therefore, throughout this work the papers or books specifically important for the topic covered are cited.

Nevertheless, at this point the most often cited references are listed separately, which are especially recommended for the interested reader:

- Cherry's book on "Continuous-Time Delta–Sigma Modulators" [3], which is also partially based on the thesis' of Gao and Shoaei, is one of the important continuous-time ΣΔ modulator references throughout the presented work. It comprehensibly covers CT ΣΔ synthesis, the theoretical treatment of clock jitter as well as compensation approaches for feedback loop delay. In addition, in Sect. 3.2 of his book a survey of important CT ΣΔ publications exists.
- Norsworthy, Schreier and Temes' extensive work on ΣΔ modulators [4] deals with a host of important subjects in discrete-time and also marginally continuous-time modulator design and will be frequently referenced.
- Medeiro's guide to switched capacitor ΣΔ modulator design [5] is one of the first works to extensively treat different nonideal behavior in order to determine the requirements of the building blocks. Additionally his papers on cascaded DT modulator design will be referenced frequently.
- Marques and Peluso's paper on the choice of "optimal parameters in ΣΔ modulators" [6] is taken as a fundamental work for successful single-bit modulator scaling.
- The books of Peluso [7] and Geerts [8], both dealing with the design of DT ΣΔ modulators, give extensive studies of optimal modulator parameters. In addition, the latter extensively deals with architectural trade-offs of single-bit vs. multi-bit, single-stage vs. multi-stage and DT vs. CT modulators and is therefore recommended to the interested reader.
- The continuous-time ΣΔ modulator works of van der Zwan [9] and Schreier [10] are often cited in the literature when the CT advantages are pointed out. In spite of their actual contribution to the explanation of advantages and nonidealities, many others exist, most of them referenced in [3].
- Tao's extensive work on the clock jitter sensitivity of ΣΔ modulators [11] together with the same topic covered by Cherry in [12] is the basis for the jitter considerations in this book. In addition, the very basic paper of Berkovitz and Rusnak [13] should be mentioned separately, whose work on different clock jitter forms has fundamental influence on the presented results.
- The works on bandpass CT ΣΔ modulators of Benabes [14–16], who belongs to the early contributors to the topic and also of Breems, who additionally considered CT ΣΔ modulators with mixing inside the ΣΔ loop [17, 18].
- The recent contributions of industrial researchers, among others Breems, van Veldhoven and Doerrer [19, 20] and [21–23] in the field of CT ΣΔ modulators.

- The papers of Oliaei especially in the field of jitter effects in CT $\Sigma\Delta$ modulators [24–26], and also the recent contributions of Hernandez in the same topic [27, 28].
- And finally the contributions of the authors themselves in the theoretical modeling, error correction as well as implementation of continuous-time $\Sigma\Delta$ modulators, beyond others [29–38].

1.4 Organization of this Book

Following this introduction, Chap. 2 introduces preliminary basics important in $\Sigma\Delta$ modulator design and understanding. Here the terminology, fundamental A/D conversion and architectural alternatives of $\Sigma\Delta$ modulators are covered.

Thereafter, Chap. 3 deals with the specialties of continuous-time $\Sigma\Delta$ implementation with attention to loop filter design and implementation, overview on advantages and drawbacks as well as a list of recently published modulator implementations. Chapters 4–6 present a detailed survey of nonidealities in continuous-time modulators, their classification and modeling. Additionally, where potential techniques for error compensation exist, these are reviewed and explained. Finally, Chap. 7 presents a low-power design strategy based on a Figure-of-Merit for continuous-time $\Sigma\Delta$ modulators. Furthermore, design examples of low-pass single-loop, $\Sigma\Delta$ modulators, employing single-bit and multi-bit quantization are shown and circuits with error suppression are reviewed.

2

Basic Understanding of ΣΔ A/D Conversion

This chapter is intended to give an introduction and overview of analog-to-digital conversion, first in general and thereafter for the ΣΔ modulator. Herein the terms and definitions used in the rest of this work will be explained. For a possible treatment of A/D converters as a linear, time-invariant system the white noise model of the quantizer and its spectral behavior will be discussed. Furthermore, the techniques of oversampling and noise shaping are introduced and the used performance measures and simulation methods are reviewed. Finally, architectural alternatives of ΣΔ modulators are discussed.

Thus, for an expert in the field of ΣΔ modulation, who is experienced with the basics and common terminology, this chapter may be read in a glance.

2.1 Basics of A/D Conversion

2.1.1 Sampling and Quantization

The conversion of analog signals to the digital domain can simply be separated into two basic operations: uniform *sampling* in time and *quantization* in amplitude.

Under the assumption that the signal information of the continuous input waveform $u(t)$ is contained in the signal band, i.e., $|f_{\mathrm{sig}}| \leq f_{\mathrm{B}}$, where f_{B} is defined as the signal bandwidth, the sampling in time is a completely invertible process. This is easily understood when considering a quantization in time as a periodization in frequency [39], which is illustrated in Fig. 2.1. There, the considered input signal is sampled at uniform time intervals T_{S}, the sampling time, or with a fixed frequency or rate, f_{S}, resulting in a periodicity of the original signal spectrum at multiples of f_{S}. (The periodization itself arises from the sampling process considered as a multiplication of the signal $u(t)$ with a train of dirac pulses spaced at $1/f_{\mathrm{S}}$ in time, which convolution of the signal spectrum with a periodic train of dirac pulses spaced at f_{S} in frequency.)

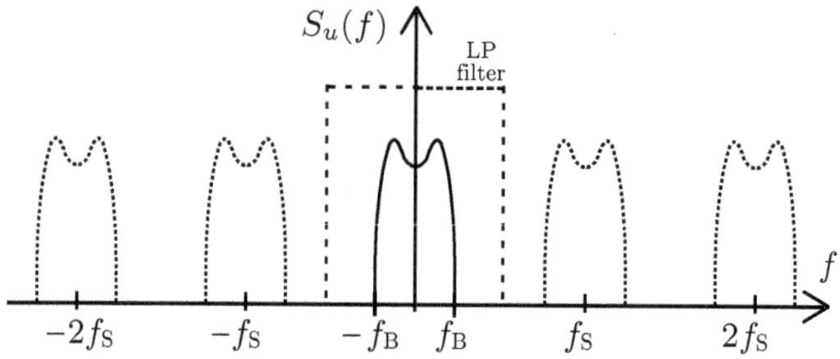

Fig. 2.1. The spectral sampling operation

From Fig. 2.1 it is obvious that by simple low-pass filtering, the original baseband spectrum can be reconstructed, provided that the sampling itself does not result in overlap or aliased regions. This is achieved when:

$$f_S \geq 2\,f_B = f_N\,, \tag{2.1}$$

which is known as the *Nyquist theorem*, where f_N is the *Nyquist frequency*. To assure a proper sampling operation, the condition in (2.1) is enforced by an analog filter preceding the sampling operation, called the *antialiasing filter* (AAF). Therewith, the basic ADC structure simplifies to Fig. 2.2. An ADC working with a sampling frequency of f_N is called a *Nyquist Rate* converter. But in real implementations, this results in a zero transition band for the filter to cut off the unwanted high frequency signals, making it hard to design. On the other hand, analog filters with a gentle roll off in their transition band are less costly, easier to design, require less power, and smaller chip area while introducing less phase distortion. Therefore, many ADCs work with sampling rates higher than f_N, and one defines:

$$\mathrm{OSR} = \frac{f_S}{2f_N} \tag{2.2}$$

as the *oversampling ratio* of the ADC.

The process of quantization in amplitude, usually referred to as the *quantization*, encodes a continuous range of analog values into a set of discrete levels. It is clear that quantization is a noninvertible process, since an infinite number of input amplitude values of the sampled analog signal $u(n)$ is mapped into a finite number of output amplitude values. Thus, even the ideal quantization

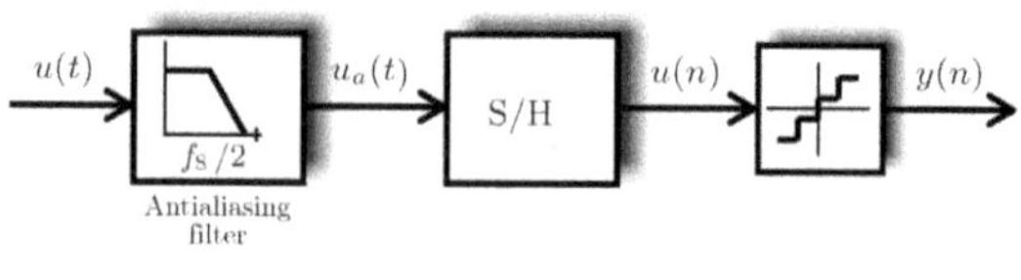

Fig. 2.2. Operation principle of an ADC

process inherently introduces errors to the signal, and the primary objective in ADC design is to limit exactly this error.

The key distinguishing characteristic of a quantizer is its number of bits B_{int}, which correlates with the number of different output levels. If the analog input is mapped into $2^{B_{\text{int}}}$ discrete levels, the quantizer is said to have B_{int}-bits of resolution. If all of these levels are equally spaced, the quantizer is uniform. The uniform spaces between two adjacent quantizer output levels are defined as the *quantizer step width*:

$$\Delta = \frac{\text{FS}}{2^{B_{\text{int}}} - 1}, \tag{2.3}$$

where FS is the *full-scale* output range. This is illustrated in Fig. 2.3, where a single-bit and a multi-bit quantizer are given with their I/O characteristic in Fig. 2.3c, d. As the quantizer input signal rises from −FS/2 to FS/2 the output

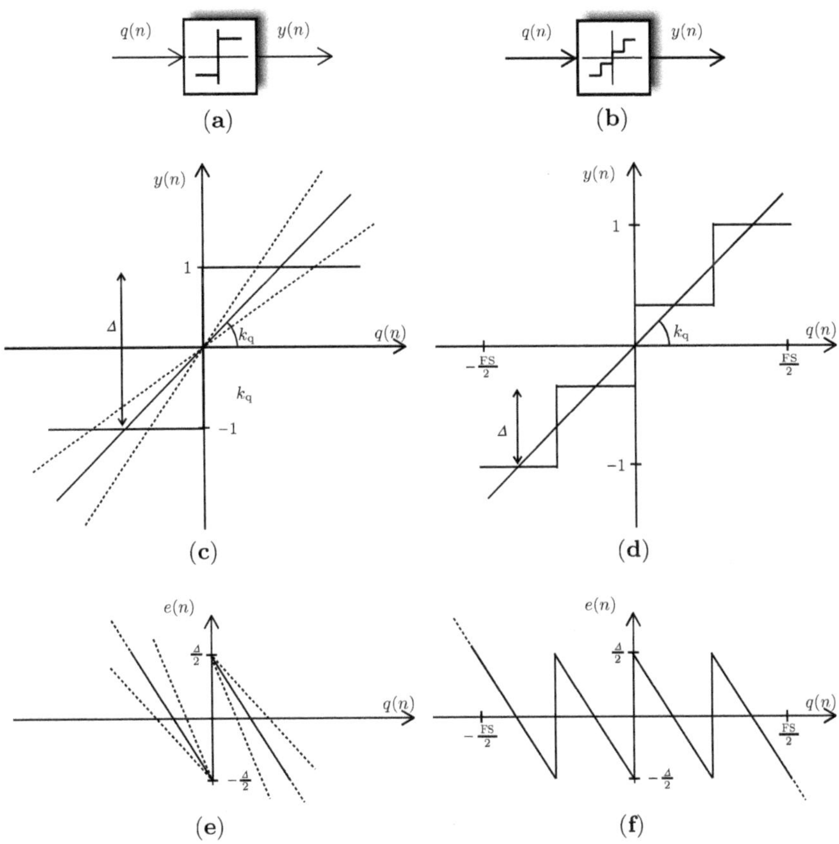

Fig. 2.3. Quantizer transfer curves and quantization error [5, 8]. (**a**) Single-bit; (**b**) multibit; (**c**) single-bit quantizer I/O; (**d**) multibit quantizer I/O; (**e**) single-bit quantizer error; (**f**) multibit quantizer error

is quantized to one of the $2^{B_{\text{int}}}$ different output levels, which are encoded into a binary digital representation. The resulting *quantization error* of that operation is defined as the difference of the quantizer input and its output, which is illustrated in Fig. 2.3e, f. Obviously, an input signal which exceeds the valid input range, results in monotonously increasing quantization error, which is commonly known as the *overload or saturation region* of a quantizer. In opposite, for input signals between −FS/2 and FS/2 the quantization error $e(n)$ is bounded within $[-\Delta/2,\ \Delta/2]$.

Since the input and output range of the quantizer are not necessarily equal, the quantizer can show a nonunity gain, which is indicated in Fig. 2.3c, d by the slope k_q. Without loss of generality, in the following it is assumed that the full scale, maximum output range is ±1, and that the input range equals the output range, giving the multibit quantizer a gain of $k_\mathrm{q} = 1$. The problem in the case of the singlebit quantizer is that its gain is not defined exactly, but can be chosen arbitrarily, as is illustrated in Fig. 2.3c. This is because the output of the singlebit quantizer only depends on the polarity but not on the magnitude of the input signal, which will be further discussed in the stability and performance analysis of ΣΔ converters [8].

2.1.2 Quantizer White Noise Model

With the above shown I/O characteristics of a quantizer, a gain k_q and a quantization error $e(n)$, the linearized quantizer model of Fig. 2.4 can be derived. Here, beside the not exactly defined gain in the case of the single-bit quantizer, also the quantization error itself is not well defined.

If one assumes that the input changes rapidly from sample to sample by amounts larger than the step size Δ, then the quantization error appears uncorrelated from sample to sample and has equal probability of lying anywhere in the range of $[-\Delta/2,\ \Delta/2]$, provided that no overload occurs. Therefore it seems to be possible to assume the quantization error to have statistical properties. This is the basis of the *additive white noise model* of the quantizer, which can be derived, if the often cited Bennett's conditions are fulfilled [40, 41]. Then, the *white noise approximation* of a quantizer yields that:

- The quantization error can be represented as a random variable
- The quantization error sequence $e(n)$ is uncorrelated with the input $q(n)$
- The probability density function of the quantization error process pdf_e is uniform over the range $[-\Delta/2,\ \Delta/2]$, Fig. 2.5a
- The power spectral density of the quantization error process $S_e(f)$ is flat

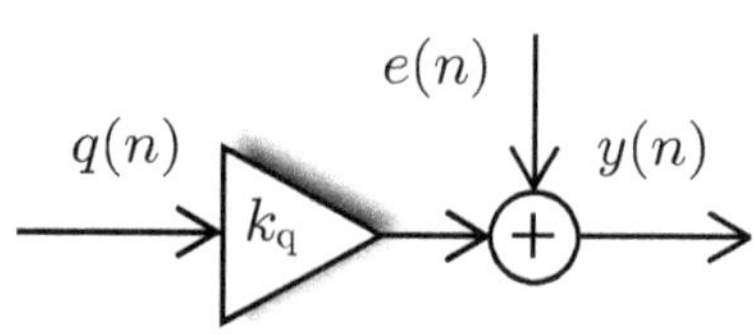

Fig. 2.4. Linear quantizer model

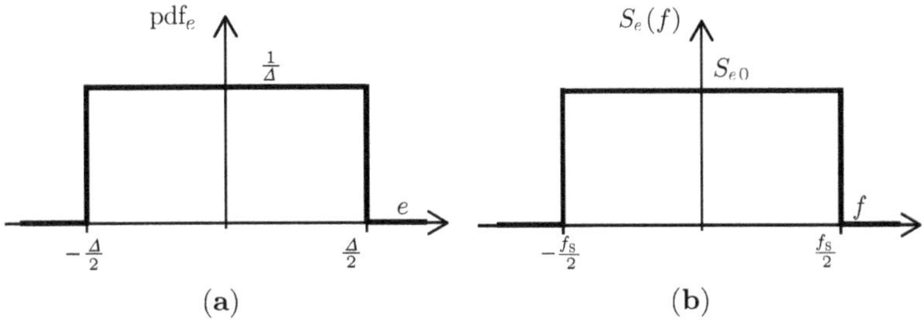

Fig. 2.5. Properties of the quantization error according to the linear model. (**a**) Probability density function; (**b**) power spectral density

This drastically simplifies the analysis of an ADC system, because a deterministic nonlinear system is replaced by a stochastic linear one, i.e., the quantization error becomes a *quantization noise*. In a practical system, this does not strictly hold true, but nevertheless over many years the assumptions and the given quantizer model yielded good results [4].

From Fig. 2.5a the total quantization noise power can be calculated as:

$$\sigma_e^2 = \int_{-\infty}^{\infty} e^2 \,\mathrm{pdf}_e \,\mathrm{d}e = \frac{\Delta^2}{12} . \tag{2.4}$$

It should be noted at this point that the total quantization error power is independent of the sampling frequency and is only determined by the quantizer resolution. Since the signals at the quantizer are sampled (discrete-time) signals, all the quantization noise power σ_e^2 is folded into the frequency range $[-f_S/2, f_S/2]$. Thus, with the white noise approximation in Fig. 2.5b the power spectral density of the quantization noise is:

$$S_e(f) = \frac{\Delta^2}{12} \frac{1}{f_S} . \tag{2.5}$$

With this white noise model, i.e., the combination of (2.5) and Fig. 2.4, together with the sampling procedure, the A/D conversion process can be approximately described.

2.2 Performance Metrics

Before introducing different ADC architectures and techniques, some methods to characterize them need to be defined. Therefore, in this section the most important specifications used in this work are briefly discussed.

ADC characteristics are classified into two categories: static and dynamic performance metrics. The former include monotonicity, offset, gain error, differential nonlinearity (DNL), and integral nonlinearity (INL), whereas the latter implies signal-to-noise ratio (SNR), total harmonic distortion (THD), signal-to-noise plus distortion ratio (SNDR), dynamic range (DR), spurious free dynamic range (SFDR), idle channel noise and the overload level (OL). A brief introduction concerning the static characteristics is given in [42].

In general, the overall performance of a ΣΔ modulator is characterized by the dynamic metrics, which can be subdivided into spectral, frequency domain, and power metrics.

2.2.1 Frequency Domain Metrics

The analysis of the power or amplitude spectral density of the ADC output signal is straightforward, because the proper functionality can be almost directly seen. Especially in the case of the ΣΔ modulator, which is a noise-shaping converter, the analysis of the frequency domain representation of the time domain output bits is of interest.

Power Spectrum Estimation

Due to its major importance for this work and the analysis of ΣΔ modulators, the calculation or estimation of the power spectrum will be first considered in more detail. In a real application, the power spectrum can only be estimated rather than calculated, for which a popular method is Welch's (averaged) periodogram [3, 39, 43]. The most common tool for its calculation is using the discrete Fourier transform, which in turn can be efficiently computed by the fast Fourier transformation (FFT), if the length n_p of the considered bitstream is a power of 2.

The resulting power spectrum is defined at $(n_\mathrm{p})/2 + 1$ uniformly spaced frequency points within $[0,\ f_\mathrm{S}/2]$. Thus, each frequency bin is of width $f_\mathrm{S}/n_\mathrm{p}$; this is why the accuracy of the estimated spectrum directly depends on the duration in time of the input or sampled signal. In this work, for the estimation of the power spectrum the standard MATLAB function psd has been used [44]. The corresponding source code is given in Appendix A.

Windowing and Signal Leakage

In practice, the spectral analysis can only be obtained from a finite sample of the infinite series of the output signal, while mathematically the Fourier transform is obtained from an infinite number of samples; this, in fact, is the same as taking the FFT of an infinite stretch of data which is multiplied by a rectangular window, therewith extracting the signal exactly in the desired time slot. In the frequency domain, this multiplication corresponds to a convolution of the exact infinite signal spectrum with the Fourier transform of the rectangle, i.e., $\sin x/x$. The resulting effect is known as *leakage*, which refers to the phenomenon by which variance, an important frequency of a strong periodicity, leaks into other adjacent frequencies in the estimated spectrum [3].

The severity of leakage can be reduced by *windowing* the data, which means its multiplication by a windowing function before taking its FFT. This results in a convolution of the exact signal spectrum with a function other than $\sin x/x$, i.e., with a sharper peak at the center and a gentle, but fast transition to zero beside it [45]. In this work, the *Blackman* window has been used. To prevent signal power mismatch, after the calculation of the periodogram the effect of windowing has been reversed as proposed in [46] and shown in Appendix A.

Nevertheless, the effect of leakage and even uncertainty remain in this analysis, and slight variations in the calculation method can result in more than a trivial difference. Thus one should not expect the decimal places of the calculation results or simulations to be meaningful [3], even if they are often given in the literature. Note that a possibility to avoid windowing and thus also the spectral signal leakage during the signal processing is possible by adjusting a simulation time as a multiple of the input signal period.

Spectral Metrics

Beside the shape of the output spectrum and therefore the possible qualitative analysis, in the frequency domain two major quantitative measures exist:

- *Total harmonic distortion* is the ratio of the sum of the signal power of all harmonic frequencies above the fundamental frequency to the power of the fundamental frequency. The xth harmonic itself is the ratio between the signal power and the power of the distortion component at the xth harmonic of the signal frequency.
- *Spurious free dynamic range* is defined as the ratio of the signal power to the power of the strongest spectral tone [47]. Its importance strongly depends on the application, since it dominates the resulting ADC linearity.

Both measures are shown in an exemplary signal spectrum in Fig. 2.6, where exemplarily a flat noise floor with a one tone signal and its harmonics are illustrated.

2.2.2 Noise and Power Metrics

These are commonly the most important measures, when $\Sigma\Delta$ ADCs are described or compared. All of these measures are obtained from the output spectrum above by integration, or by interpretation of these:

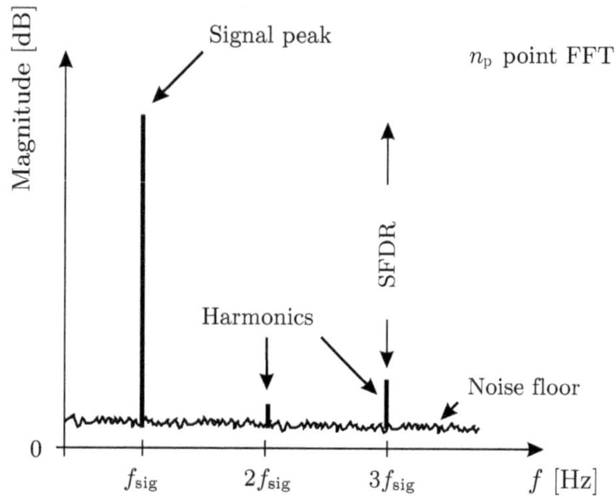

Fig. 2.6. Typical spectrum with signal peak and harmonics

- *In-band noise.* The integrated in-band noise (IBN) is a common converter measure, because it gives the total output noise in the band of interest and thus the minimum resolvable signal power. In spite of its little importance as a real ADC performance metric in terms of number of bits, it gives direct insight into the influence of architectural alternatives, nonideal behavior and so on. Therefore, it will be most often used throughout this work. Additionally it should be noted that the IBN contains all in-band power including noise, distortion or other tones. Only the signal itself is subtracted.
- *Signal-to-noise ratio.* The SNR of a converter is the ratio of the signal power to the noise power at the output of the converter, specified for a certain input amplitude. Most often, the maximum SNR, the peak $\mathrm{SNR_p}$, is pointed out. In this work, the SNR is obtained by applying a signal at a frequency $f_{\mathrm{sig}} > f_\mathrm{B}/2$, i.e., no signal distortion will appear in the band of interest (when no intermodulation occurs).
- *Signal-to-noise and distortion ratio.* The SNDR is the ratio of the signal power to the noise and all distortion power components. Thus, the corresponding spectra are obtained by applying a signal at $f_{\mathrm{sig}} \leq f_\mathrm{B}/3$ to include at least the second and third harmonic inside the band of interest.
- *Overload level* also known as maximum stable amplitude is defined as the maximum input sinusoidal signal, for which the structure still operates correctly. This level can almost be arbitrarily chosen, but most publications regard the ADC to operate correctly, until the SNR falls 6 dB below the peak value $\mathrm{SNR_p}$ for higher amplitudes.
- *Dynamic range.* The DR is defined as the rms value of the maximum amplitude input sinusoidal signal, for which the structure still operates

correctly, to the rms value of the smallest detectable input sinusoidal signal.
- *FOM.* The definition of a figure of merit is often taken into consideration to be able to compare different modulator implementations with each other. The most commonly used FOM takes the relation of power consumption over modulator resolution and bandwidth [48].

Usually, for $\Sigma\Delta$ converters most of these power measures are collected in a plot as shown in Fig. 2.7, where the SNR/SNDR is given vs. the *relative* input signal power [7], and both are given in *decibel* (dB). This unit on the other hand is a measure for the relation of two values. This is easy for almost all the performance specifications given above, because SNR, SNDR, and DR are relative expressions. But the IBN and the signal power in Fig. 2.7 are not related to any other value. Therefore it is important to note that these values have to be set in relation to something, and usually they are set relative to a converter's full scale amplitude [3], even if this is not explicitly mentioned in most publications.[1]

2.2.3 Used Tools and Program Code

Throughout this work various simulations will be performed, most of them giving the spectral behavior of an ADC, its SNR similar to Fig. 2.7 or the IBN plotted vs. some nonideal variable. Therein the tools, with which the DT output bit stream is generated, strongly depend on the depth of the analysis.

In the literature plenty of different methods have been proposed to perform simulations on $\Sigma\Delta$ ADCs, some of them in special programs for DT (Midas, Tosca, Switcap, and others [5]) or CT modulators [15, 49], some using c-Code modeling [3] or high-level simulators as MATLAB with SIMULINK [44], and a plenty of circuit simulators for the design of integrated circuits.

In this work the usage is limited to MATLAB and SIMULINK for the high-level synthesis and simulation of ideal and nonideal performance, while for low-level simulations a circuit simulator (SPECTRE) under the CADENCE DESIGN FRAMEWORK has been used, partially fully on transistor level, partially using some AHDL or VERILOGA block models. This restriction has been chosen to give this work a fully comprehensible style, while the results remain reproducible for a large audience due to the wide spread of these tools. In Appendix A, the MATLAB program code is explained, which has been used to obtain the performance measures from the discrete-time output bitstream of the ADC. The given code has been partially chosen according to [46].

[1] In the case of the $\Sigma\Delta$ modulator full scale amplitude (FS/2) is commonly defined as the input, whose magnitude equals the maximum magnitude of the quantizer feedback (relative to the common mode level), i.e., the *reference voltage* $V_{\rm ref}$ of the modulator. An input larger than that inherently overloads the converter

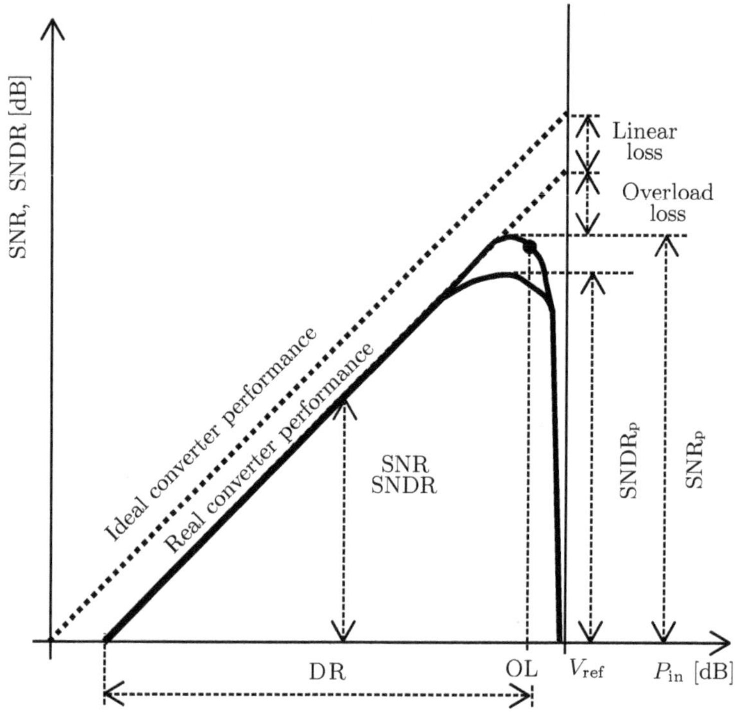

Fig. 2.7. Typical performance characteristic of a ΣΔ converter [8]

2.3 Performance of Nyquist Rate Converters

With the above found performance metrics for ADCs, the quantizer model derived in Sect. 2.1.2 and the spectral behavior of a quantizer in combination with a sampled signal, from Fig. 2.5b and (2.5) the dynamic range of Nyquist ADCs can be derived. Herein, the complete quantization noise power lies in the band of interest and contributes to the performance loss.

Suppose the multibit quantizer gain k_q is equal to 1, from Fig. 2.3d it can be seen that the amplitude of a full-scale sinusoid (before the A/D overloads) equals

$$\frac{\mathrm{FS}}{2} = \frac{\Delta}{2}\left(2^{B_{\mathrm{int}}} - 1\right) = \frac{\Delta}{2}\left(2^{B_{\mathrm{int}}}\right) \tag{2.6}$$

and the maximum SNR becomes the ratio of the corresponding input signal power ($P_{\mathrm{FS}/2}$) and the total quantization noise in (2.4):

$$\begin{aligned} \mathrm{SNR_p}|_{\mathrm{Nyquist}} &= 10\log_{10}\left(\frac{P_{\mathrm{FS}/2}}{\sigma_e^2}\right) = 10\log_{10}\left(\frac{\Delta^2(2^{(2B_{\mathrm{int}}-3)})}{\Delta^2/12}\right) \\ &= 6.02\ B_{\mathrm{int}} + 1.76\,[\mathrm{dB}]\,, \end{aligned} \tag{2.7}$$

which is the well-known signal-to-noise ratio of an ideal quantizer, where each additional bit results in approximately 6 dB increase in $\mathrm{SNR_p}$ [40]. The main benefit of (2.7) is the possibility to compare every ADC architecture to a simple Nyquist rate quantizer. Therefore from a specific performance plot as in Fig. 2.7 the *effective number of bits* can be obtained by approximately dividing the peak $\mathrm{SNR_p}$ by 6 dB.

While there are many Nyquist-rate converter architectures, especially for high speed applications, they all generally require operations such as comparison, amplification, etc., which have to be performed with the accuracy of the overall precision of the converter; this sets a tough requirement for the intrinsic precision of the integrated circuits, which in practice is rarely achieved with better than 12-bits of accuracy [50]. Therefore, Nyquist rate converters mostly rely on correction or calibration techniques for higher resolutions.

2.4 Performance of Oversampled Converters

A converter with a sampling frequency f_S much higher than the Nyquist frequency f_N is called an *oversampling converter*, (2.2). The benefit of a high oversampling ratio is twofold: first, as stated in Sect. 2.1.1 the implementation of an antialiasing filter is much easier, because the transition band increases enormously for $\mathrm{OSR} \gg 1$. This is because no frequency components between $[f_\mathrm{B}, f_\mathrm{S} - f_\mathrm{B}]$ can alias into the band of interest $[0, f_\mathrm{B}]$ during the sampling procedure, as can be seen from Fig. 2.1. But additionally, and this is even more important, the total quantization noise is independent from the sampling frequency. Therefore, as could be seen in (2.5), the power spectral density of the quantization noise reduces proportionally with increasing sampling frequency, while the integrated part in the band of interest decreases alike.

This is illustrated in Fig. 2.8. If a digital low-pass filter follows the oversampled converter, all parts outside the in-band can be eliminated leaving only a small part of the total quantization noise, shown as hatched or cross hatched areas in Fig. 2.8. The integrated in-band quantization noise becomes:

$$\mathrm{IBN_{OSR}} = \int_{-f_\mathrm{B}}^{f_\mathrm{B}} S_e(f)\mathrm{d}f = \frac{\Delta^2}{12}\frac{2f_\mathrm{B}}{f_\mathrm{S}} = \frac{\Delta^2}{12}\frac{1}{\mathrm{OSR}}, \tag{2.8}$$

thus increasing the maximum $\mathrm{SNR_p}$ equivalently to (2.7) into:

$$\begin{aligned}\mathrm{SNR_p}|_\mathrm{OSR} &= 10\log_{10}\left(\frac{P_{\mathrm{FS}/2}}{\mathrm{IBN_{OSR}}}\right) \\ &= 6.02\ B_\mathrm{int} + 10\log_{10}\mathrm{OSR} + 1.76\,[\mathrm{dB}]\,.\end{aligned} \tag{2.9}$$

Therefore, every doubling of the oversampling ratio increases the resolution by 3 dB or approximately 0.5-bit. The above mentioned digital low-pass filter

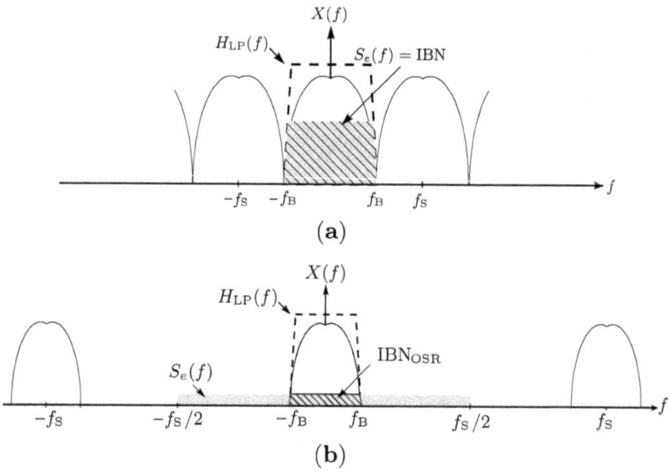

Fig. 2.8. Illustration of spectral effect of oversampling. (**a**) Quantization noise and in-band noise for OSR = 1; (**b**) quantization noise and in-band noise for OSR = 4

is usually called the *decimation filter* or simply the *decimator*. Beyond the elimination of high frequency noise components its task is also to reduce the sample rate at the overall output to the Nyquist rate. This can be done without aliasing due to the cancellation of all frequency components in $[f_B, f_S - f_B]$ by the low-pass filtering. Thus the decimator converts a high rate (f_S) B_{int}-bit word into a higher bit word at a low rate (f_N).

Nevertheless, the pure adoption of oversampling is not really profitable, because the enhancement of resolution is poor compared to the necessary increase of the sampling rate, which corresponds to a loss in conversion speed or an increase in power consumption.

2.5 Oversampled Noise-Shaping Converters: ΣΔ ADC

From a system point of view, the ADCs discussed up to now are pure open loop controls, i.e., an output signal is steered by an input signal (command variable) and affected by noise (disturbance variable). Therefore, the idea is self evident, to introduce a feedback path in order to achieve a closed control loop. Thus, with an additional controller in the forward or feedback path, different transfer functions can be obtained for the disturbing noise on the one hand and the desired signal on the other hand, defining a *signal transfer function* (STF) and a *noise-transfer function* (NTF).

An optimal configuration for the STF and NTF is the complete extinction of the noise inside the desired signal bandwidth (for a low-pass converter $[0, f_B]$), while herein the signal itself is not affected. Outside this in-band region the NTF gain can be high, while the STF can be low instead, thus defining an ideal NTF being the inverse of an ideal STF:

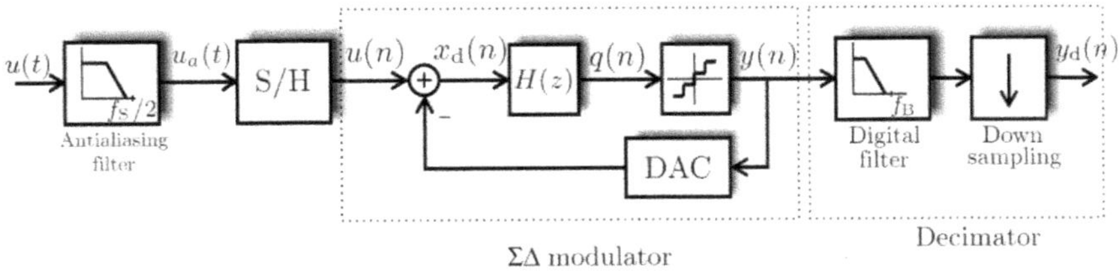

Fig. 2.9. Block diagram of a ΣΔ ADC

$$\mathrm{NTF_{Opt}} = \overline{\mathrm{STF_{Opt}}} = \begin{cases} 0\,, & -f_\mathrm{B} \le f \le f_\mathrm{B} \\ 1\,, & \text{else}\,. \end{cases} \tag{2.10}$$

This corresponds to an ideal low-pass filter for the STF and to an ideal high-pass filter for the NTF, both with a cut-off frequency of at least the signal bandwidth f_B. Thus, the quantization noise is shaped away from the desired frequency range, leading to a *noise-shaping converter*. The architectural principle of the ΣΔ ADC is shown in Fig. 2.9 [1], where the converter consists of:

- *Antialiasing filter* (AAF), which eliminates spectral components above half the sampling frequency from the input signal, so that the modulator input signal is band-limited and the subsequent sampling operation does not alias input signals from higher frequencies into the band of interest.
- ΣΔ *modulator*, which performs the actual A/D conversion by means of sampling and quantizing the band-limited input signal as well as by filtering the quantization error from the internal quantizer out of the in-band. The quantizer is typically of low resolution; implementations up to 6-bit wide internal quantization are reported [51]. The internal feedback DAC is commonly implemented with the same low resolution as the internal quantizer and thus does not introduce an additional quantization error.
- *Decimation filter*, performing low-pass filtering and down sampling, just as it was used for the pure oversampling ADC in Sect. 2.4. The former operation eliminates all the noisy out of band components from the spectrum whereas the latter reduces the output rate down to the Nyquist frequency with an output bit width, appropriate to the overall ADC resolution.

When replacing the quantizer in Fig. 2.9 by the model in Fig. 2.4, the linearized ΣΔ modulator in Fig. 2.10 is obtained. Its STF and NTF yield:

$$Y(z) = \mathrm{STF}(z)U(z) + \mathrm{NTF}(z)E(z)\,, \tag{2.11}$$
$$\mathrm{STF}(z) = \frac{1}{\frac{1}{H(z)k_\mathrm{q}} + 1}\,, \quad \mathrm{NTF}(z) = \frac{1}{1 + H(z)k_\mathrm{q}}\,.$$

In the above calculation the DAC is assumed to be ideal, while $U(z)$, $Y(z)$, and $E(z)$ are the $\mathcal{Z}$-domain representations of the input and output signal and the quantization error. By comparing (2.11) with the NTF

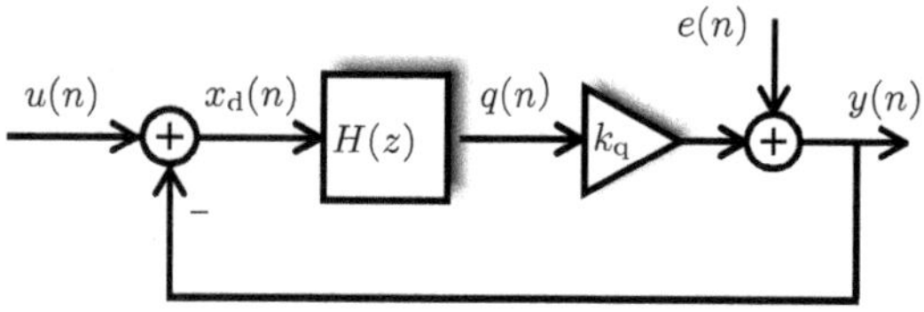

Fig. 2.10. Linearized model of a ΣΔ modulator

in (2.10), it reveals that the loop filter $H(z)$ should show a large gain within the in-band, while its gain may decrease outside the desired frequency bandwidth.

2.5.1 The First-Order ΣΔ Modulator

The easiest filter $H(z)$ is a simple first-order integrator:

$$I(z) = \frac{z^{-1}}{1 - z^{-1}}, \tag{2.12}$$

and the resulting architecture is the first-order ΣΔ modulator. For low frequencies (2.11) yields:

$$\mathrm{STF}(z) = z^{-1}, \quad \mathrm{NTF}(z) \approx \frac{1 - z^{-1}}{k_{\mathrm{q}}}, \tag{2.13}$$

where the NTF tends to realize a high-pass filter function for low frequencies. The effect of such noise shaping is illustrated in Fig. 2.11.

Obviously, in addition to the reduction of the quantization noise due to oversampling, the effect of noise shaping shifts the majority of the remaining noise power to higher frequencies. This leaves only little noise power in the in-band. With $z = \mathrm{e}^{(sT_{\mathrm{S}})} = \mathrm{e}^{(\mathrm{j}2\pi f/f_{\mathrm{S}})}$, where s is the Laplace-domain variable, the power spectral density of the shaped quantization noise yields [5]:

$$\begin{aligned} S_{\mathrm{Q}}(f) &= S_e(f)\, |\mathrm{NTF}(f)|^2 \\ &= S_e(f) \frac{1}{k_{\mathrm{q}}^2} \left| 1 - \exp\left(-\mathrm{j}2\pi \frac{f}{f_{\mathrm{S}}}\right) \right|^2 = S_e(f) \frac{4}{k_{\mathrm{q}}^2} \sin^2\left(\pi \frac{f}{f_{\mathrm{S}}}\right). \end{aligned} \tag{2.14}$$

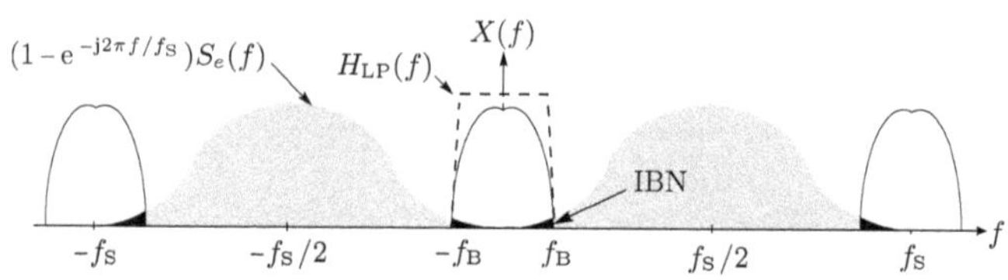

Fig. 2.11. Illustration of spectral effect of oversampling with noise shaping

By approximating the sine-function for small $f/f_S \approx 0$ (i.e., OSR $\gg 1$), $\sin(\pi f/f_S) \approx \pi f/f_S$ or, equivalently, in (2.13) $z = e^{(s/f_S)} \approx 1 + s/f_S$, the integrated in-band quantization noise becomes:

$$\mathrm{IBN}_1 \approx \int_{-f_B}^{f_B} \frac{\Delta^2}{12\, f_S} \frac{4\pi^2}{k_q^2} \left(\frac{f}{f_S}\right)^2 \mathrm{d}f = \frac{\Delta^2}{12} \frac{\pi^2}{3} \frac{1}{k_q^2\, \mathrm{OSR}^3} \,. \tag{2.15}$$

An increase of the oversampling ratio reduces the in-band noise power by 9 dB per octave. The corresponding, maximum $\mathrm{SNR_p}$ could be equivalently derived to (2.7) and (2.9), but in the case of the ΣΔ modulator the full scale amplitude does not necessarily correspond with the maximum SNR, as was illustrated in Fig. 2.7. For a general input amplitude one derives:

$$\mathrm{SNR}_1 = 10 \log_{10} \left(\frac{P_{\mathrm{sig}}}{\mathrm{IBN}_1}\right) \sim 10 \log_{10} \mathrm{OSR}^3 \, [\mathrm{dB}] \,, \tag{2.16}$$

which corresponds to 1.5-bit increase in resolution for a doubling of the OSR. Thus it is evident that a combination of noise shaping and oversampling is a favorable concept for achieving high resolution ADCs.

Finally, an illustrative explanation of the ΣΔ modulator in Fig. 2.10 together with a filter as in (2.12) can be given as follows: The integrator accumulates the actuating variable $x_d(n)$, i.e., the difference between the feedback and the input signal. The resulting output is quantized and fed back. The negative feedback control loop forces the actuating variable $x_d(n)$ to be zero, (which is approximately true for a higher internal quantizer bit-width B_{int}). Thus, the control loop tries to equalize the input and feedback signal. In the case of a multibit internal quantizer, the output $y(n)$ will follow the input signal with an error mostly smaller than the stepwidth Δ. In contrast, for a single-bit internal quantizer the output appears as a pulse code modulated signal [52] and tracks the input signal over time.

2.5.2 Pattern-Noise and Dithering in ΣΔ Modulators

A main drawback of the first-order ΣΔ modulator described above is the generation of tones and pattern noise appearing in the output [4], which is in strong disagreement with the theoretical behavior following the linearized model in Fig. 2.10. This is due to the assumption of the lack of a correlation between the quantizer input signal $q(n)$ and the quantization error $e(n)$ in Bennett's theorems in Sect. 2.1.2 [40]. Thus, when the input to the quantizer is not a random signal, the quantization noise will no longer be white. In a first order modulator, the input signal and the quantizer input are strongly correlated, while the output contains strong in-band frequency tones. This is even worse, if an additional dc-signal is applied. As has been shown in [5, 53], for dc-input signals the first-order ΣΔ modulator runs into repetitive patterns, trying to equal the input level on average.

A possible solution to this problem is to include a (pseudo) random noise source at the quantizer input (or at the modulator input) in order to add some nonperiodic, random signals [4], which are called *dither*. Therewith a partial decorrelation of the quantization error from the input signal can be achieved in low order modulators.

This is shown in Fig. 2.12b, where the first-order ΣΔ modulator in Fig. 2.10 with $H(z) = 1/(z-1)$ has been simulated with a dither-signal at the input of the quantizer. In comparison, Fig. 2.12a shows the same simulation without the adaption of a dither signal. The cancellation of spectral tones through dithering is obviously seen. Nevertheless, the introduction of dithering increases the complexity of the modulator and furthermore, real dithering is never totally random, thus again tones are seen due to the colored dither signal. Consequently, the usage of first-order modulators is nowadays mostly avoided. As will be seen, in higher order ΣΔ modulators this effect is much less visible, which is achieved through the reduced correlation of the input signal and the quantization error.

Throughout this book, in some MATLAB simulations dithering has been introduced to smooth the spectrum, especially when simulating second-order modulators; but this dither has been added in amounts, which are comparable to the thermal circuit noise of an actual implementation.

2.6 Performance Increase in ΣΔ Modulators

During the last 25 years a large number of ΣΔ modulator architectures has been proposed in the literature [2, 9, 54–58], all of which are intended to further

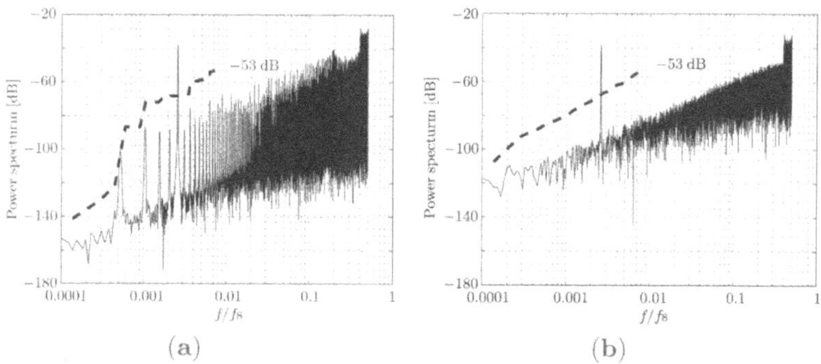

Fig. 2.12. Simulated power spectrum of a DT first-order, single-bit ΣΔ modulator. $\Delta = 2$, OSR $= 64$, $P_{\text{sig}} = -20\,\text{dB}$ at $f_{\text{sig}} = f_{\text{B}}/3$ and $P_{\text{dc}} = -20\,\text{dB}$. *Dashed*: integrated noise power. (**a**) First-order modulator; (**b**) first-order modulator with dither

reduce the in-band quantization noise. In principle, all possibilities can be reduced to the following strategies:

2.6.1 High OSR ΣΔ Modulators

Increasing the oversampling ratio of a ΣΔ modulator is twofold beneficial: first, the required front-end antialiasing filter can be easier implemented, because the transition band of that filter increases enormously; thus, the required power and area consumption decrease. Additionally, for ΣΔ modulators the achievable performance increases exponentially with the OSR, as can be seen e.g., in (2.15).

Nevertheless, if the signal bandwidth of a modulator is kept constant, for higher OSR the sampling frequency has to be increased, which obviously results in faster circuits and a drastically increasing power consumption [59]. Additionally, the incorporation of the digital part of the ΣΔ converter on the same chip and also timing requirements in switched circuits present a non-negligible challenge when drastically increasing the sampling frequency.

Finally, there exists nowadays an increasing demand for very high conversion rates, e.g., in communication electronics and wireless (e.g., UMTS) or wireline (e.g., xDSL) telecom applications, driving the development of modulators well beyond the 1 MHz baseband limit at similar resolutions as before [60–64]. Therefore, in modern high-speed applications the increase of the OSR is strongly limited and usually kept as low as possible. In [8] it was shown that a minimum OSR ≈ 4 should be maintained to profit from the ΣΔ implementation at all, which is therefore the lower limit for the oversampling.

2.6.2 Higher Order ΣΔ Modulators

The operating principle of ΣΔ modulators is based on shaping the quantization noise away from the in-band to higher frequencies by means of a loop filter. Therefore it is obvious to use a more aggressive filter function as in (2.12). An ideal, Nth order modulator should show a noise-transfer function, as it has been derived for the first-order modulator in (2.13), but with a higher order:

$$\mathrm{NTF}_N|_{\mathrm{ideal}} = (1 - z^{-1})^N . \tag{2.17}$$

A plot of this ideal higher order NTF is shown in Fig. 2.13, where the magnitude of (2.17) is given with $z = \mathrm{e}^{(-\mathrm{j}2\pi f/f_\mathrm{S})}$. The possible decrease of the in-band noise through higher order filtering is seen. Out of this ideal NTF, similar to (2.15) the performance of an ideal Nth order ΣΔ modulator could be derived. But actually, the agreement of calculation and simulation results would be poor and not transferable to real implementations.

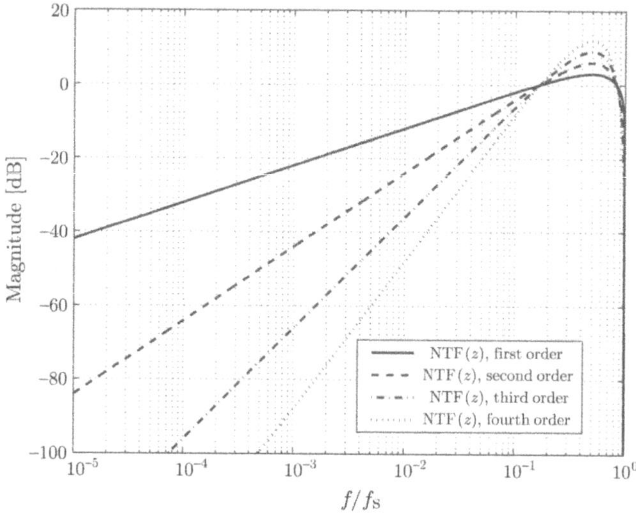

Fig. 2.13. NTF(z) for an ideal Nth order ΣΔ modulator

The reason for this is twofold: First, due to the required modeling of the quantizer as gain and additive white noise source as in Sect. 2.1.2, where the gain was not yet specified in detail for single-bit quantizers. Additionally, as can be seen in Fig. 2.13, the out-of-band gain of the higher order NTF increases rapidly, therewith resulting in instability through overloading the quantizer. Thus, as generally done in control systems, in order to increase the stability the loop gain is commonly reduced by scaling, as will be discussed more detailed in Sect. 2.7.4.

2.6.3 Multibit ΣΔ Modulators

Another approach is using an internal multibit quantization in Fig. 2.9 [4, 65]. Here, the intrinsic resolution is increased proportionally to $(2^{B_{\text{int}}}-1)^2$, because with increasing quantizer resolution in accordance to Fig. 2.3d and (2.4) the quantizer step width and therewith the quantization noise power proportionally decrease. Furthermore, the incorporation of multibit internal quantization tends to make higher order modulators more stable, the internal quantizer gain can be approximated to be unity, and the requirement for loop gain scaling is reduced. Thus almost ideal performance is obtained for medium modulator orders [8].

But while the implementation of the internal ADC is rather simple, because any errors are subject to the maximum noise shaping [66], the feedback DAC requires a linearity better than the overall modulator [5]. This is due to the DAC errors, which are directly fed into the input of the modulator as can be seen from Fig. 2.9. Therefore, even if the internal quantizer and DAC have a low resolution, the linearity of the DAC has to be as good as the accuracy of the overall modulator.

For a single-bit internal quantizer this problem does not arise, because a two level DAC is intrinsically linear. Here, only offset and gain errors remain, which a ΣΔ modualtor is highly tolerant to [5]. Actually, there have been many publications dealing with techniques to achieve the linearity requirements, based on design consideration [67] as well as digital or analog correction techniques [4, 8, 68–70]. However, all of these possibilities make use of rather complex circuitry, causing an increase of area and power consumption as well as switching activity.

2.7 Single-Loop, Single-Bit, Higher Order ΣΔ Modulators

Single-loop, single-bit architectures are widespread, due to the fact that they are least prone to circuit imperfections [34, 71], which is commonly accepted as one major advantage of ΣΔ ADCs. The loop filters of such ΣΔ modulators can be implemented in different ways. The following list shows today's most commonly realized structures:

- Chain of integrators with distributed feedback [71, 72]
- Chain of integrators with distributed feedback and distributed feedforward [73]
- Chain of integrators with weighted feedforward summation [17, 56, 74]
- Inclusion of local feedback loops

Nonetheless, also other single-loop structures have been used [75].

2.7.1 Distributed Feedback Topology

Even before Candy's paper on the second-order modulator [2], in [76] the incorporation of a higher order loop filter had been proposed, by means of a cascade of several integrators in the forward path of the ΣΔ modulator loop, with each integrator receiving a weighted feedback path from the DAC, where the coefficients a_i, $i = 1, \ldots, N$ are called *integrator scalings or weights*, Fig. 2.14. This topology has been widely used [34, 63, 71, 72] and optimal scaling coefficients were presented in [6]. In Fig. 2.14 the STF and NTF yield:

$$\mathrm{STF} = \frac{k_\mathrm{q} \prod_{i=1}^{N} a_i H(z)}{1 + k_\mathrm{q} \sum_{i=1}^{N} \prod_{k=i}^{N} a_k H^{N-i+1}(z)}, \tag{2.18}$$

$$\mathrm{NTF} = \frac{1}{1 + k_\mathrm{q} \sum_{i=1}^{N} \prod_{k=i}^{N} a_k H^{N-i+1}(z)}. \tag{2.19}$$

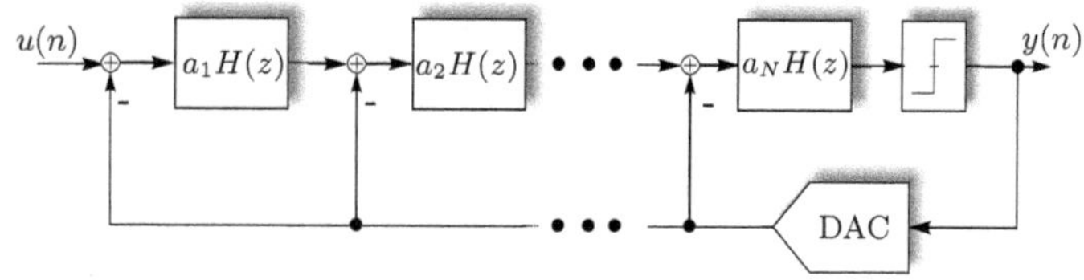

Fig. 2.14. Nth order single-bit ΣΔ modulator consisting of a chain of integrators with distributed feedback a_i

When using DT integrators as filters $H(z) = I(z)$, the NTF is approximated for low frequencies [6]:

$$|\mathrm{NTF}| \approx \frac{\left|1 - z^{-1}\right|^N}{k_\mathrm{q} \prod_{i=1}^{N} a_i} . \tag{2.20}$$

Obviously, the complete scaling of the outermost feedback branch dominates the noise-shaping behavior. Similar to (2.14, 2.15), the in-band noise of an Nth order modulator as in Fig. 2.14 yields:

$$\mathrm{IBN}_N \approx \frac{\Delta^2}{12} \frac{\pi^{2N}}{2N+1} \frac{1}{k_\mathrm{q}^2 \prod_{i=1}^{N} a_i^2} \frac{1}{\mathrm{OSR}^{2N+1}} . \tag{2.21}$$

Here, the major advantage and disadvantage of higher order, single-loop ΣΔ modulators reveals: first, the increase of the exponent over OSR and thus the enhanced exponential reduction of the in-band noise, but second the increase of the in-band noise through the scaling of the outermost feedback loop, i.e., $\prod_{i=1}^{N} a_i$ in Fig. 2.14, which becomes smaller for higher orders.

A major drawback of this loop filter implementation is that beside the filtered quantization noise, the integrator outputs also contain a significant amount of the input amplitude [4], which can be understood as follows: since the integrator input signals are controlled to be zero over time, the summed feedback 1-bit DAC signal and the integrator outputs are equal over time. Because in a ΣΔ modulator the DAC signal tracks the input signal over time, the outputs of the integrators have to carry a significant amount of input signal. Thus, the integrators both require significant swing capabilities and, in order to avoid clipping or even overload, the scaling coefficients need to be low or special circuit techniques need to be applied, e.g., [77]. Small scaling coefficients on the other hand yield discrete-time SC integrators with small sampling capacitor or continuous-time integrators with small transconductance/large resistance, both causing higher circuit noise contribution. This

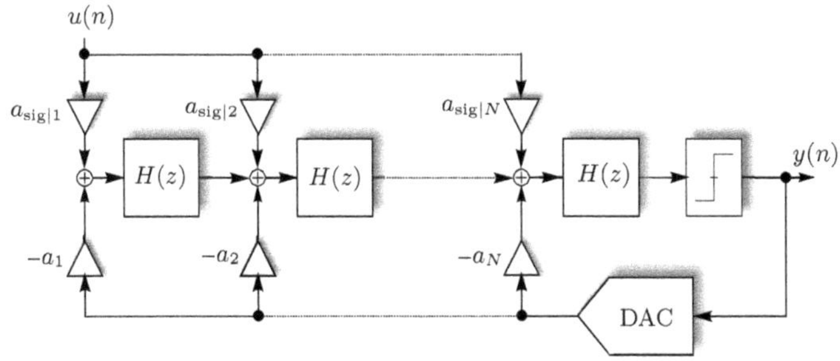

Fig. 2.15. Nth order single-bit ΣΔ modulator consisting of a chain of integrators with distributed feedback k_i and feed-forward $a_{\text{sig}|i}$

issue will be shortly discussed in Sect. 5.6.3, also with respect to linearity requirements.

In addition, as seen in (2.18, 2.19) and shown in [4], the loop filter for the signal and noise are essentially identical. Thus, if an optimal noise-shaping function has been determined, the STF is fixed. One possibility to overcome this drawback is to include also feed-forward coefficients from the signal input to every integrator [73]; the corresponding architecture is known as a chain of integrators with weighted feedback and distributed feed-forward and is shown in Fig. 2.15. Therewith, the STF becomes somewhat independent from the NTF and both can be separately optimized [4].

2.7.2 Feed-Forward Topology

Another style to implement the loop filter of an Nth order ΣΔ modulator is the chain of integrators with feed-forward summation, which is illustrated in Fig. 2.16. Also this loop filter implementation is commonly used [17, 78, 79]. For a quantizer model as in Fig. 2.4, the corresponding signal and noise-transfer functions are given in (2.22, 2.23):

$$\text{STF} = \frac{a_{\text{sig}} k_{\text{q}} \sum_{i=1}^{N} a_{N-i+1} H^i(z)}{1 + k_{\text{q}} \sum_{i=1}^{N} a_{N-i+1} H^i(z)}, \tag{2.22}$$

$$\text{NTF} = \frac{1}{1 + k_{\text{q}} \sum_{i=1}^{N} a_{N-i+1} H^i(z)}, \tag{2.23}$$

which is obviously the same structure of an NTF as in (2.19), thus also the IBN yields (2.21). The major advantage of the topology in Fig. 2.16 is that the integrator outputs do not contain the significant part of the input signal as seen above for the feedback architecture. Thus, the necessity for scaling and also the requirements on integrator dynamics are much more relaxed.

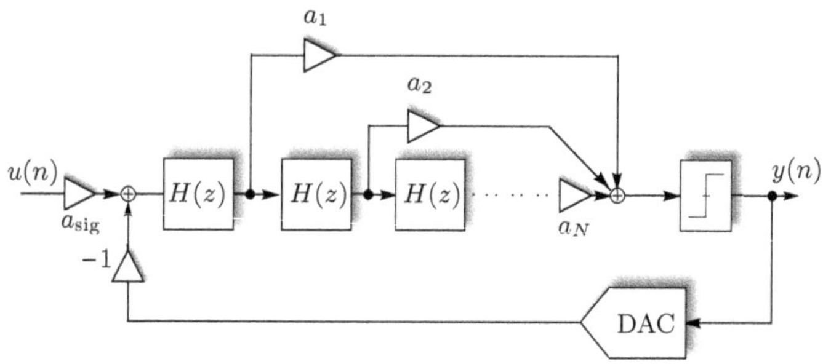

Fig. 2.16. Implementation of a Nth order single-bit ΣΔ modulator consisting of a chain of integrators with weighted feed-forward a_i summation

Nonetheless, as was the case for the pure feedback architecture, for a desired noise-shaping function, the STF is also chosen. In addition, the STF of the feed-forward architecture in Fig. 2.16 contains peaking at high frequencies due to the resulting filter characteristic; with input signals at these frequencies a modulator with feedforward summation could overload, thus precautions have to be taken [4].

2.7.3 Local Feedback Loops

One common characteristic of the architectures shown in Figs. 2.14–2.16 is the placement of the poles of the loop filter and thus of the zeros of the NTF. If a low-pass ΣΔ modulator is considered with integrators as filter $H(z)$, then all zeros of the NTF are placed at dc, which corresponds to a simple Nth order high-pass. Thus, as also seen in Fig. 2.13, the noise shaping is effective only at very low frequencies, while next to the base bandwidth f_B the noise quickly increases.

By adding local feedback loops around pairs of integrators, as illustrated in Fig. 2.17, the NTF zeros can be moved away from dc and spread over the signal band. As a consequence, the IBN and the DR will be improved. A detailed analysis concerning the performance gain as well as the optimal zero placements for modulators up to the eighth order is given in [4]. The optimal

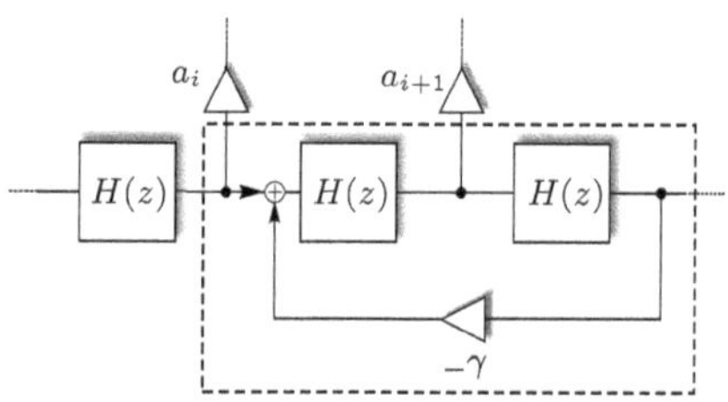

Fig. 2.17. Implementation of a local feedback loop

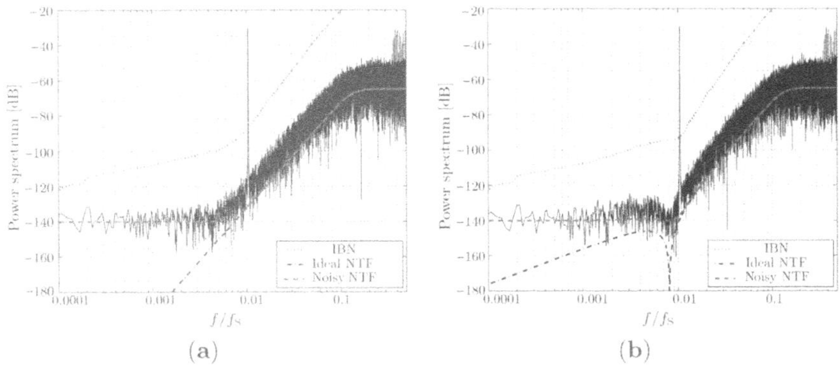

Fig. 2.18. Power spectrum, IBN and calculated NTF of the feed-forward structure with and without optimized zeros. $P_{\mathrm{sig}} = -15\,\mathrm{dB}$, $f_{\mathrm{sig}} = f_{\mathrm{B}}$ and OSR = 48. IBN ≈ −87 dB, $\mathrm{IBN_{Opt}} \approx -93\,\mathrm{dB}$. (**a**) Without optimized zero; (**b**) with optimized zero

zero placement can be found by minimizing the integrated IBN with respect to the feedback coefficient γ:

$$\mathrm{Min}\left\{\int_{-f_{\mathrm{B}}}^{f_{\mathrm{B}}} |\mathrm{NTF}(f,\gamma)|^2 \mathrm{d}f\right\}. \tag{2.24}$$

Note that by using local feedback loops, also the drawback of high-frequency peaking of the feed-forward architecture is alleviated.

Unfortunately, it turns out that the local feedback coefficients γ scale with OSR^{-2}. As a result, they rapidly decrease with the OSR and the technique is more feasible for low oversampling, because for a significant performance increase the relative placement of the zeros within the signal band has to be rather exact. This is important if the integrator gains can not be precisely realized as in the case of continuous-time implementations. Figure 2.18 exemplarily shows the simulated output spectrum, the integrated IBN as well as the calculated NTF of a third-order, feed-forward ΣΔ modulator, with and without optimal zero placement through a local feedback loop. In order to show the effectiveness of the zero spread even in the presence of realistically dominant circuit noise, the input signal was affected with a large white noise! Obviously, even if the optimized NTF notch is completely filled with circuit noise, the shift of the optimized NTF near the baseband frequency still yields 6 dB IBN improvement! This is why this technique is commonly implemented in all most recent designs.

2.7.4 ΣΔ Modulator Loop Filter Stability and Scaling

The above presented architectures provide the possibility to implement a noise-transfer and signal-transfer function with a certain filter characteristic.

But together with considering about the loop filter implementation, the finding of such loop filter is of major interest.

The well-known drawback of single-loop single-bit ΣΔ modulators with orders higher than two is their tendency to instability [80]. Thereby, stability is defined as a modulator condition, where all internal *state variables*, which are the integrator outputs, remain bounded over time [81]. As mentioned in Sect. 2.6.2, one measures to ensure stable operation in a possibly unstable control loop is the reduction of the loop-gain by appropriate filter scaling. The need for this scaling can be also seen from Fig. 2.13: the out-of-band gain of the ideal noise-transfer function increases heavily for increasing loop filter order N. Consequently, it starts to overload the quantizer input, which yields a significantly decreased effective quantizer gain k_{q}.

Thus, after having chosen a certain filter characteristic through one specific architecture, and after having designed an optimized noise-transfer function, the analysis of stability is of major importance. Therefore, several methods can be used, beyond others simulation or calculation. While simulations are surely the final prove for performance, stability, and other effects like pattern noise (see Sect. 2.5.2), calculations give the starting point for the loop filter and can even provide a closer insight into the behavior of the chosen ΣΔ modulator: in [82] the method of root-locus plots has been adopted for that purpose in single-loop, single-bit ΣΔ modulators. Therefore, the unknown quantizer gain k_{q} has been used as the variable gain of the root locus, which was found to be not defined in the case of a single-bit quantizer, (see Sect. 2.1.1).

In the following, the STF of a third-order modulator with distributed feedback as in Fig. 2.14 with $H(z) = I(z) = 1/(z-1)$ is exemplarily considered: the open loop transfer function, which is needed for the root locus calculation, is the denominator in (2.18) minus 1. From this open loop transfer function, the root-locus plot can be derived and is shown in Fig. 2.19. The scaling coefficients have been illustratively varied from no scaling toward an optimal set of scaling coefficients for a third-order, single-bit ΣΔ modulator according to [6].

Obviously, three poles start at dc ($z = \{1, 0\}$) for $k_{\mathrm{q}} = 0$. The real pole tends to $-\infty$, but actually it will always be on the real axis inside the unit circle. This is due to the effect of a stable limit cycle as descriptively shown in [8]. The possible instability arises from the conjugate complex pair of poles, which leave the unit circle for small quantizer gains k_{q} and enter it at a *critical gain* $k_{\mathrm{q,crit}}$: without scaling ($a_i = 1$), this pole pair never gets into the stable area, i.e., the modulator will not work. For smaller scaling coefficients, the branches of the root-locus plot are bent into the unit circle, and a conditional stability is achieved, if the quantizer gain is larger than the critical gain. On the other hand, a high single-bit quantizer gain corresponds to small quantizer input signals and thus to a limitation of the quantizer input, since the output is fixed to the reference $\pm\Delta/2$, as can be seen in Fig. 2.3c.

A feasible enhancement of this linear examination is the calculation of the quantizer gain k_{q} in the case of zero input signal, i.e., k_{q0}. It has been

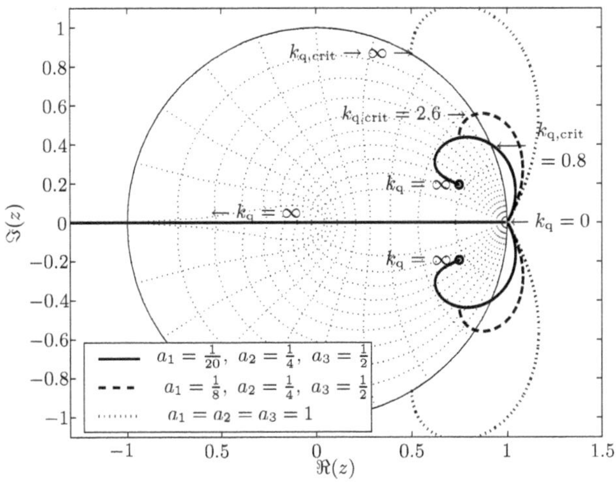

Fig. 2.19. Root-locus of a third-order single-bit ΣΔ modulator with different sets of scaling coefficients

shown in [83] that this calculation can only be formally done for orders $N \leq 2$, while for higher order loop filters numerical methods have to be used. The determination of k_{q0} together with the critical gain $k_{q,crit}$ then yields an estimate, if the modulator is conditionally or unconditionally stable or even completely unstable. The modulator will be:

- Completely unstable, if one of its poles is out of the unit circle for $k_q = k_{q0}$, i.e., if the quantizer gain for zero input signal causes poles outside the unit circle
- Unconditionally stable, if the locus of its poles is completely inside the unit circle for $k_q < k_{q0}$, i.e., if for all input amplitudes larger than zero all poles remain inside the unit circle
- Conditionally stable in any other case, i.e., for zero input signal all poles are inside the unit circle, but for increasing input amplitude at least one pole shifts into instability

The given root-locus method is a linear approach to a strongly nonlinear system. Thus, this conditional stability and the finding of optimal scaling coefficients must be confirmed by behavioral simulations, and is usually based on two requirements [80]: first, the modulator input signal has to be bounded into a specific interval, and second, the scaling has to prevent the noise-transfer function from peaking at high frequencies. As a rule of thumb, in [4] it has been proposed to limit the out-of-band gain of the noise-transfer function of single-bit, single-loop modulators to 1.5. More detailed analysis about stability considerations and simulations was presented in [81].

In practice, there exist tools [84], cook-book like design procedures [4] and a wide range of behavioral simulations [6–8], all of which propose some kind of optimized modulator design. Nonetheless, the design of ΣΔ loop filters is still a matter of hands-on design and optimization for a specific implementation. In Chap. 7, design examples will be given, which present different procedures and also trade-offs for circuit implementations, while in the rest of this book most simulations are based on the loop filters proposed in [6]. There scaling coefficients have been derived providing stable operation for second-order to fourth-order distributed feedback architectures, which were found with extensive simulation and optimization efforts.

2.7.5 Effective Quantizer Gain in ΣΔ Modulators

Above, the quantizer gain of the linear quantizer model has been left unspecified in calculations or even changed over a large range in order to analyze stability. Nonetheless, for a quantitative analysis of the shape of the noise-transfer function and thus of the performance of a ΣΔ modulator a feasible estimate for this quantizer gain has to be determined.

For multibit internal quantizers, the gain can be safely modeled with $k_{\mathrm{q}} \approx 1$, because many transition steps define a straight conversion line. As indicated in Fig. 2.3c, the gain of a single-bit quantizer can be chosen to an arbitrary value, and many approaches exist to find a good approximation: The approach used in multibit, i.e., the assumption of a unity gain of the quanitzer, has been chosen in [85], while a more complicated possibility has been introduced in [86] using a describing function analysis and assuming different quanitzer models for the signal and the quantization noise path. In [4] an extensive overview of published attempts is presented. Nevertheless, many of these methods either strongly deviate from the reality and simulation results, or become rather complex and loose the possibility to give an easy insight.

In contrast to such sophisticated, but practically too extensive methods, in [87] and in [6] methods were proposed, which are easy to adopt and commonly lead to accurate models for a rough analytical analysis. Both methods were proposed for the commonly used higher order ΣΔ modulator architecture with distributed feedback shown in Fig. 2.14, where N DT integrators are cascaded and scaled with different coefficients, which can be drawn in the feedback or in the forward path in front of the integrators [6]:

- *LF unity gain approximation.* In [47, 87] the quantizer in first-order and second-order modulators has been modeled using the loop filter unity gain approximation. This is to assume a quantizer gain k_{q}, such that the product of the loop filter gains and the quantizer gain around the outermost feedback loop of a ΣΔ modulator as in Fig. 2.14 is equal to 1. It should be noted that the only justification for this approximation is the good agreement of analytical and simulation results, which were stated in [47, 87]. But it is to mention that in both works, only first-order and second-order loops were considered.

- *Effective quantizer gain "k_{eff}."* In [6–8] the product of the quantizer gain k_{q} and the scaling coefficient of the last integrator a_N of a chain of integrators have been combined to an *effective quantizer gain* k_{eff}. This can be done, because the output of a single-bit quantizer (comparator) only depends on the polarity of the quantizer input. Therefore the scaling of the last integrator is rather unimportant for the functionality. In [6], a choice of $k_{\mathrm{eff}} = 2$ has been found to achieve a good agreement of calculations and simulations of single-bit ΣΔ modulators, while here only higher order modulators ($N = 2, \ldots, 4$) were considered.

Simulations reveal that for single-bit, first-order modulators the *unity gain approximation* gives the best results, while for higher order modulators the assumption of the *effective quantizer gain* $k_{\mathrm{eff}} = 2$ results in the best agreement. For a multibit quantizer, as stated above and in [7, 8], a quantizer gain equal to unity has been chosen. Consequently, our assumption for the quantizer gain is:

$$k_{\mathrm{q}} = \begin{cases} 1/a_1, & \text{single-bit, first-order modulator} \\ 2/a_N, & \text{single-bit, } N\text{th order modulator} \\ 1, & \text{multibit internal quantizer}. \end{cases} \tag{2.25}$$

A further note should be mentioned at this place: if the input signal on a multibit ΣΔ modulator becomes very low, the input to the quantizer decreases simultaneously. For low input values to the quantizer, its output toggles only between ±1 LSB, i.e., the multibit quantizer behaves as a single-bit comparator and thus the effective quantizer gain has to be adopted. This feature will be illustratively shown in Sect. 5.3 in Fig. 5.3c.

In addition to this easy modeling of the quantizer gain for the architecture in Fig. 2.14, the model proposed in [88] should be mentioned, which is a bit more sophisticated: It relies on the assumption that the single-bit quantizer output power is constant, since its output level is constant and only changes its sign. Therewith, slightly more complex but still practical expressions can be derived, which have been proven to match also for architectures other than the distributed feedback.

2.8 Multiloop, Cascaded ΣΔ Modulators

The augmentation of the loop filter to orders higher than two in single-stage modulators results in architectures which are prone to instability and thus they suffer from reduced scaling coefficients and reduced performance than an ideal higher order filter. As an alternative, *cascaded or* MASH (multistage noise shaping [89]) topologies can combine high order noise shaping with the stability of low order single-stage modulators. They typically consist of several

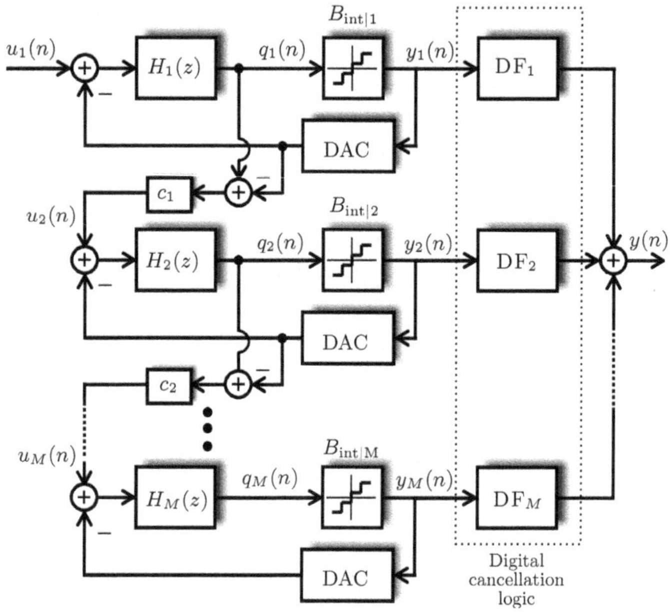

Fig. 2.20. Illustrative general structure of a cascaded ΣΔ modulator

stages of lower-order ΣΔ modualtors, where each higher stage receives a representation of the quantization noise of the previous stage as its input. With an appropriate recombination of all digital outputs the quantization noise of all but the last stage can be canceled. An illustrative implementation of a general cascaded ΣΔ modulator with M stages is shown in Fig. 2.20. There, $H_i(z)$ are loop filters of order N_i, the quantizers have a $B_{\text{int}|i}$-bit resolution and the DF_i are the digital recombination logic.

The operation principle is easily understood as follows: by modeling the quantizer of the ith stage as additive noise $e_i(n)$ as in Fig. 2.10, the outputs of an exemplary two-stage cascaded modulator become:

$$y_1(n) = \text{STF}_1\, u_1(n) + \text{NTF}_1\, e_1(n)\,, \tag{2.26}$$

$$y_2(n) = \text{STF}_2\, u_2(n) + \text{NTF}_2\, e_2(n)\,. \tag{2.27}$$

For the input of the second-stage follows from Fig. 2.20:

$$u_2(n) = -c_1 e_1(n)\,. \tag{2.28}$$

Thus, the overall output yields:

$$\begin{aligned} y(n) &= y_1(n)\,\text{DF}_1 + y_2(n)\,\text{DF}_2 && (2.29)\\ &= \text{STF}_1\, u_1(n)\,\text{DF}_1 + \text{NTF}_1\, e_1(n)\,\text{DF}_1 && (2.30)\\ &\quad + c_1\text{STF}_2\,(-e_1(n))\,\text{DF}_2 + c_1\text{NTF}_2\, e_2(n)\,\text{DF}_2\,. && (2.31) \end{aligned}$$

The first objective of the digital cancellation logic is to eliminate the quantization noise of the first stage $e_1(n)$, which yields the condition:

$$\mathrm{DF}_1\,\mathrm{NTF}_1 = c_2\,\mathrm{DF}_2\,\mathrm{STF}_2 \tag{2.32}$$

$$\rightarrow \mathrm{DF}_1 = \mathrm{STF}_2|_\mathrm{d}, \quad \mathrm{DF}_2 = \frac{1}{c_1}\mathrm{NTF}_1|_\mathrm{d}\,, \tag{2.33}$$

where the suffix $|_\mathrm{d}$ stands for an equivalent digital implementation of an analog transfer function. Doing so, the output of the cascaded modulator becomes:

$$y(n) = \mathrm{STF}_1\,\mathrm{STF}_2|_\mathrm{d}\;u_1(n) + \frac{1}{c_1}\mathrm{NTF}_1|_\mathrm{d}\;\mathrm{NTF}_2\,e_2(n)\,. \tag{2.34}$$

Interestingly, apart from the interstage coupling coefficient c_1, (2.34) shows the ideal noise-shaping of second-stage cascaded ΣΔ modulator, because through the digital cancellation logic all but the last quantization noise is canceled completely. The quantizer error of the last stage is attenuated by a noise-shaping function of an order equal to the number of integrators N_casc in the overall cascade. Therefore, the resulting noise-transfer function almost follows the ideal one in (2.17):

$$\mathrm{NTF}_\mathrm{casc} = \frac{(1-z^{-1})^{N_\mathrm{casc}}}{\prod_{i=1}^{M} c_i}\,, \qquad N_\mathrm{casc} = \sum_{i=1}^{M} N_i\,. \tag{2.35}$$

Thus, only the interstage coupling coefficients c_i, which prevent a possible overload of the input of the higher stages, decrease the performance below the ideal value. From (2.35), similar to (2.20, 2.21) the integrated in-band noise of cascaded ΣΔ modulators can be derived to:

$$\mathrm{IBN}_\mathrm{casc} = \frac{\pi^{2N_\mathrm{casc}}}{2N_\mathrm{casc}+1}\,\frac{\Delta_M^2}{12}\,\frac{1}{\mathrm{OSR}^{2N_\mathrm{casc}+1}\prod_{i=1}^{M-1} c_i^2}\,, \tag{2.36}$$

where M is the total number of cascaded stages, N_i the order of the ith stage, c_i the interstage connection scalings, while Δ_M is the quantizer step width of the last stage.

Principally, any stable single-loop ΣΔ modulator can be used as one stage in a cascaded modulator. But in practice there exist some restrictions: first, due to their intrinsic, unconditional stability, mostly first-order and second-order single-loop modulators are employed; by that, first stage first order modulators are avoided due to the tonal problems (Sect. 2.5.2) and an increased sensitivity to nonidealities [87, 90, 91].

Over the last two decades, there have been publications of numerous different architectures for the implementation of cascaded ΣΔ modulators. The first two-stage, second-order modulator, 1–1, has been reported in [89], and in the following years modulators of different orders with different numbers of stages have been implemented: third-order, third-stage, 1–1–1 in [92], third-order, second-stage modulators, 2–1 or SOFO (Second-order First-order) in [93] and fourth-order, third-stage, 2–1–1 in [94]. Finally in [5, 61, 64, 95] the possibility of using multibit (mb) quantization in higher stages has been implemented in a fourth-order 2–1–1_{mb} modulator incorporating a 3-bit last stage quantizer. Based on these works dozens of publications were published, presenting theoretical work and extensions of the basic architectures. A general overview of the most common discrete-time cascaded modulators is given in [6, 8], while a comprehensible way for the top down design of switched-capacitor cascaded ΣΔ modulators is given in [5].

Recently, even continuous-time realizations of cascaded ΣΔ modulators were rediscovered [20, 96, 97], after being proposed already in [98]. These promising architectures will be considered more extensively in the subsequent chapters.

2.9 Specialized Architectures

Beyond the architectures for implementing ΣΔ loop filters discussed above, there were numerous other proposals over the last decades, many of which were proposed to overcome special nonidealities of other architectures or tried to improve the ΣΔ performance in a special way. It is out of the scope of this book trying to cover all of these proposals, nonetheless some further techniques shall be mentioned in a glance:

- *Multirate, single-loop architectures.* In contrast to the loop filters shown above, it is obviously possible to use different oversampling ratios in different feedback loops. This enables the possibility, to chose low oversampling ratio and thus low sampling frequency in the power-hungry first integrator, while the latter filters and the quantizer work at the high sampling frequency. Consequently, the power consumption can be reduced, while the modulator performance can be kept rather high [99, 100].
- *Multirate, multiloop architectures.* Cascaded ΣΔ modulators use additional stages to measure and subsequently cancel the quantization noise of the previous stages. In order to increase their effectiveness, multibit quantization was employed in higher stages, which decreases the overall in-band noise, while the requirements on the multibit DAC in the higher stages is drastically relaxed. In the same manner, higher stages can employ higher oversampling ratios, which exponentially increases the resulting performance. Concurrently, the expected higher speed requirements of active circuitry is drastically relaxed compared to the first stage, because errors are suppressed in these higher loops. Even if this technique has been proposed both for DT [100] as well as for CT cascaded modulators [101], they are not widely used up to now, since their drawback is the higher sampling

rate of the subsequent digital cancellation logic as well as the decimation filter.

- *Swing reduction concepts.* For deep submicron processing, the almost linear reduction of the supply voltage, but the nonlinear (slower) decrease of the threshold voltages causes the available voltage swing to rapidly decrease, thus requiring different circuit techniques to overcome this lag. A simple scaling of the integrator outputs reduces the signal amplitude while the circuit noise maintains constant [90]. Therefore, the SNR is degraded. In [102] and [103, 104] two different integrator output swing reduction schemes are proposed, which overcome this effect. The aim of the former technique is the swing-reduction of the first integrator, whereas the latter reduces the signal swing of the last integrator.

2.10 Loop Filters with Bandpass Characteristic

In the exemplarily shown architectures above, all filters and corresponding calculations were presented for low-pass characteristic of the $\Sigma\Delta$ modulator. Nonetheless, there is no reason why the loop filter could not show bandpass filter behavior, thus reducing the quantization noise in a band around a center frequency f_0.

While for LP modulators, the filters $H(z)$ show high gain in the baseband, i.e., around $f = 0$, and thus quantization noise is suppressed here, in bandpass $\Sigma\Delta$ modulators $H(z)$ shows this high gain around $f_0 > 0$. The subsequent decimation filter is then designed to remove the noise content outside the signal band, this time around the center frequency $f_0 > 0$. Concurrently, the sampling rate is reduced and the signal translated into the baseband for further digital processing. The simplest way to synthesize bandpass $\Sigma\Delta$ loop filters is to start the design of a LP filter as described above, and then transforming the LP to a bandpass characteristic. For example, applying the transformation

$$z \to -z^2 \tag{2.37}$$

maps all zeros of a low-pass $\mathrm{NTF}(z)$ from dc to $\pm\pi/2$; thus, this simple transformation places the center frequency of the resulting bandpass NTF to $f_0 = f_S/4$, and the structure is called a $f_S/4$ bandpass modulator [4]. The corresponding architecture behaves exactly as its LP original concerning dynamics, stability, SNR, etc. But note that bandpass modulators need an order twice as high as their LP equivalent, while the order of the NTF and the resolution remain the same.

In practice, the mentioned transformation was commonly used, but more custom-designed loop filters are nowadays widely seen. Especially, if the loop filter has to show special characteristic, as for example suppression of adjacent radio channels or if partial mixing is placed inside the $\Sigma\Delta$ modulator, hands-on design becomes necessary and the interested reader is referred to the publications specialized on this topic [3, 18, 105, 106].

3

Continuous-Time ΣΔ Modulators

This chapter illustrates the general analogies and differences as well as the advantages and drawbacks of continuous-time and discrete-time ΣΔ modulators. Various synthesis procedures are explained for single-loop modulators which are based on several possibilities found throughout the literature. Thereafter, they are extended to cascaded, multistage implementations. The differences of both architectures considering the signal and noise transfer function as well as the implicit antialiasing filter are emphasized. Finally, the most commonly used CT loop filter realizations are discussed.

3.1 CT ΣΔ Modulator Issues

The introduction into the theory of ΣΔ A/D conversion presented in the previous chapter is based on purely discrete-time (DT) circuitry, i.e., the analog input to the modulator, $u(n)$ is a sampled signal, which has been subject to antialiasing filtering before. The loop filter $H(z)$ consists of DT, $\mathcal{Z}$-domain filters, while the actual DT loop filter transfer function is based on architectural design and scaling coefficients, as could be exemplarily seen for example in Fig. 2.14. In the last two decades, the majority of published ΣΔ modulators throughout the literature have actually been realized in discrete-time circuits, implemented using the switched capacitor (SC), e.g., [107], or the switched current (SI), e.g., [108, 109], technique. This was primarily due to the emerging ease with which monolithic SC filters can be designed, the high degree of linearity of the resulting circuits and finally the natural allegory between the mathematics of the ΣΔ system and the DT circuit-level implementation. Therefore, also most of the theoretical research and inventions focussed on these DT ΣΔ modulators, e.g., [2, 5, 6, 8, 89, 91–93, 110], and nowadays there exists a great deal of software, tools, and publications to support the synthesis and to perform the design of DT ΣΔ loop filters for a given specification.

Beyond this method of DT modulator design, the implementation of the loop filter can also be performed using continuous-time (CT) circuits, which

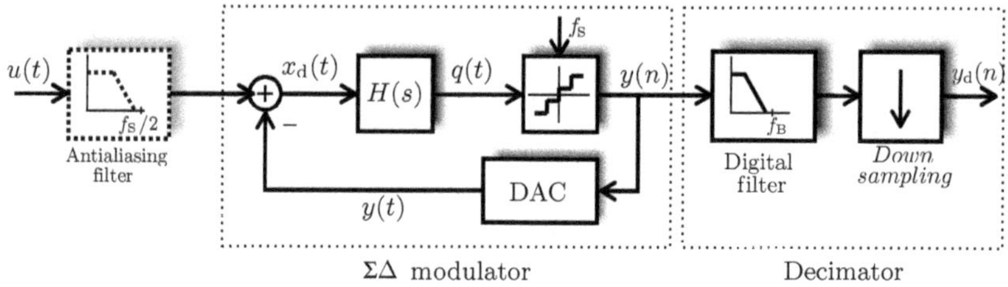

Fig. 3.1. Block diagram of a CT ΣΔ A/D converter

is also the historical origin of ΣΔ modulation: First mentioned in [1] this technique has been used early for example in [54, 111, 112]. An illustrative schematic of a CT ΣΔ converter implementation is shown in Fig. 3.1. Here, the input signal is subject to antialiasing filtering or, which is noteworthy, feeds directly into the CT modulator without preceding AAF, as will be elaborated later. The loop filter $H(s)$ now consists of CT filters, which can be active *RC*-filters using OpAmps, operational transconductance amplifiers (OTA) [34], *gmC*-filters [9, 113] or even LC-resonator structures [3]. Finally, the internal quantizer is a sampled or latched circuit, which is clocked at the modulator's sampling frequency f_S. The comparison of the working principle of a DT and a CT modulator and therewith the origin of their assets and drawbacks can be attributed to three major differences.

3.1.1 Sampling Operation

Probably the key advantage of CT modulators over their DT competitors is that the sampling operation takes place inside the ΣΔ loop, in contrast to DT modulators, where a S/H circuit is placed at the input of the converter (Figs. 2.9 and 3.1). Thus, every error of this critical block adds to the input signal in the DT case; in the CT modulator all nonidealities of the sampling process are subject to noise-shaping. In particular, if a large conversion bandwidth is aimed, high-performance DT ΣΔ modulators are more difficult to design because of the stringent requirements on the fast, high-precision sample and hold building block [114–116].

Beyond this favorable feature of noise-shaped S/H errors, the shift of the sampling operation behind a continuous-time filter in the forward, signal path results in some degree of *implicit antialiasing filtering* [9, 10, 105]. This attribute can heavily reduce the required specifications of a front-end AAF, and even more, makes it sometimes unnecessary. Especially in high-speed circuits or architectures with very low oversampling ratios, this can be the most emphasizing argument for choosing a continuous-time ΣΔ implementation. This topic will be discussed more in detail in Sect. 3.5.

Another performance limitation arises from the fact that the noise generated by the *nonzero on-resistance* of the switches and that of the amplifier is sampled together with the input signal [5]. Considering that the overall accuracy of a switched-capacitor ΣΔ modulator is given by the settling accuracy of the charge-transfer characteristic [60, 117], which recommends a small time constant compared to the clock period, a noise cut-off frequency several times larger than the sampling frequency is the result. Accordingly, a large amount of thermal noise is fold back into the signal band, constituting a fundamental resolution limit in SC ΣΔ modulators [5, 63]. Further performance limitations are introduced by the finite and signal-swing dependant on-resistance of the switches, glitch-induced errors and the generated digital switching noise [103, 118].

Timing errors of the sampling clock have a significant impact on the DT modulator sampling operation: this is obvious from the fact that if the sampling clock is jittered, the front-end S/H circuit will sample the input at the wrong time instant, producing an equivalent amplitude error that degrades the SNR of the converter [11]. Thus, the front-end S/H sets an upper limit on the performance of the entire DT modulator. In opposite, the filters in SC circuits are usually designed as to safely settle within half a sampling period, and thus clock jitter is of minor importance here; this alleviates if high speed of low power SC circuits are considered [119].

Consequently, a common benefit that comes along with the sampled nature of the DT counterpart is its relaxed sensitivity to timing variations or delays within the feedback path, both of which are a strong performance limitation in CT modulators. This is further discussed in Sect. 3.1.4 as well as in Chap. 4.

3.1.2 Filter Realization

The loop filters in DT and CT modulators are integrators or resonators, depending on the filter-transfer function, which has to be realized. The former are commonly implemented in switched-capacitor circuits, while the latter are continuous-time integrators.

The signal representation in DT circuits are quickly changing pulses; thus, ΣΔ modulators employing the SC technique have a maximum clock rate, which is limited by the OTA bandwidth and the fact that the entire charge transfer needs several time constants to settle to the required accuracy [63, 120]. In contrast, in CT modulators, all signals are represented by analog, continuous-time waveforms; therefore the OpAmp speed restrictions are drastically relaxed, and in theory CT ΣΔ modulators could be clocked an order of magnitude faster in the same technology as their DT counterparts [10]. Actually, in most implementations the factor tends to lie between three and five [9, 38, 120–122].

Another interesting aspect is the linearity behavior concerning the virtual ground node: while in DT modulators large glitches appear at the virtual ground node due to the fast SC pulses, this node can be usually kept quiet in CT modulators due to the continuously changing signals, which has been used for a high resolution modulator in [123]. In contrast, in [8] it has been stated that the virtual ground node is not as important in DT modulators, because it only matters that the finally settled value of the signals shows a sufficient accuracy, while it does not matter how this value is achieved. In opposite, in CT modulators the virtual ground node has to be kept quiet all the time [124], because the integration of a continuously changing waveform requires permanent linearity. Therefore it can be said that this feature is still a matter of an actual implementation, rather than being a real advantage for one or another architecture.

Furthermore, the filter implementation with discrete or continuous-time integrators heavily affects the actual integrated filter transfer function. While in SC circuits, the DT integrator gain is determined by a capacitor ratio, in a CT architecture, this gain depends on an *RC* or *gmC* product. The former is defined very precisely, since absolute mismatch of both capacitors does not affect the integrator gain. Accordingly, only the relative mismatch remains, which can be better than 0.1% [58], thus the modulator transfer function is accurately determined. In contrast, the CT integrator gain is subject to large process dependent variations, which usually lead to mismatched integrated functionality compared to the designed [121, 125] or in worst case to an unstable system [126].

Another issue is the voltage dependency (nonlinearity) of the used passive and active components. This imperfection leads to harmonic distortion at the modulator output. In general, the performance of CT ΣΔ modulators is limited by the linearity of the integrated resistors or transconductance stages of the respective integrators [9, 127]. In particular, the first transconductor must have the overall accuracy of the system [35, 128]. Unlike, the linearity of an SC integrator is usually one order of magnitude better than that of the resistors. These characteristics will be addressed in Chap. 5.

3.1.3 Quantizer Realization

Common to both implementations of ΣΔ modulators is that all nonidealities of the quantization process are subject to noise shaping, since the quantizer resides within the modulator's feedback loop after the respective filter. The offset voltage of the internal comparator or the multibit quantizer is attenuated by the integrator dc gains. In the same way, the ADC's integral and differential nonlinearity errors are suppressed.

In contrast, the decision time of the internal quantizer and its signal dependent variation (metastability [3]) have a significantly different influence: while the sampled DT systems grants the quantization process half of sampling period for the decision, a CT ΣΔ modulator ideally needs infinitely fast quantization, since the result is needed immediately to generate the continuous-time feedback signal. Therefrom, severe performance limitation can arise. This and possible countermeasures are discussed in a glance in Sect. 4.2 and Chap. 6.

Recently, techniques have been published to implement the internal quantizer in a CT ΣΔ modulator by a low-resolution ADC different from the usually adopted flash! The motivation for this is the exponentially increasing number of comparators for increasing number of bits in the internal quantizer, and thus the exponentially increasing power and area consumption of this circuit. Additionally, the capacitive load seen by the last integrator in the loop filter exponentially increases, which again imposes tougher requirements on its drive capability and therefore its power consumption [126]. Therefore, [129] proposed the usage of a low resolution successive approximation ADC as internal quantizer, which is advantageous concerning the number of comparators, decreasing to only one in this case. The requirements on the DAC of the successive approximation quantizer are low, since it only adds to the nonidealities at the ΣΔ modulator quantizer, which are highly suppressed through the entire loop filter gain. Nonetheless, this technique has the obvious disadvantage that it requires an even higher sampling frequency within the successive approximation quantizer as is used in the overall ΣΔ modulator. This increases on the one hand again the power consumption of the single quantizer, and even more, it prevents the adoption of this technique for high-speed modulators, where the sampling frequency f_S itself is already high!

Consequently, [22] proposed to use a tracking ADC as quantizer in the ΣΔ modulator loop. In a multibit ΣΔ modulator the feedback signal is usually changing only by one LSB at every sampling instant. Thus, not the full quantization range of the internal quantizer is needed, but only the decision at a certain interval within the full-scale range. The same technique is adopted in a tracking ADC. Since in irregular instances also two LSB steps may occur, not only one but three comparators have been used within the tracking ADC in order to improve the stability of the implemented multibit ΣΔ modulator [22]. Nonetheless, it still has to be assured that the input signal to the tracking ADC within the modulator loop is not sloping to fast, since otherwise the ADC looses the input signal and again the modulator becomes unstable. Consequently, both techniques impose their own advantages and drawbacks; in the following chapters, when considering multibit internal quantizers always the standard flash quantizer is taken into account!

3.1.4 Feedback Realization

The signals in DT modulators are represented by SC-pulses which are based on charge transfer operation. Thus, in the DT system the feedback signal is applied by charging a capacitor to a reference voltage and discharging it onto the integrating capacitance. In contrast, the analog continuous-time feedback waveform is integrated over time and thus the ΣΔ modulator is sensitive to every deviation from the exact, ideal waveform of that feedback signal. From this, some severe nonidealities result, as will be shown in Chap. 4.

A survey of commonly used and further possible feedback DAC waveforms is given in Figs. 3.2a–3.3d. The nomenclature used for the feedback waveforms is explained in Fig. 3.4, where a typical rectangular feedback pulse form is

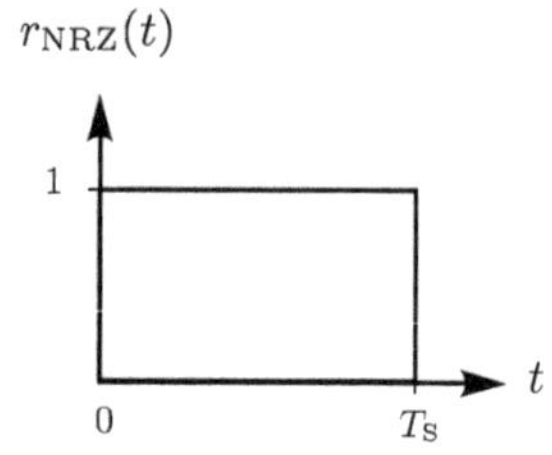

(**a**)

$$
\begin{aligned}
r_{\mathrm{NRZ}}(t) &= \begin{cases} 1 \,, & 0 \le t < T_{\mathrm{S}} \\ 0 \,, & \text{otherwise} \end{cases} \\
R_{\mathrm{NRZ}}(s) &= \frac{1 - \mathrm{e}^{-sT_{\mathrm{S}}}}{s}
\end{aligned}
\tag{3.1}
$$

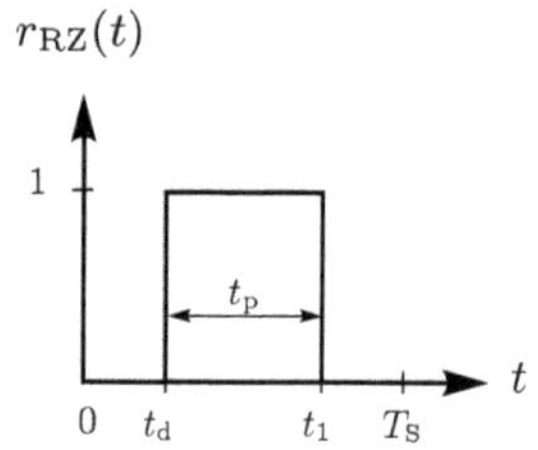

(**b**)

$$
\begin{aligned}
r_{\mathrm{RZ}}(t) &= \begin{cases} 1 \,, & t_{\mathrm{d}} \le t < t_1 \\ 0 \,, & \text{otherwise} \end{cases} \\
R_{\mathrm{RZ}}(s) &= \frac{\mathrm{e}^{-st_{\mathrm{d}}}(1 - \mathrm{e}^{-st_{\mathrm{p}}})}{s}
\end{aligned}
\tag{3.2}
$$

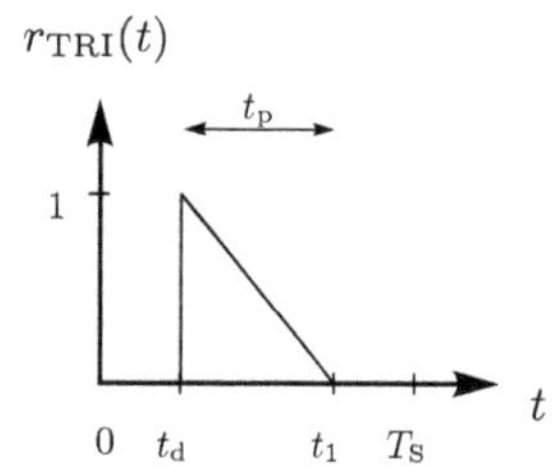

(**c**)

$$
\begin{aligned}
r_{\mathrm{TRI}}(t) &= \begin{cases} 1 - \frac{(t-t_{\mathrm{d}})}{t_{\mathrm{p}}} \,, & t_{\mathrm{d}} \le t < t_1 \\ 0 \,, & \text{otherwise} \end{cases} \\
R_{\mathrm{TRI}}(s) &= \frac{\mathrm{e}^{-st_{\mathrm{d}}}}{s} \left(1 - \frac{(1 - \mathrm{e}^{-st_{\mathrm{p}}})}{st_{\mathrm{p}}} \right)
\end{aligned}
\tag{3.3}
$$

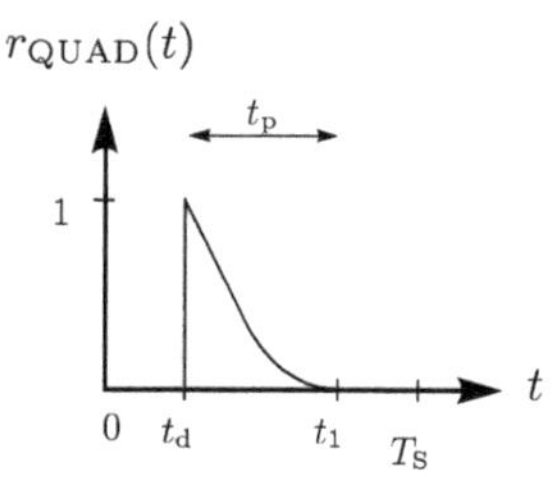

(**d**)

$$
\begin{aligned}
r_{\mathrm{QUAD}}(t) &= \begin{cases} \left(1 - \frac{(t-t_{\mathrm{d}})}{t_{\mathrm{p}}}\right)^2 \,, & t_{\mathrm{d}} \le t < t_1 \\ 0 \,, & \text{otherwise} \end{cases} \\
R_{\mathrm{QUAD}}(s) &= \frac{\mathrm{e}^{-st_{\mathrm{d}}}}{s} \left(1 - \frac{2}{st_{\mathrm{p}}} + \frac{2(1 - \mathrm{e}^{-st_{\mathrm{p}}})}{s^2 t_{\mathrm{p}}^2} \right)
\end{aligned}
\tag{3.4}
$$

Fig. 3.2. Common DAC feedback impulse responses $r_{\mathrm{DAC}}(t)$ and the corresponding Laplace transform $R_{\mathrm{DAC}}(s)$. The normalized time instants α and β yield $t_{\mathrm{d}}/T_{\mathrm{S}}$ and $(t_{\mathrm{d}} + t_{\mathrm{p}})/T_{\mathrm{S}}$, respectively. (**a**) NRZ-DAC; (**b**) RZ-DAC; (**c**) linear-decaying DAC; (**d**) quadratic-decaying DAC

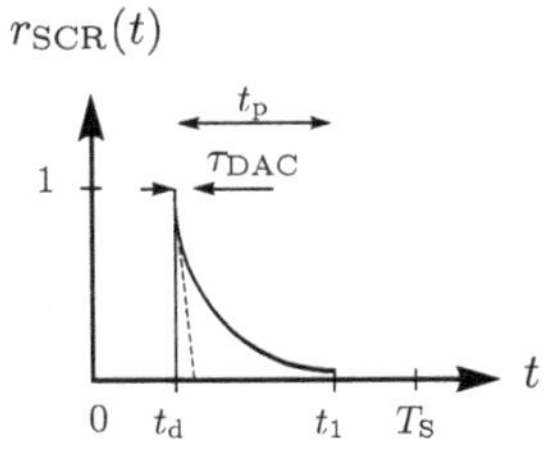

(**a**)

$$r_{\mathrm{SCR}}(t) = \begin{cases} \mathrm{e}^{-(t-t_\mathrm{d})/\tau_{\mathrm{DAC}}} & , \ t_\mathrm{d} \le t < t_1 \\ 0 & , \ \text{otherwise} \end{cases}$$
$$R_{\mathrm{SCR}}(s) = \frac{\mathrm{e}^{-st_\mathrm{d}}\left(1-\mathrm{e}^{-t_\mathrm{p}(s+1/\tau_{\mathrm{DAC}})}\right)}{s+\frac{1}{\tau_{\mathrm{DAC}}}} \tag{3.5}$$

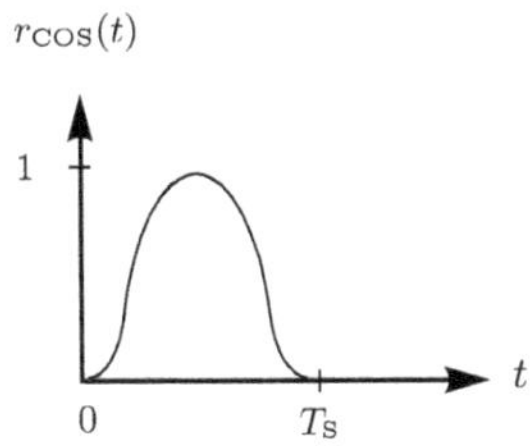

(**b**)

$$r_{\mathrm{COS}}(t) = \begin{cases} \left(1-\frac{(t-t_\mathrm{d})}{t_\mathrm{p}}\right)^2 & , \ 0 \le t < t_1 \\ 0 & , \ \text{otherwise} \end{cases}$$
$$R_{\mathrm{COS}}(s) = \frac{\mathrm{e}^{-st_\mathrm{d}}}{s}\left(1-\frac{2}{st_\mathrm{p}}+\frac{2(1-\mathrm{e}^{-st_\mathrm{p}})}{s^2t_\mathrm{p}^2}\right) \tag{3.6}$$

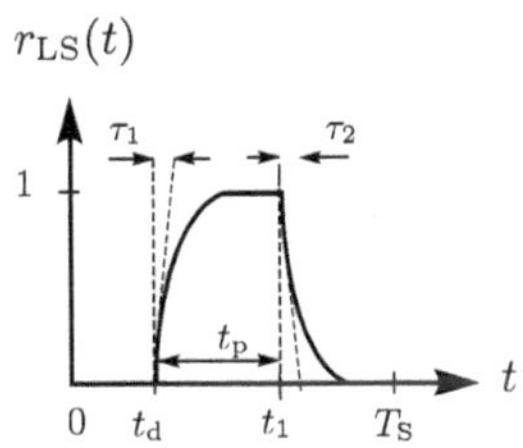

(**c**)

$$r_{\mathrm{LS}}(t) = \begin{cases} 0 & , \ 0 \le t < t_\mathrm{d} \\ 1-\mathrm{e}^{-(t-t_\mathrm{d})/\tau_1} & , \ t_\mathrm{d} \le t < t_1 \\ \mathrm{e}^{-(t-t_\mathrm{d}-t_\mathrm{p})/\tau_2} & , \ t_1 \le t < T_\mathrm{S} \end{cases}$$
$$R_{\mathrm{LS}}(s) = \frac{\mathrm{e}^{-st_\mathrm{d}}(1-\mathrm{e}^{-st_\mathrm{p}})}{s(s\tau_1+1)(s\tau_2+1)} + \frac{\mathrm{e}^{-st_\mathrm{d}}(\tau_2-\tau_1\mathrm{e}^{-st_\mathrm{p}})}{(s\tau_1+1)(s\tau_2+1)} \tag{3.7}$$

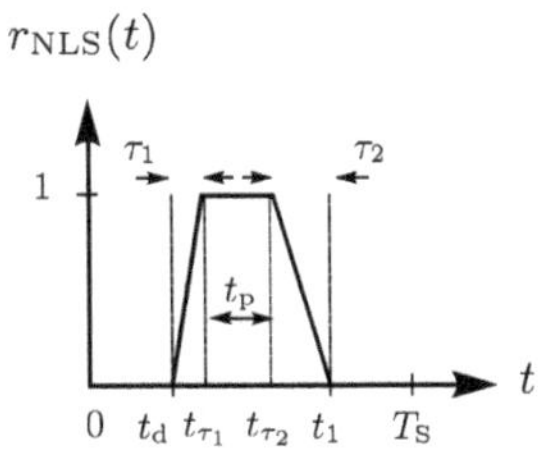

(**d**)

$$r_{\mathrm{NLS}}(t) = \begin{cases} \frac{t-t_\mathrm{d}}{\tau_1} & , \ t_\mathrm{d} \le t < t_{\tau_1} \\ 1 & , \ t_{\tau_1} \le t < t_{\tau_2} \\ \frac{\tau_2-(t-t_{\tau_2})}{\tau_2} & , \ t_{\tau_2} \le t < t_1 \\ 0 & , \ \text{otherwise} \end{cases}$$
$$R_{\mathrm{NLS}}(s) = \frac{\mathrm{e}^{-st_\mathrm{d}}\tau_2(1-\mathrm{e}^{-s\tau_1})}{\tau_1\tau_2 s^2} - \frac{\mathrm{e}^{-st_\mathrm{d}}\tau_1\mathrm{e}^{-s(t_\mathrm{p}+\tau_1)}(1-\mathrm{e}^{-s\tau_2})}{\tau_1\tau_2 s^2} \tag{3.8}$$

Fig. 3.3. Common DAC feedback impulse responses $r_{\mathrm{DAC}}(t)$ and the corresponding Laplace transform $R_{\mathrm{DAC}}(s)$. The normalized time instants α and β yield $t_\mathrm{d}/T_\mathrm{S}$ and $(t_\mathrm{d}+t_\mathrm{p})/T_\mathrm{S}$, respectively. (**a**) SCR-DAC; (**b**) cosine-DAC; (**c**) linear-slope DAC; (**d**) nonlinear-slope DAC

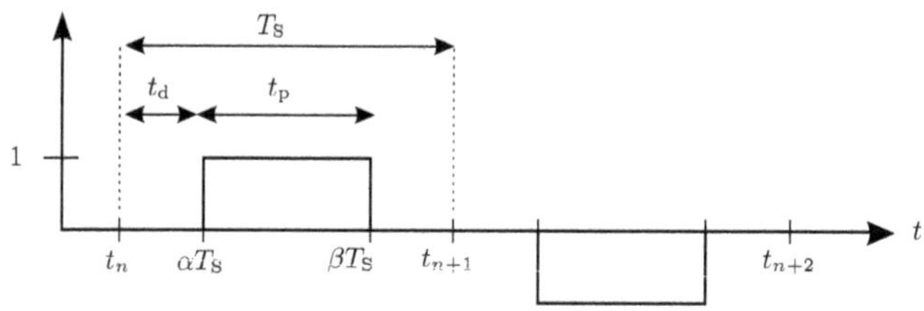

Fig. 3.4. Illustrative rectangular feedback pulse form after [31]

given in detail. Here, ideally every sample should provide a feedback signal exactly within $t_d \leq t \leq t_1$, or relative to T_S within $\{\alpha T_S, \beta T_S\}$, where

$$\alpha = t_d/T_S\,, \qquad (3.9)$$
$$\beta = t_1/T_S\,.$$

Additionally, the pulse or clock interval limits t_n should be ideally multiples of the sampling time T_S of the modulator, i.e., $t_n = nT_S$, or on the other hand the length of the nth period T_n yields $T_n = t_{n+1} - t_n = T_S$.

Over many years, continuous-time ΣΔ modulators only incorporated rectangular feedback pulses in a nonreturn-to-zero (NRZ, Fig. 3.2a) or return-to-zero (RZ, Fig. 3.2b) implementation [3, 54, 112, 113, 123]. These are particularly advantageous, because their waveforms can be generated by simply switching the feedback current or voltage sources on and off by using the system clock and its edges [36]. On the other hand, from the rectangular feedback realization one major nonideality arises, i.e., the purity of the sampling clock: any timing jitter in the sampling clock modulates directly the ADC decision point as well as the rising and falling edges of the DAC output and thus the quantity of the DAC feedback signals. While every error introduced by the ADC is noise shaped, the resulting DAC error is directly introduced into the modulator input, which is why clock jitter can limit the overall ΣΔ modulator performance in terms of noise. Therefore, in recent years more and more the usage of shaped feedback pulses was discussed, some of them shown in Figs. 3.2 and 3.3 [19, 31, 33, 130]. The influence of clock jitter on CT ΣΔ modulators as well as possible countermeasures are discussed in Sect. 4.3.

Additionally in practice, the quantizer latch and the feedback DAC cannot switch simultaneously [3]. Thus a delay between the quantizer clock and DAC current pulse exists. The delay causes a shift of the DAC feedback pulse which affects the entire CT loop filter characteristic. This effect is mathematically analyzed in Sect. 4.2, where also circuit techniques to mitigate the impact of loop delay are discussed.

In practical realizations of CT ΣΔ modulators, the implemented feedback DAC also exhibits slew rate limitations as well as different rise and fall times. Both nonidealities tend to degrade the performance, unless these errors or their effects on the overall modulator are made small enough. Especially in CT ΣΔ modulators employing a NRZ-DAC, the impact of both errors causes

Table 3.1. Benefits that are obtained by using a CT and a DT loop filter

Pro-CT	Pro-DT
Implicit antialiasing filter	Low sensitivity to clock jitter
Attenuated (noise shaped) S/H errors	Low sensitivity to excess loop delay
Higher sampling frequency possible	Low sensitivity to DAC waveform
Relaxed OpAmp speed requirements	Accurately defined integrator gains and transfer functions
Reduced supply and ground noise impact	Highly linear SC integrator
Less glitch sensitive	Lower simulation time (high level)
Less digital switching noise	Capacitive loads only
Breadboard prototyping possible	Compatible with VLSI CMOS processes
Lower simulation time (circuit level)	

intersymbol interference (ISI) due to the fact that the resulting errors depend on the output sequence of the modulator, which causes additional noise and tones within the spectrum that fold into the baseband. Analysis and improvements regarding the maximal tolerable DAC asymmetry maintaining the overall modulator performance are given in Sect. 4.4.

As known from DT ΣΔ modulator implementations, the usage of multibit internal quantization usually requires linearization techniques in the feedback DAC. While a feedback DAC with only two conversion levels is inherently linear, any error in the level spacing (nonlinearity) of a multibit quantizer is directly input-referred. Techniques to improve the matching dynamically or by calibration are shortly mentioned in Sect. 4.5. But in contrast to DT modulators, in CT ΣΔ implementations, the delays encountered with the linearization techniques are of major importance, because also here loop-delay can arise [21].

3.1.5 DT/CT Modulators Trade-offs

As a summary, Table 3.1 gives the main advantageous points that distinguish DT and CT ΣΔ modulators.

3.2 DT-to-CT Conversion of ΣΔ Modulators

In the past, most work on ΣΔ modulators is focused on DT implementations. As a result, a great deal of software, tools, and innovative architectures have been presented in a vast of publications over the last two decades. Thus, a straight forward design procedure of a CT ΣΔ modulator loop filter might start with a DT loop filter $H(z)$. This way, design and simulation of the ideal CT ΣΔ modulator can be done in discrete-time domain to speed-up the overall design procedure [4, 6, 8, 105, 131, 132]. Consequently, it is highly recommended to start the CT modulator design with the synthesis of a DT modulator showing the required performance and to proceed with a DT-to-CT conversion in order to obtain the equivalent CT modulator [25].

The most common methods for this transformation are: the modified $\mathcal{Z}$-transformation [105, 133–135] or the impulse-invariant transformation [3, 136]. Beside, also the state-space method has been used for designing a CT ΣΔ modulator [4, 10], but this technique is rather uncommon in most recent publications. Finally, there exist even some build in functions in MATLAB to perform transformations between the DT and CT domain, but these are afflicted with some nontolerable restrictions for a general use [3, 4, 81].

3.2.1 The Impulse-Invariant Transformation

Principally the DT-to-CT approach is based on the clocked internal quantizer of the CT modulator, which makes it a kind of DT system at this point. Following [105, 132] a DT-to-CT equivalence is achieved in Fig. 3.5, if the input to both quantizers $x(t)$ and $x(n)$ are the same at the sampling instants.

$$q(n) \stackrel{!}{=} q(t)|_{t=nT_S}\,. \tag{3.10}$$

Thus, the output bitstreams of both modulators and therefore the noise-performance are identical. In Fig. 3.5, it is important to realize that the digital-to-analog converter in the feedback of the CT modulator acts as a discrete-to-continuous converter: while its input is a DT sample $y(n)$, its output is a continuous waveform $y(t)$, whose shape depends on the DAC transfer function

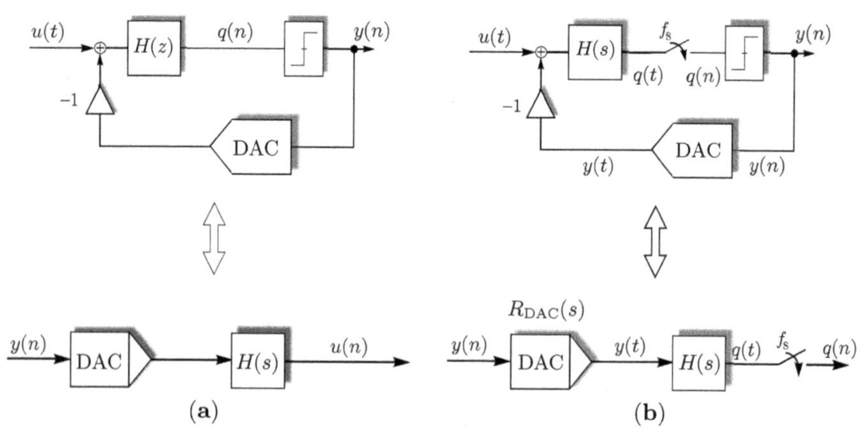

Fig. 3.5. Loop filter representation for DT and CT modulators, after [3, 31]. (**a**) DT modulator; (**b**) CT modulator

$R_{\mathrm{DAC}}(s)$. The above condition for the loop filter equivalence translates directly into:

$$\mathcal{Z}^{-1}\{H(z)\} = \mathcal{L}^{-1}\{R_{\mathrm{DAC}}(s)H(s)\}|_{t=nT_{\mathrm{S}}} \,. \tag{3.11}$$

In the time domain this leads to the condition [105]:

$$h(n) = [r_{\mathrm{DAC}}(t) * h(t)]|_{t=nT_{\mathrm{S}}} = \int_{-\infty}^{+\infty} r_{\mathrm{DAC}}(\tau)h(t-\tau)\,\mathrm{d}\tau|_{t=nT_{\mathrm{S}}} \tag{3.12}$$

with $r_{\mathrm{DAC}}(t)$ as the impulse response of the specific DAC (concerning different impulse responses see Figs. 3.2 and 3.3). This transformation between DT and CT domain is called the *impulse-invariant transformation* [137], because it makes the open-loop impulse responses equal at the sampling times. Alternatively to (3.11, 3.12), also the expressions derived in e.g., [14] are commonly used.

This transformation now enables the design of a CT loop filter $H(s)$, which together with a specific DAC transfer function $R_{\mathrm{DAC}}(s)$ exactly matches the noise-shaping behavior of a DT loop filter $H(z)$. For a given architecture, the actual transformation can be done using (3.12) for a specific loop filter arrangement [8]. In order to provide a more general approach, in [3, 132] the transformation has been performed for basic loop filter-poles $1/(z-z_k)^i$, $i = 1.3$ for a rectangular DAC transfer function as presented in Fig. 3.4. Therewith it is possible, to transform every DT loop filter to a CT equivalent, by simply dividing it into its partial fractions and adopting the basic equivalents, while incorporating a NRZ or RZ feedback DAC.

Due to the intention of also transforming cascaded modulators of orders higher than 3, Table 3.2 reviews the equivalents presented in [3] and extends the list to poles up to the order of 4. The calculations were done using the symbolic math program MAPLE [138].

DT-to-CT Conversion of a Second-Order Modulator Using the Impulse-Invariant Transformation

To show the transformation procedure, which will be used extensively through this book, as an example the second-order modulator of Fig. 3.6a incorporating DT integrators $I(z) = z^{-1}/(1-z^{-1})$ with scaling coefficients as proposed in [6] is converted to the CT domain. According to the open loop in Fig. 3.5, the complete loop filter of Fig. 3.6a results in:

$$\mathrm{LF}(z) = -a_2 I(z) - a_1 a_2 I^2(z) = -\frac{a_2}{z-1} - \frac{a_1 a_2}{(z-1)^2} \,. \tag{3.13}$$

Table 3.2. CT equivalents for rectangular feedback DACs for the first to fourth-order DT low-pass loop filter poles, after [3, 32]

$\mathcal{Z}$-domain	$\mathcal{S}$-domain equivalents with f_S (Hz) $= 1/T_S$
$\frac{1}{(z-1)}$	$\frac{w_0}{s}, \quad w_0 = \frac{f_S}{\beta-\alpha}$
$\frac{1}{(z-1)^2}$	$\frac{w_1 s + w_0}{s^2}, \quad w_0 = \frac{f_S^2}{\beta-\alpha}, \quad w_1 = \frac{1}{2}\frac{f_S(\alpha+\beta-2)}{\beta-\alpha}$
$\frac{1}{(z-1)^3}$	$\frac{w_2 s^2 + w_1 s + w_0}{s^3}, \quad w_0 = \frac{f_S^3}{\beta-\alpha}, \quad w_1 = \frac{1}{2}\frac{f_S^2(\alpha+\beta-3)}{\beta-\alpha},$ $w_2 = \frac{1}{12}\frac{f_S[\beta(\beta-9)+\alpha(\alpha-9)+4\alpha\beta+12]}{\beta-\alpha}$
$\frac{1}{(z-1)^4}$	$\frac{w_3 s^3 + w_2 s^2 + w_1 s + w_0}{s^4}, \quad w_0 = \frac{f_S^4}{\beta-\alpha}, \quad w_1 = \frac{f_S^3}{2}\frac{\beta+\alpha-4}{\beta-\alpha},$ $w_2 = \frac{f_S^2}{12}\frac{(\beta-\alpha)^2+2\beta\alpha-12(\beta+\alpha)+22}{\beta-\alpha},$ $w_3 = \frac{f_S}{12}\frac{\beta^2(\alpha-2)+\alpha^2(\beta-2)-8\alpha\beta+11(\beta+\alpha)-12}{\beta-\alpha}$

Adopting rows 1 and 2 of Table 3.2 yields:

$$\mathrm{LF}(s)|_{\mathrm{DT-CT}} = -a_2\frac{\frac{f_S}{\beta-\alpha}}{s} - a_1 a_2\frac{\frac{1}{2}\frac{f_S(\alpha+\beta-2)}{\beta-\alpha}s + \frac{f_S^2}{\beta-\alpha}}{s^2}\,. \tag{3.14}$$

Using an NRZ feedback DAC pulse, i.e., $\{\alpha = 0,\ \beta = 1\}$, one obtains:

$$\mathrm{LF}(s)|_{\mathrm{DT-CT}} = -\left(a_2 - \frac{a_1 a_2}{2}\right)\frac{f_S}{s} - a_1 a_2\frac{f_S^2}{s^2}\,. \tag{3.15}$$

Equation (3.15) represents the DT-to-CT converted loop filter transfer function. Consequently, this loop filter now has to be implemented in a CT architecture. One possibility is again the common feedback architecture including two CT integrators similar to the DT modulator presented in Fig. 3.6a. A corresponding block diagram is given in Fig. 3.6b. Here, the *CT integrator transfer functions* are $I(s) = \omega_I/s$, where ω_I is the respective unscaled integrator corner frequency, and k_i are the continuous-time modulator scaling coefficients. Without loss of generality, the CT scaling coefficients in Fig. 3.6a are moved into the feedback path [34, 36], which can be used in order to account for some nonideal behavior, as will be further discussed in Chap. 5. The discrimination of a signal scaling coefficient k_{sig} and a first feedback scaling

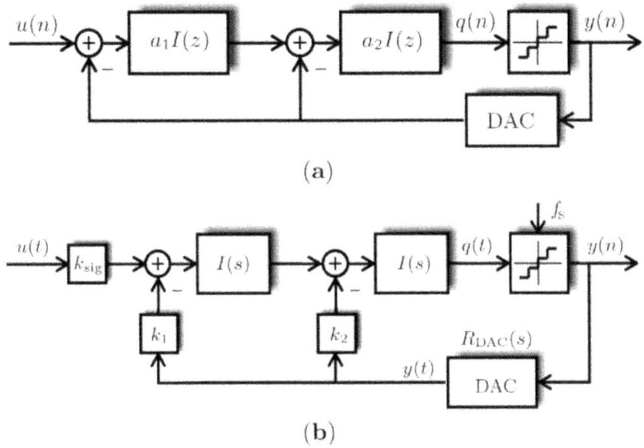

Fig. 3.6. Block diagram of (**a**) a second-order, single-stage DT ΣΔ modulator; (**b**) a second-order, single-stage CT ΣΔ modulator with distributed feedback

coefficient k_1 accounts for changing feedback coefficients, when adopting a DAC other than NRZ: the input scaling coefficient remains constant and independent from the feedback waveform:

$$k_{\text{sig}} = k_{1|\text{NRZ}} . \tag{3.16}$$

The loop filter of the second-order CT modulator in Fig. 3.6b is derived to:

$$\text{LF}(s) = -k_2 I(s) - k_1 I^2(s) = -k_2 \frac{\omega_\text{I}}{s} - k_1 \frac{\omega_\text{I}^2}{s^2} , \tag{3.17}$$

where ω_I are the unscaled integrator corner frequencies. The integrator transfer function is set to:

$$I(s) = \frac{\omega_\text{I}}{s} \overset{!}{=} \frac{f_\text{S}}{s} . \tag{3.18}$$

Concerning the choice of this nomenclature for the integrator transfer function and the dependencies of f_S and ω_I, please see also the illustrative explanation in Appendix C. With (3.18), the CT loop filter of (3.17) becomes:

$$H(s) = -k_2 \frac{f_\text{S}}{s} - k_1 \frac{f_\text{S}^2}{s^2} . \tag{3.19}$$

By comparison of this loop filter with the DT-to-CT converted one in (3.15), the CT scaling coefficients $\{k_1,\ k_2\}$ can be calculated dependent on the DT scalings $\{a_1,\ a_2\}$, resulting in a scaled second-order CT modulator, which behaves exactly as its DT counterpart:

$$k_{1|_{\text{Mod2-NRZ}}} = a_1 a_2, \quad k_{2|_{\text{Mod2-NRZ}}} = a_2 - \frac{a_1 a_2}{2}. \tag{3.20}$$

For completeness and due to a regular use of it throughout this book, (3.21, 3.22) give the CT modulator scaling coefficients for a general rectangular feedback DAC for a second-order modulator as presented above in Fig. 3.6b and additionally for an equivalently implemented third-order modulator.

$$k_{1|_{\text{Mod2}}} = \frac{a_1 a_2}{\beta - \alpha}, \quad k_{2|_{\text{Mod2}}} = \frac{a_1 a_2}{2} \frac{\frac{2}{a_1} + \alpha + \beta - 2}{\beta - \alpha}, \tag{3.21}$$

$$k_{1|_{\text{Mod3}}} = \frac{a_1 a_2 a_3}{\beta - \alpha}, \quad k_{2|_{\text{Mod3}}} = \frac{a_1 a_2 a_3}{2} \frac{\frac{2}{a_1} + \alpha + \beta - 3}{\beta - \alpha}, \tag{3.22}$$

$$k_{3|_{\text{Mod3}}} = \frac{a_1 a_2 a_3}{12} \frac{\alpha(\alpha - 9) + \beta(\beta - 9) + 4\alpha\beta + \frac{6}{a_1}(\beta + \alpha) + \frac{12}{a_1}(\frac{1}{a_2} + a_1 - 1)}{\beta - \alpha}.$$

The signal scaling coefficients k_{sig} of either modulator are calculated by the feedback coefficient of the outermost loop with NRZ implementation. The above presented method is one possibility for the synthesis of continuous-time ΣΔ modulators; while due to presentability reasons the equivalents given in Table 3.2 are only useful for low-pass modulators, i.e., the basic filter poles are at dc ($z = 1,\ s = 0$), the original table in [132] is given for general poles (for modulator orders of 1–3) and can therefore be used also for bandpass modulators [3, 105]. Finally, the use of these tables of equivalent poles is easy and could also be automated, and furthermore extended to different waveforms: Appendix B exemplarily shows the method of finding the equivalent poles for the exponential feedback waveform illustrated in Fig. 3.3a as shown in [31].

3.2.2 Modified $\mathcal{Z}$-Transform

Another method for establishing the CT loop filter is the modified $\mathcal{Z}$-transform [35, 105, 139, 140]. This approach is based on the general $\mathcal{Z}$-transform [141, p. 649] but extended, such that the discrete system behavior can be calculated at all instants of time, which is particularly important for mixed-signal, sampled-data systems with delay or multirate sampled systems.

In order to determine the equivalent CT loop filter for a certain modulator architecture with a certain feedback DAC pulse shape, the DT loop transfer function is computed and compared with the original DT loop filter function, in the same way as shown during the impulse invariant transform. $H(s)$ will be multiplied with the desired DAC impulse response $R_{\text{DAC}}(s)$ and according to (3.23) the modified $\mathcal{Z}$-transform is adopted on it.

$$H(z) = \sum_i \mathcal{Z}_{m_i}\{H(s) R_{\text{DAC}}(s)\}. \tag{3.23}$$

The variable delay factor m_i is the key parameter of the modified $\mathcal{Z}$-transform. The value m_i is normalized to the sampling period and bounded between $0 < m_i < 1$, whereas the extremes 0 confirm to the previous sample instant and 1 to the next. In general, for every time instance, in which the CT loop filter function changes its behavior, an additional delay factor is introduced. For example that means for an ideal NRZ-DAC pulse:

- The first time instant is the rising edge of the DAC pulse at $t = 0$, which results in $m_1 = 1 - 0/T_S = 1$.
- The second time instant is the falling edge at $t = T_S$. Thus m_2 yields $m_2 = 1 - T_S/T_S = 0$.

Next, each loop filter term is transferred with respect to all time instances m_i according to Table 3.3. An extension of Table 3.3 up to the seventh order is presented in [38]. Finally, a coefficient comparison with the original DT loop filter function leads to the wanted CT coefficients k_i. This approach will be exemplary shown on a second-order modulator with SCR-DAC feedback in the following.

DT-to-CT Conversion of a Second-Order Modulator Using the Modified $\mathcal{Z}$-Transformation

Figure 3.3a illustrates the DAC impulse response of the SCR feedback scheme, which is given by

$$R_{\mathrm{DAC}}(s) = \frac{\mathrm{e}^{-st_{\mathrm{d}}}\left(1 - \mathrm{e}^{-t_{\mathrm{p}}(s+1/\tau_{\mathrm{DAC}})}\right)}{s + \frac{1}{\tau_{\mathrm{DAC}}}} \tag{3.24}$$

$$= \frac{\mathrm{e}^{-st_{\mathrm{d}}}}{s + \frac{1}{\tau_{\mathrm{DAC}}}} - \frac{\mathrm{e}^{-s(t_{\mathrm{d}}+t_{\mathrm{p}})}}{(s + \frac{1}{\tau_{\mathrm{DAC}}})\mathrm{e}^{t_{\mathrm{p}}/\tau_{\mathrm{DAC}}}} . \tag{3.25}$$

Table 3.3. Modified $\mathcal{Z}$-transform for the corresponding loop filter order [139]

$\mathcal{S}$-domain	$\mathcal{Z}_m$-domain equivalents
$\left(\frac{1}{s^2}\right)$	$\frac{mT_S}{z-1} + \frac{T_S}{(z-1)^2}$
$\left(\frac{1}{s(s+s_k)}\right)$	$\frac{1}{s_k}\left(\frac{1}{z-1} - \frac{\mathrm{e}^{-s_k mT_S}}{z-\mathrm{e}^{-s_k T_S}}\right)$
$\left(\frac{1}{s^3}\right)$	$\frac{T_S^2}{2}\left(\frac{m^2}{z-1} + \frac{2m+1}{(z-1)^2} + \frac{2}{(z-1)^3}\right)$
$\left(\frac{1}{s^2(s+s_k)}\right)$	$\frac{1}{s_k^2}\left(\frac{s_k mT_S - 1}{z-1} + \frac{s_k T_S}{(z-1)^2} + \frac{\mathrm{e}^{-s_k mT_S}}{(z-\mathrm{e}^{-s_k T_S})}\right)$
$\left(\frac{1}{s^4}\right)$	$\frac{T_S^3}{6}\left(\frac{m^3}{z-1} + \frac{3m^2+3m+1}{(z-1)^2} + \frac{6m+6}{(z-1)^3} + \frac{6}{(z-1)^4}\right)$
$\left(\frac{1}{s^3(s+s_k)}\right)$	$\frac{1}{s_k^3}\left(\frac{s_k^2 m^2 T_S^2/2 - s_k mT_S + 1}{z-1} + \frac{s_k^2 T_S^2(m+1/2) - s_k T_S}{(z-1)^2} + \frac{s_k^2 T_S^2}{(z-1)^3} - \frac{\mathrm{e}^{-s_k mT_S}}{(z-\mathrm{e}^{-s_k T_S})}\right)$

The delay factors for the SCR-DAC feedback are $m_1 = 1 - t_\mathrm{d}/T_\mathrm{S}$ and $m_2 = 1 - t_\mathrm{d}/T_\mathrm{S} - t_\mathrm{p}/T_\mathrm{S}$. Note that $\mathrm{e}^{t_\mathrm{p}/\tau_\mathrm{DAC}}$ is a constant factor, since it is independent of s. The loop filter in the special case of a second-order modulator results in:

$$\mathrm{LF}(s) = \frac{-k_2}{T_\mathrm{S}s} + \frac{-k_1}{T_\mathrm{S}^2 s^2}\,. \tag{3.26}$$

Thus, using (3.25) and (3.26), the expression in (3.23) yields

$$\mathrm{LF}(z)|_\mathrm{CT-DT} = \mathcal{Z}_{m_1}\left\{\frac{\frac{-k_1}{T_\mathrm{S}s} + \frac{-k_2}{T_\mathrm{S}^2 s^2}}{s + \frac{1}{\tau_\mathrm{DAC}}}\right\} - \mathcal{Z}_{m_2}\left\{\frac{\frac{-k_1}{T_\mathrm{S}s} + \frac{-k_2}{T_\mathrm{S}^2 s^2}}{(s + \frac{1}{\tau_\mathrm{DAC}})\mathrm{e}^{t_\mathrm{p}/\tau_\mathrm{DAC}}}\right\}.$$

After some simplifications one obtains

$$\mathrm{LF}(z)|_\mathrm{CT-DT} = -\frac{k_1}{T_\mathrm{S}}\mathcal{Z}_{m_1}\left\{\frac{1}{s(s + \frac{1}{\tau_\mathrm{DAC}})}\right\} - \frac{k_2}{T_\mathrm{S}^2}\mathcal{Z}_{m_1}\left\{\frac{1}{s^2(s + \frac{1}{\tau_\mathrm{DAC}})}\right\}$$

$$+\frac{k_1}{T_\mathrm{S}\,\mathrm{e}^{t_\mathrm{p}/\tau_\mathrm{DAC}}}\mathcal{Z}_{m_2}\left\{\frac{1}{s(s + \frac{1}{\tau_\mathrm{DAC}})}\right\} + \frac{k_2}{T_\mathrm{S}^2\,\mathrm{e}^{t_\mathrm{p}/\tau_\mathrm{DAC}}}\mathcal{Z}_{m_2}\left\{\frac{1}{s^2(s + \frac{1}{\tau_\mathrm{DAC}})}\right\}.$$

If the particular CT loop filter terms are substituted now into the DT by using Table 3.3 (second and fourth row) and by considering a SCR-DAC feedback scheme with $t_\mathrm{d} = T_\mathrm{S}/2$, $t_\mathrm{p} = T_\mathrm{S}/2$, this yields $m_1 = 0.5$, $m_2 = 0$ and $s_k = 1/\tau_\mathrm{DAC}$. Finally, a coefficient comparison with the original DT modulator yields the corresponding CT coefficients k_i.

$$k_1|_\mathrm{Mod2\text{-}SCR} = \frac{a_1 a_2 T}{\tau_\mathrm{DAC}\left(1 - \mathrm{e}^{-T/2/\tau_\mathrm{DAC}}\right)}\,, \tag{3.27}$$

$$k_2|_\mathrm{Mod2\text{-}SCR} = \frac{a_2\left(-a_1 T_\mathrm{S} + (2a_1\tau_\mathrm{DAC} + 2T_\mathrm{S})\left(1 - \mathrm{e}^{-T/2/\tau_\mathrm{DAC}}\right)\right)}{2\tau_\mathrm{DAC}\left(1 - \mathrm{e}^{-T/2/\tau_\mathrm{DAC}}\right)^2}\,. \tag{3.28}$$

If τ_DAC is set e.g., to $0.05T_\mathrm{S}$ and $\{a_1,\ a_2\} = \{0.5,\ 0.5\}$ the CT loop filter coefficients yield $k_1 = 5$ and $k_2 = 7.75$. The signal scaling coefficient is again given by the coefficient of the outermost loop with NRZ-DAC (see (3.21)), which results here in $k_\mathrm{sig} = a_1 a_2 = 0.25$.

3.2.3 Differences of the Two Transformations

The modified $\mathcal{Z}$-transform is particularly useful in modeling and computing multirate-sampling systems [101, 142]. However, it possesses indeed application-specific differences. This means that the resulting loop filter remains not the same, if – for example – an excess loop delay is taken into account. This can be reduced to the fact that the modified $\mathcal{Z}$-transform does not accurately model the effect of excess loop delay compared to the impulse invariant approach. The transformation causes the delay to appear at the loop filter output, whereas in fact the delay happens after the quantizer sample instant and the resulting DAC pulse decision [140, 143].

Nevertheless, the resulting differences are marginal as calculations predict. In particular, these differences are negligible for frequencies $f \leq f_S/10$, and since ΣΔ modulators employ oversampling to achieve the desired performance, this condition is almost always satisfied.

3.2.4 DT-to-CT Conversion of Cascaded ΣΔ Modulators

Cascaded continuous-time ΣΔ modulators have not been popular in the past, but recently more and more publications arise covering this interesting topic. Principally, in order to implement cascades of CT modulators and to calculate a correct working recombination logic, the CT waveforms at various nodes inside the analog circuitry have to be determined. Furthermore, this theoretical determination has to match with the real, integrated circuit almost perfectly.

From this, two challenges arise for the design of CT cascaded modulators, i.e., an easy synthesis procedure, in spite of the required connection of a CT analog circuitry to a DT, digital noise cancellation circuitry. Furthermore, a solution has to be found for noncanceled, leaking errors of the first modulator stage in the output. Here, especially the typically large CT integrator gain errors cause the resolution to drop as low as that of the first stage alone. The former of both tasks will be the focus of the next sections, while the latter will be addressed separately in Sect. 5.3.6.

The difference of synthesizing a CT in contrast to a DT cascaded modulator is once more that in the DT, $\mathcal{Z}$-domain the analog part shows a natural allegory between the mathematics and circuit level and the corresponding compatibility to digital $\mathcal{Z}$-domain operation. Therefore the DT, digital noise-cancellation filter can be easily calculated [29]. In contrast, in the CT case the analog part is represented by $\mathcal{S}$-domain transfer-functions. In the literature, there exist three ways to perform the synthesis of a cascaded CT ΣΔ modulator:

- Early implementations did not really take the above problem of analog–digital connection or modulator scaling into consideration [144]. They simply adopted the known digital recombination logic of DT cascaded modulators to CT ΣΔ modulators of the same order. Despite the functionality shown in [144], the comparison with the theoretical performance is rather poor.
- In [4] and [145] the outputs $y(n)$ of all CT modulator stages have been calculated in the frequency domain and consequently matched to equivalent

$\mathcal{Z}$-domain functions using the bilinear transform [145], equivalent to the state space-based method mentioned at the beginning of Sect. 3.2. This method was also adopted in [20]. The main disadvantage of this procedure is the lack of generality for different (feedback) waveforms as shown in [3]. Additionally, by designing CT ΣΔ modulators completely new in the analog and the digital part, the preferable feature is missed that the large amount of work on DT-cascaded modulators in the literature can be used and adopted.

- Finally, in [96] the method presented above for DT-to-CT conversion using the impulse-invariant transformation has been adopted to synthesize the analog part of cascaded CT ΣΔ modulators. This is done in a way that exactly matches them to their DT originals with regard to the discrete-time output bitstreams $y_i(n)$ of the different stages in Fig. 2.20. Therewith, the design of cascaded CT modulators can directly profit from existing DT architectures as well as DT modulator scaling and the digital recombination logic can be reused. Additionally, feedback waveforms different from NRZ can be chosen by simply adopting the DAC transfer function in Table 3.2. This technique will be shown in detail in the following.

To be able to use the time-invariant transformation for cascaded structures, first the constitution of all loop filters has to be considered. In single-loop modulators this is by definition the transfer-function from the modulator output to the input of the quantizer. In the multiloop case of a cascaded modulator as in Fig. 2.20, *connecting loop filters* resulting from the connection network between the different stages have to be additionally taken into account [96]. The output of the modulator of preceding stages exhibits a transfer function to the input of the quantizers of higher stages, which is nothing but a loop filter by definition. In the following, the DT-to-CT conversion of a 2–1 and 2–1–1 modulator architectures will be exemplarily shown.

2–1 (SOFO) CT Modulator

For the DT SOFO modulator in Fig. 3.7a the following LFs can be found:

$$\begin{aligned}
\mathrm{LF}_1|_{\mathrm{DT}}(z) &= \frac{q_1}{y_1} = -a_1a_2I(z)^2 - a_2I(z) = -\frac{a_1a_2}{(z-1)^2} - \frac{a_2}{z-1} \\
\mathrm{LF}_2|_{\mathrm{DT}}(z) &= \frac{q_2}{y_2} = -a_3I(z) = \frac{a_3}{z-1} \\
\mathrm{LF}_{12}|_{\mathrm{DT}}(z) &= \frac{q_2}{y_1} = (g_1\mathrm{LF}_1 - b_1)\,c_1a_3I(z) \\
&= -\frac{c_1a_3}{(z-1)^3} - \frac{c_1a_3}{a_1(z-1)^2} - \frac{c_1a_3b_1}{z-1}\,.
\end{aligned} \tag{3.29}$$

Replacing the poles $1/(z-1)^i$ in (3.29) by the CT equivalents given in Table 3.2 results in the corresponding CT LFs given in (3.30). Due to

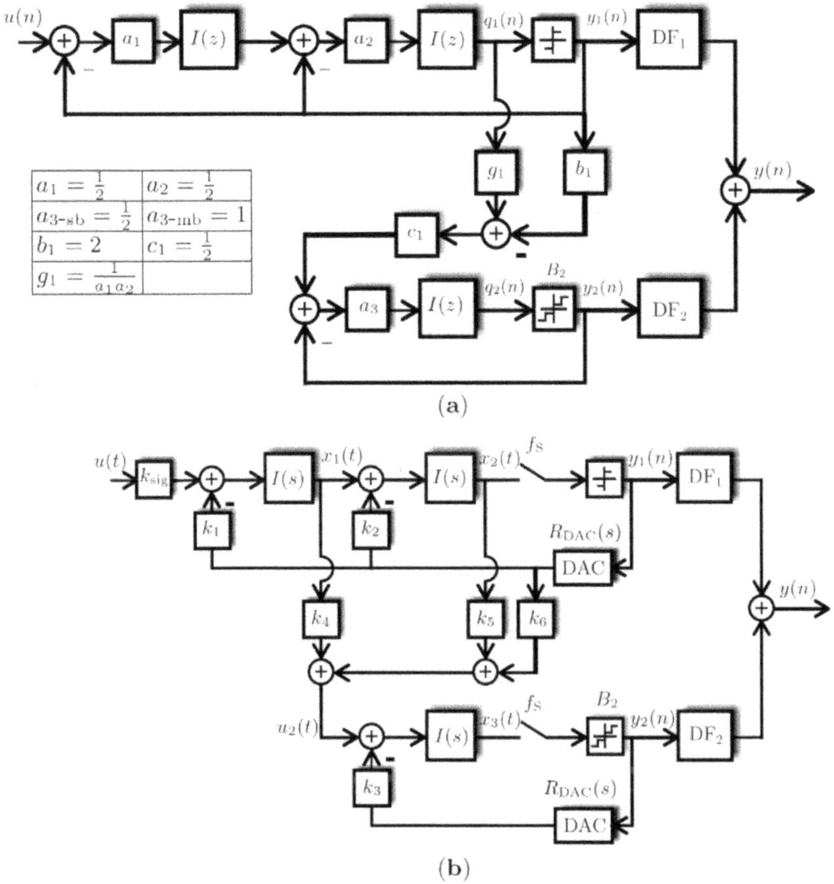

Fig. 3.7. DT and CT 2–1 (SOFO) modulators. (**a**) DT 2–1 modulator [6]; (**b**) CT 2–1 modulator, after [96]

simplicity reasons, the calculations are only presented for an NRZ feedback pulse ($\alpha = 0,\ \beta = 1$):

$$
\begin{aligned}
\mathrm{LF}_1|_{\mathrm{DT-CT}}(s) &\overset{\mathrm{NRZ}}{=} -\frac{a_1 a_2 f_S^2}{s^2} - \frac{1}{2}\frac{a_2(2-a_1) f_S}{s}, \\
\mathrm{LF}_2|_{\mathrm{DT-CT}}(s) &\overset{\mathrm{NRZ}}{=} -\frac{a_3 f_S}{s}, \\
\mathrm{LF}_{12}|_{\mathrm{DT-CT}}(s) &\overset{\mathrm{NRZ}}{=} -\frac{c_1 a_3 f_S^3}{s^3} - \frac{c_1 a_3 (\frac{1}{a_1} - 1) f_S^2}{s^2} - \frac{c_1 a_3 (b_1 + \frac{1}{3} - \frac{1}{2a_1}) f_S}{s}.
\end{aligned}
\tag{3.30}
$$

The objective now is to find a continuous-time structure which implements these loop filters thus making it equivalent to the analog part of the cascaded, discrete-time structure in Fig. 3.7a. The most obvious approach would be to use a CT structure similar to the DT one, i.e., a cascaded second-order and first-order modulator with a connection network as simple as in the DT approach. But it turns out that the LFs in (3.30) cannot be represented by this simple system. This is obvious at first sight, because in the DT-cascaded modulator in Fig. 3.7a, there exist five independent variables: two integrator scaling coefficients in the first stage $\{a_1, a_2\}$, one integrator scaling coefficient in the second stage $\{a_3\}$ as well as two coefficients in the connection network $\{b_1c_1,\ g_1c_1\}$. An equivalent CT modulator of the same structure therefore shows also only five independent variables. In contrast, in (3.30) there exist six state variables (different powers of s), all of which have to be controlled by one variable in an architectural representation of that CT cascaded SOFO modulator. Therefore an extension has to be made: To realize the connecting loop filter from lower to higher stages in cascaded CT ΣΔ modulators, which results from a DT-to-CT conversion, it is necessary to provide at every integrator input of the later stages connections from every (weighted) CT state variable as well as the feedback DAC output of all previous stages [25, 96].

A corresponding architecture of the CT SOFO modulator is shown in Fig. 3.7b, where the uncommon part is the feed-forward coefficient k_4, connecting the state variable x_1 to the input of the second stage as stated above. With a CT integrator transfer function chosen as in (3.18), the loop filters of this structure are found to be:

$$\mathrm{LF}_1|_{\mathrm{CT}}(s) = \frac{x_2}{y_1} = -\frac{k_1 f_\mathrm{S}^2}{s^2} - \frac{k_2 f_\mathrm{S}}{s}, \quad \mathrm{LF}_2|_{\mathrm{CT}}(s) = \frac{x_3}{y_2} = -\frac{k_3 f_\mathrm{S}}{s}, \tag{3.31}$$

$$\mathrm{LF}_3|_{\mathrm{CT}}(s) = \frac{x_3}{y_1} = -\frac{k_1 k_5 f_\mathrm{S}^3}{s^3} + \frac{(k_1 k_4 - k_2 k_5) f_\mathrm{S}^2}{s^2} + \frac{k_6 f_\mathrm{S}}{s}.$$

By comparing (3.30) and (3.31), the CT scaling coefficients can be calculated, and are given in Table 3.4 for a general rectangular feedback DAC, where α and β determine the beginning and the end time point of the rectangular feedback pulse form, while a_i are the DT modulator scaling coefficients as in Fig. 3.7a. The so-derived CT modulator behaves equivalently to its DT counterpart with regard to the quantizer outputs $y_i(n)$. Therefore the digital

Table 3.4. CT 2–1 (SOFO) modulator scaling coefficients [96]

$k_\mathrm{sig} = a_1 a_2$	$k_1 = \dfrac{a_1 a_2}{\beta - \alpha}$	$k_2 = \dfrac{\frac{1}{2} a_1 a_2 \left(\beta + \alpha - 2 + \frac{2}{a_1}\right)}{\beta - \alpha}$
$k_3 = \dfrac{a_3}{\beta - \alpha}$	$k_4 = -\dfrac{\frac{1}{2} c_1 a_3}{a_1 a_2}$	$k_5 = \dfrac{c_1 a_3}{a_1 a_2}$

$$k_6 = -\frac{\frac{1}{12} c_1 a_3 \left(\beta(\beta - 9) + \alpha(\alpha - 9) + 4\alpha\beta + 12(b_1 + 1) + \frac{6}{a_1}(\alpha + \beta - 2)\right)}{\beta - \alpha}$$

cancellation logic derived for discrete-time modulators in [5, 6] can be adopted without modification.

The need of the additional forward paths can be generalized to a structure as in Fig. 2.20. There, M different stages with orders N_i are connected through two independent paths, i.e., the input and output of the quantizer of the respective preceding stage. Therefore, in the DT domain one has exactly $\sum_{i=1}^{M}(N_i+2)$ independent variables.

In the CT structure, the loop filter orders remain constant and with it the number of coefficients inside each loop. In contrast, the connecting loop filters from the ith to the kth stage are of orders $N_{ik} = \sum_{r=i}^{k} N_r$. Therefore, the number of states in the DT-to-CT converted loop filters resulting from the general structure in Fig. 2.20 will be:

$$\sum_{i=1}^{M} N_i + \sum_{\substack{i=1,\ldots,M-1\\ k=2,\ldots,M}} N_{ik} \text{ with } N_{ik} = \sum_{r=i}^{k} N_r\,. \tag{3.32}$$

This number of states, which has to be controlled by independent scaling coefficients, is always higher, than the number of DT coefficients above. It has been found that by providing at every integrator input of the later stages connections from every CT state variable as well as the feedback DAC output of all previous stages, this number will always be met. Furthermore, it will mostly be exceeded and therefore some of the coefficients can result in zero [25, 97].

2–1–1 CT Modulator

Following this method also the equivalent CT architecture and scaling coefficients of the DT 2–1–1 modulator in Fig. 3.8a are derived, and are shown in Fig. 3.8b with Table 3.5. For simplicity, the scaling coefficients are given with DT coefficients a_i substituted from Fig. 3.8a for a single-bit rectangular feedback DAC $\{\alpha, \beta\}$.

Table 3.5. CT 2–1–1 modulator scaling coefficients for rectangular feedback DAC $(\alpha,\ \beta)$ and a DT modulator scaling as in Fig. 3.8a [6], after [32]

$k_1 = \frac{1}{4}\frac{1}{\beta-\alpha}$	$k_2 = \frac{1}{8}\frac{2+\alpha+\beta}{\beta-\alpha}$	$k_3 = \frac{1}{2}\frac{1}{\beta-\alpha}$	$k_4 = \frac{1}{2}\frac{1}{\beta-\alpha}$
$k_{51} = -\frac{1}{2}$	$k_{52} = \frac{1}{3}$	$k_{61} = 1$	$k_{62} = -\frac{1}{2}$
$k_{71} = -\frac{1}{48}\frac{\alpha(\alpha+3)+\beta(\beta+3)+4\alpha\beta+12}{\beta-\alpha}$		$k_{72} = -\frac{1}{48}\frac{\beta(\alpha^2+5)+\alpha(\beta^2+5)-12)}{\beta-\alpha}$	
$k_{82} = 1$	$k_{92} = -\frac{1}{4}\frac{\alpha+\beta}{\beta-\alpha}$		$k_{\text{sig}} = \frac{1}{4}$

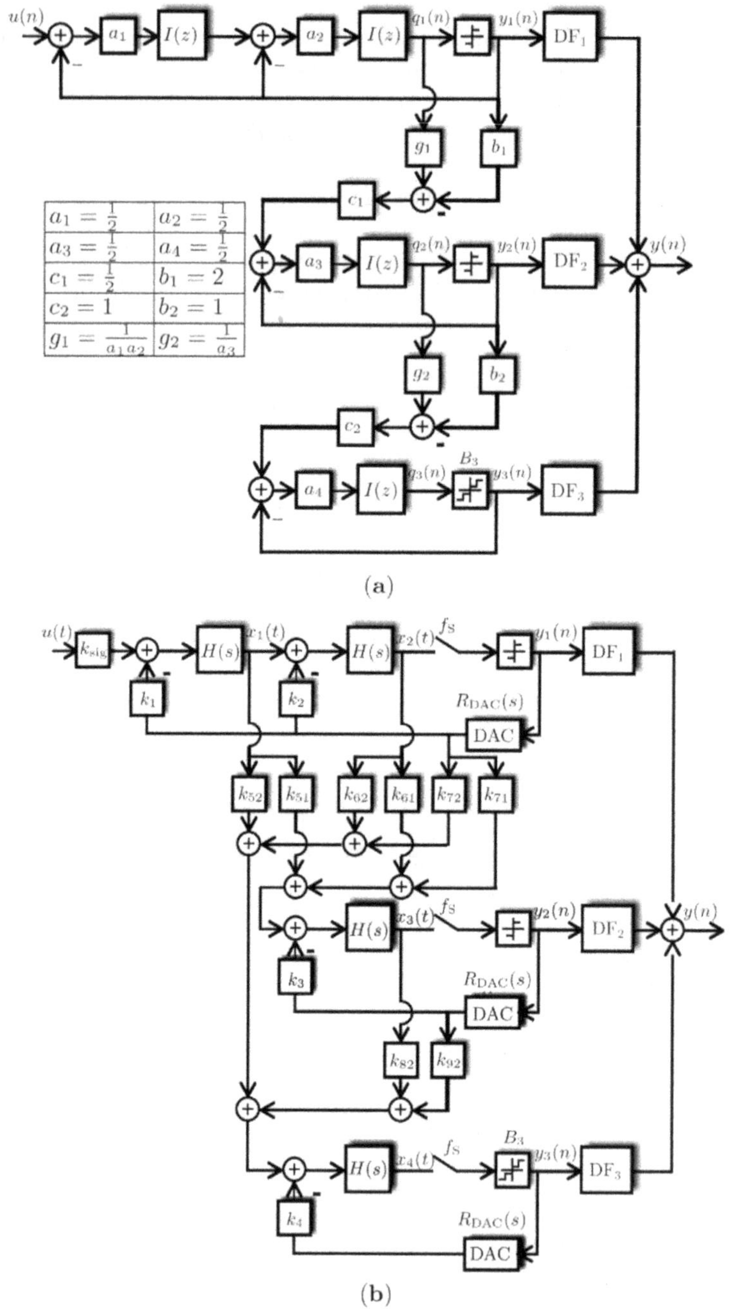

Fig. 3.8. DT and CT 2–1–1 modulators. (**a**) DT 2–1–1 modulator [6]; (**b**) CT 2–1–1 modulator, after [97]

Again note that the connecting, interstage loop filters as proposed in [96], always have to be regarded, when dealing with the cascaded CT $\Sigma\Delta$ modulators; but the additional forward paths are only needed, if the CT modulator is derived from a DT one with a DT–CT conversion [146].

3.3 Direct Filter Synthesis

A method to directly design a CT loop filter from its desired noise-transfer function prototype is carried out in detail in the following. Here, the loop filter is obtained by reverse calculations as proposed in [81] for the DT counterpart. Originate from a rational $\mathcal{S}$-domain prototype $\mathrm{NTF}(s) = A(s)/B(s) = 1/(1 + k_{\mathrm{q}}\mathrm{LF}(s))$ the CT loop filter $\mathrm{LF}(s)$ yields

$$\mathrm{LF}(s) = \frac{1}{k_{\mathrm{q}}} \frac{B(s) - A(s)}{A(s)} . \tag{3.33}$$

For reasons of simplicity, k_{q} can be arbitrarily set to unity [81]. Subsequently, the high-pass NTF prototype is designed with the base-band as stop-band in order to obtain a high base-band noise suppression. Increasing the base-band rejection of the NTF prototype, leads necessarily to an increased pass-band gain. As a consequence the maximal stable amplitude is reduced [3, 81], which is why a fundamental trade-off exists between the $\Sigma\Delta$ modulator noise attenuation and stable amplitude range [4].

A suitable synthesis tool for the filter transfer function is, e.g., MATLAB. The signal processing toolbox includes digital and analog filter synthesis options. In the following, a CT Chebyshev type II is designed using MATLAB and thereafter compared with a real $\Sigma\Delta$ modulator. The example considers a third-order CT Chebyshev type II, using a NRZ-DAC and a 3-bit internal ADC. Here it is important to note that the explained method is only able to synthesize CT loop filter functions employing NRZ-DACs. The MATLAB code synthesizing a CT Chebyshev type II NTF is:

```
[B,A] = cheby2(order,R,Wn,'high','s'); %CT NTF synthesis
```

where `order=3` defines the NTF order, with the stopband ripple `R=50 dB` decibels down and stopband edge frequency `Wn=12e6` in [rad/s] and `'high','s'` designs a CT high-pass filter. Next, the received polynomial coefficients of the loop filter numerator $B(s) - A(s)$ and denominator $A(s)$ have to be mapped to the aimed modulator topology. As an example, the above estimated loop filter is mapped to a third-order $\Sigma\Delta$ modulator with distributed feedback, as shown in Fig. 3.9 [36]. Doing so, the scaling coefficients k_i yield $\{k_1, k_2, k_3, \gamma\} = \{0.51, 0.97, 1.95, 0.04/(k_2 k_3)\}$. γ specifies the coefficient for the local feedback loop, as introduced in Sect. 2.7.3.

The corresponding simulation results regarding the spectrum and SNDR use a sample frequency of $f_{\mathrm{S}} = 52\,\mathrm{MHz}$ and a signal bandwidth of $f_{\mathrm{B}} = 2\,\mathrm{MHz}$

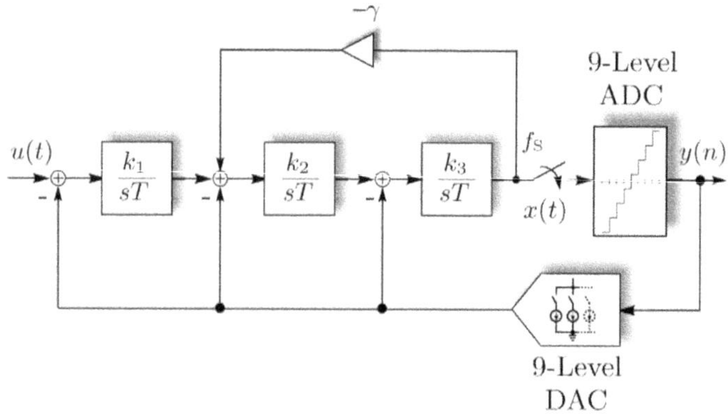

Fig. 3.9. Exemplary third-order modulator architecture with Chebyshev loop filter

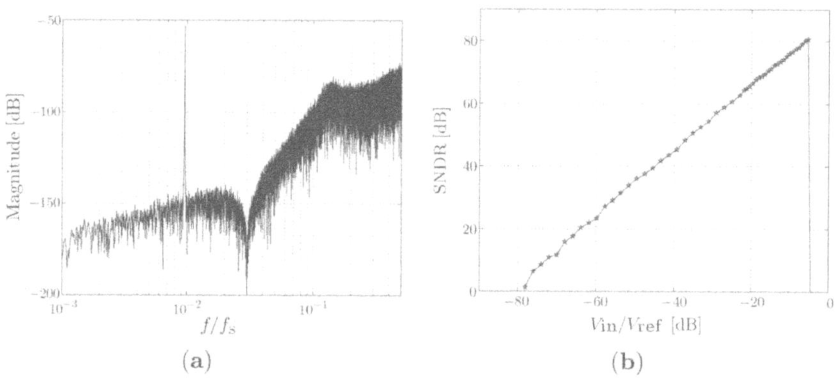

Fig. 3.10. Spectrum and SNDR of the directly synthesized CT Chebyshev type II NTF. (**a**) Simulated spectrum; (**b**) signal–noise ratio

are shown in Fig. 3.10. With a nine-level internal quantizer the obtained in-band noise and peak SNDR are −86 dB and 80 dB, respectively, while the maximal stable amplitude (the overload level) is at −5 dB.

Finally, a short note should be made about the direct synthesis of cascaded continuous-time ΣΔ modulators. While the DT–CT conversion in Sect. 3.2.4 required the introduction of additional paths to provide the state variables of previous stages at all later stages, this can be avoided by directly synthesizing the CT loop filters and designing new digital recombination logic for the CT cascade. This features some advantages, since the CT loop filters can be optimized separately [20]; but nonetheless note that it is mandatory to include the connecting, interstage loop filters as introduced in [96] and explained in Sect. 3.2.4 into the direct design considerations [146].

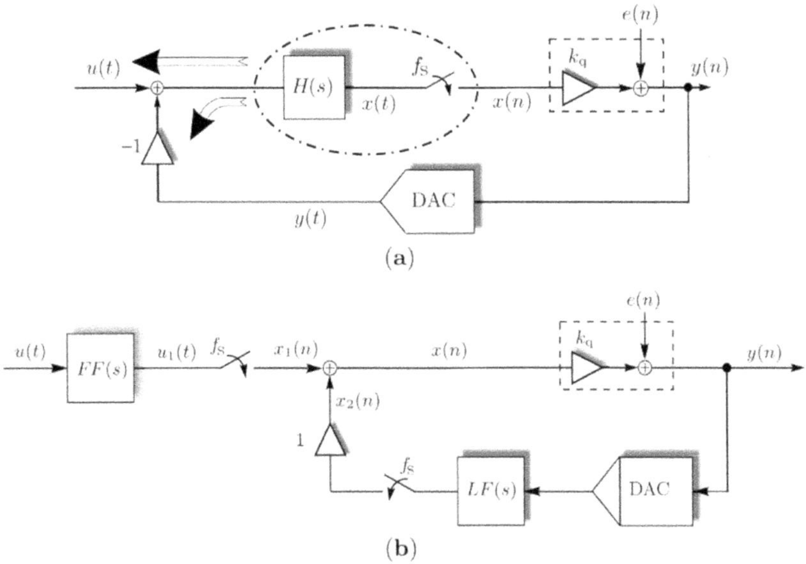

Fig. 3.11. Equivalent representation of a CT ΣΔ modulator. (**a**) General CT ΣΔ modulator architecture; (**b**) modified representation of a CT ΣΔ modulator

3.4 Signal and Noise Transfer Function in CT ΣΔ Modulators

In CT ΣΔ modulators, the input signal $u(t)$ is continuous-time and the output $y(n)$ is discrete-time, see Fig. 3.11a. Accordingly, it is not possible to define a straight $\mathcal{Z}$-domain or $\mathcal{S}$-domain signal transfer function in a CT modulator [105]. In order to be able to calculate the signal and noise transfer functions, some system theoretical modifications as shown in Fig. 3.11a have to accomplished. For that purpose, the loop filter and the implicit sampler are relocated across the summation point and placed in front of the CT ΣΔ modulator and in the feedback path. Here, FF represents the resulting forward filter and LF the feedback loop filter. Please note that the two filters are not necessarily identical as will be shown later on.

The presented simple mathematical manipulation reveals that the CT modulator behaves still like a DT one due to the sampling operation in the feedback path, but with the exception of the additional CT forward filter FF just in front of the sampler. An essential item is the position of the feedforward filter FF in front of the sampler. Keep in mind that both filters (FF, LF) feature a low pass characteristic, which is why the aliased spectrum introduced by the succeeding sampling operation is heavily suppressed [105]. This antialiasing effect is more extensively analyzed in Sect. 3.5. Furthermore, it is evident from Fig. 3.11b that the sampler in the feedback loop ensures a DT behavior as already mentioned above. Thus, the NTF remains

$$\mathrm{NTF}(z = \mathrm{e}^{\mathrm{j}\omega T_\mathrm{S}}) = \frac{1}{1 - \mathrm{LF}(z = \mathrm{e}^{\mathrm{j}\omega T_\mathrm{S}})} \,. \tag{3.34}$$

Next, the STF is calculated regarding Fig. 3.11b. Accordingly the STF of the CT modulator can be determined [105]:

$$\mathrm{STF}(\omega) \approx \frac{k_\mathrm{q}\,\mathrm{FF}(\mathrm{j}\omega)}{1 - k_\mathrm{q}\mathrm{LF}(\mathrm{e}^{\mathrm{j}\omega T_\mathrm{S}})} \,, \tag{3.35}$$

provided that the continuous-time front-end filter ($\mathrm{FF}(\mathrm{j}\omega)$) sufficiently attenuates the replica spectrum of the input signal at higher frequencies [105]. Thus the higher order replicated spectral terms are neglected. Furthermore, the sinc term associated with the pulse shape sampling is present in both systems (DT and CT), which is why this term is neglected if a comparison between both architectures is aimed.

As an example the feed-forward and loop filter (FF and LF) as well as the STF and NTF of a third-order CT modulator with distributed feedback is calculated. Therefore, the CT third-order modulator shown in Fig. 3.12a is modified according to Fig. 3.11b. Thus, from Fig. 3.12b the resulting feed-forward and loop filter (FF and LF) are calculated to be $\mathrm{FF}(s) = k_\mathrm{sig} I_1(s) I_2(s) I_3(s)$ and $\mathrm{LF}(s) = -k_1 I_1(s) I_2(s) I_3(s) - k_2 I_2(s) I_3(s) - k_3 I_3(s)$. As mentioned above both filters are not identical for the distributed feedback structure. More generalized, the filters yield for a modulator order N $\mathrm{FF}(s) = k_\mathrm{sig} \prod_{k=1}^{N} I^k(s)$ and $\mathrm{LF}(s) = -\sum_{i=1}^{N} (\prod_{k=i}^{N} k_k\, I^{N-i+1}(s))$. Beyond the distributed feedback

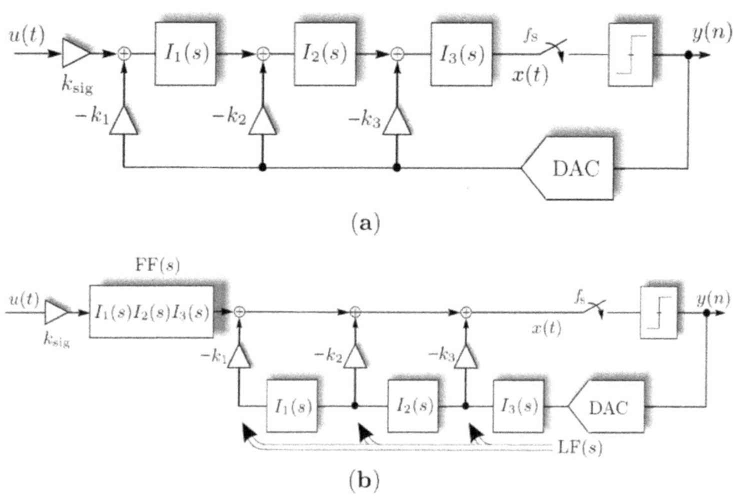

Fig. 3.12. (**a**) Third-order CT ΣΔ modulator consisting of a chain of integrators with weighted feedback k_i, after [37]; (**b**) its modified representation

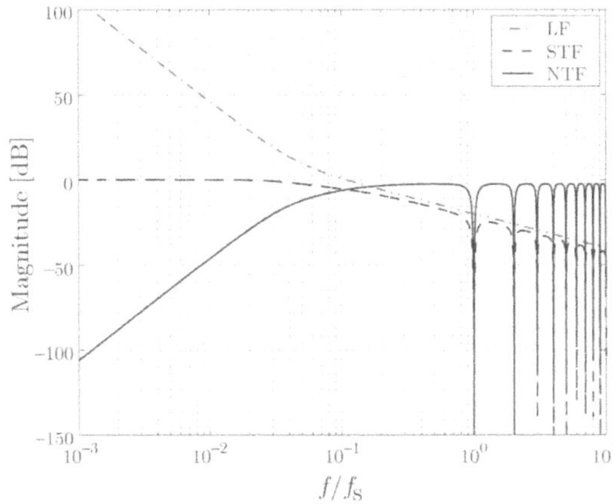

Fig. 3.13. Signal and noise-transfer function as well as the loop filter function of a third-order CT ΣΔ modulator with distributed feedback

structure, by using signal feed-forward pathes or even a feed-forward topology as shown in Sect. 2.7, the FF and LF filter are altered [147]. For example, the feed-forward topology (Fig.2.16) uses only a single feedback to the first integrator; consequently, both filters are identical (FF = −LF).

A graphical representation of the former calculated loop filter LF(s) and both transfer functions (NTF, STF) of the third-order CT modulator with distributed feedback ($\{k_{\text{sig}}, k_1, k_2, k_3\} = \{0.05, 0.05, 0.3, 0.641\}$) are shown in Fig. 3.13. As demanded, the NTF shows a strong noise attenuation while the STF shows a flat frequency response within the band of interested.

3.5 Implicit Antialiasing Filter in CT ΣΔ Modulators

As early as in [2] the inherent, implicit antialiasing filter in CT modulators has been reported, which was found to show a sinc characteristic signal-transfer function. For a first-order modulator obtained from Fig. 3.1 ($H(s) = I(s)$ and without additional AAF) this is easily understood: The input to the quantizer at the sampling instant depends on the feedback signal and additionally on

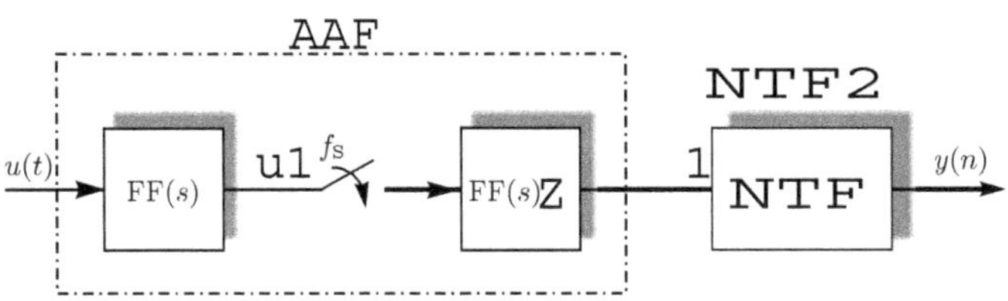

Fig. 3.14. Equivalent representation of a CT ΣΔmodulator based on Fig. 3.11b considering its implicit AAF

an integrated version of the input signal $\int_{nT_S}^{(n+1)T_S} u(t)dt$. This integration, referred to as a *boxcar integration* [105], can be expressed by a convolution of the input signal with a rectangular pulse, and therefore in the frequency domain as a multiplication of the input spectrum with a sinc($\pi f/f_S$)-function. This sinc-function attenuates the signal spectrum exactly at multiples of the sampling frequency and therefore shows the function of an AAF. Thus, one can expect that higher order modulation and larger loop-gain results in higher order antialiasing property.

Generally, the implicit input signal AAF of a CT ΣΔ modulator is presented in Fig. 3.14. Obviously, the antialiasing filter can be approximately expressed by [105]:

$$F_{\mathrm{AAF}}(\omega) = \frac{\mathrm{FF}(s = \mathrm{j}\omega)}{\mathrm{FF}(z = \mathrm{e}^{(\mathrm{j}\omega T_S)})}, \tag{3.36}$$

where FF(s) is the feed-forward filter of a CT ΣΔ modulator and FF(z) its DT equivalent. Equation (3.36) shows the AAF-behavior for different loop filter characteristics, i.e., low-pass, band-pass, all poles at dc or optimally spread inside the in-band [105].

3.5.1 Implicit AAF of the CT Third-Order Modulator

As an example, the aliasing behavior of the ideal third-order CT modulator shown in Fig. 3.12a is discussed. The original CT feed-forward filter $\mathrm{FF}_{\mathrm{CT,Mod3}}$ yields

$$\mathrm{FF}_{\mathrm{CT,Mod3}}(s) = k_{\mathrm{sig}} \left(\frac{f_S}{s}\right)^3 \tag{3.37}$$

with $k_{\mathrm{sig}} = 1/20$, see (3.22). The loop filter of the original equivalent DT modulator yields [6]:

$$\begin{aligned} \mathrm{FF}_{\mathrm{CT-DT,Mod3}} &= a_1 a_2 a_3 \left(\frac{z^{-1}}{1 - z^{-1}}\right)^3 \\ \text{with} \quad a_1 a_2 a_3 &= k_{\mathrm{sig}}. \end{aligned} \tag{3.38}$$

Inserting $\mathrm{FF}_{\mathrm{CT,Mod3}}$ and $\mathrm{FF}_{\mathrm{CT-DT,Mod3}}$ into (3.36) and after a few trigonometrical and mathematical simplifications one obtains:

$$\begin{aligned} |F_{\mathrm{AAF}}(\omega)|_{\mathrm{Mod3}} &= \left|\frac{\mathrm{FF}_{\mathrm{CT,Mod3}}(\mathrm{j}\omega)}{\mathrm{FF}_{\mathrm{CT-DT,Mod3}}(\mathrm{e}^{\mathrm{j}\omega T_S})}\right| = \left|\left(\frac{\frac{f_S}{s}}{\frac{z^{-1}}{1-z^{-1}}}\right)^3\right| \\ &\approx \mathrm{sinc}^3\left(\frac{\pi f}{f_S}\right). \end{aligned} \tag{3.39}$$

As discussed above for the first-order CT modulator, also for the third-order modulator a sinc-function appears as part of the signal transfer function in

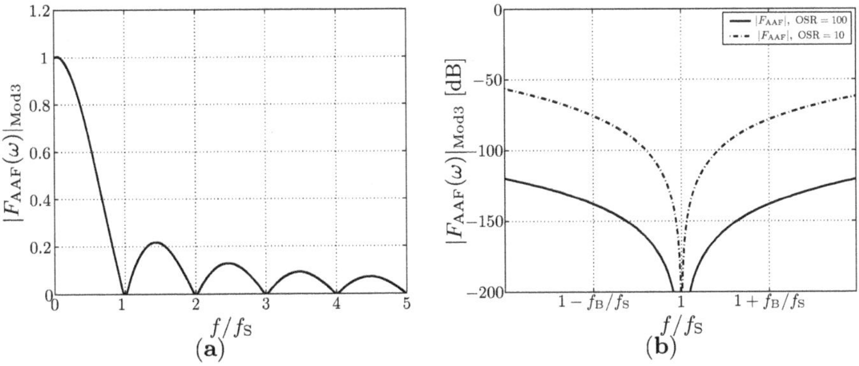

Fig. 3.15. (**a**) Antialiasing filter effect of the third-order CT modulator (see also Sect. 7.3); (**b**) aliasing attenuation for an OSR of 10 and 100

the case of a CT implementation. Fig. 3.15a illustrates the implicit antialiasing filter effect of the third-order modulator graphically. As expected, the antialiasing function $|F_{\mathrm{AAF}}(\omega)|_{\mathrm{Mod3}}$ is zero at nf_{S}. Furthermore, Fig. 3.15b illustrates the attenuation of $|F_{\mathrm{AAF}}(\omega)|_{\mathrm{Mod3}}$ at the passband edge for two different oversampling ratios (OSR = 10 and 100).

Frequency components, which fold down into the baseband due to the sampling process are located in the range of $\pm f_{\mathrm{B}}$ (baseband) around multiples of the sampling frequency nf_{S}. Due to the sinc-behavior, worst case suppression occurs at the lower edge of these frequency bands. In order to determine the resulting attenuation (3.39) is used, where f is equal to $f = f_{\mathrm{S}}(1-0.5\,\mathrm{OSR}^{-1})$ [147]. Thus, the calculated worst case error suppression at the passband edge yields −76 dB and −137 dB for oversampling ratios of 10 and 100 (compare with Fig. 3.15b). An even improved antialiasing performance is obtained by using spread NTF zeros, as exemplarily presented in Fig. 3.16 for a third-order

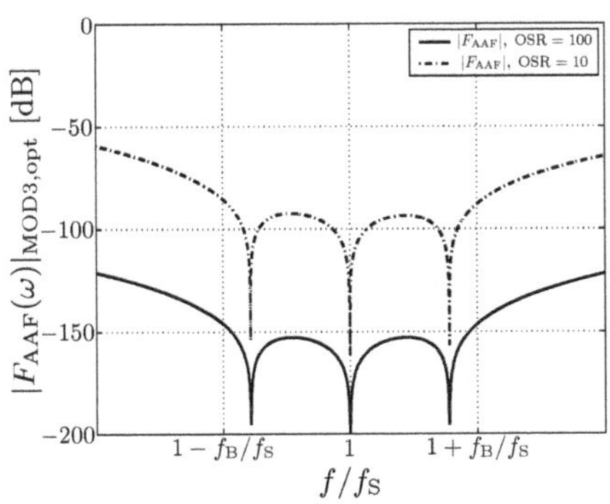

Fig. 3.16. Antialiasing filter of the third-order CT modulator with optimized NTF zeros

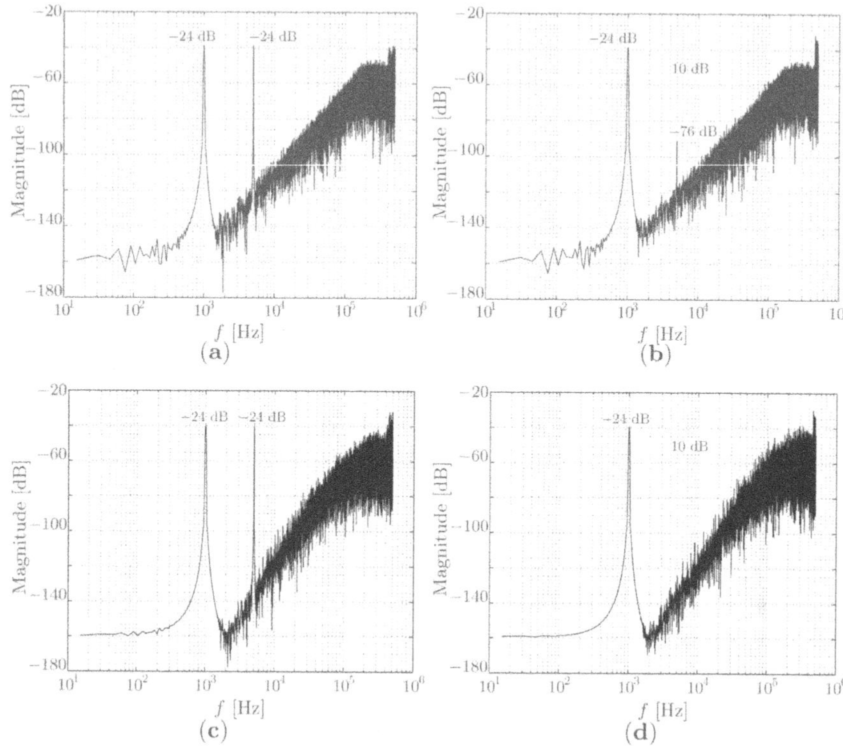

Fig. 3.17. Simulated PSD of a second-order and third-order, single-bit DT and CT ΣΔ modulator concerning AAF. $f_S = 1\,\mathrm{MHz}$, $P_{sig} = -24\,\mathrm{dB}$ at $f_{sig} = 1\,\mathrm{kHz}$. $P_{sig|_{DT}} = -24\,\mathrm{dB}$ at $f_{sig} = f_S - 5\,\mathrm{kHz}$ and $P_{sig|_{CT}} = +10\,\mathrm{dB}$ at $f_{sig} = f_S - 5\,\mathrm{kHz}$. (**a**) DT second-order modulator; (**b**) CT second-order modulator; (**c**) DT third-order modulator; (**d**) CT third-order modulator

CT modulator. As a result, additionally 5 dB–6 dB of error attenuation is gained.

These results can be qualitatively verified by the simulation of the second-order and third-order DT and CT modulators of Figs. 3.6a–b and 3.12a. (Note, the second-order modulator simulation is shown since the AAF in the third-order CT modulator does not show the aliased component any more in the simulation above the quantization noise floor!). Here, two input signals were applied to both modulators: a low-frequency, in-band sine wave signal with $P_{sig} = -24\,\mathrm{dB}$ at 1 kHz and additionally a high-frequency, out-of-band sine wave signal at $f_S - 5\,\mathrm{kHz}$, while the sampling frequency has been chosen to $f_S = 1\,\mathrm{MHz}$. Through aliasing this out-of-band signal is folded into the baseband. The magnitude of the out-of-band signal has been chosen to $P_{sig} = -24\,\mathrm{dB}$ at $f_S - 5\,\mathrm{kHz}$ for the DT modulator, but had to be increased to a very large $P_{sig} = +10\,\mathrm{dB}$ at $f_S - 5\,\mathrm{kHz}$ for the CT modulator to clearly see an aliased component. The output power spectra are shown in Fig. 3.17a–d.

It can be seen clearly that in the case of the second-order and third-order DT modulator, the out-of-band signal is completely folded into the baseband without attenuation. In contrast, in the second-order CT modulator the aliased signal is suppressed by almost 90 dB. As expected, an even higher attenuation is obtained by the third-order CT modulator, where the aliased signal disappears within the shaped quantization noise.

3.5.2 Implicit AAF of the CT SOFO Modulator

When considering the antialiasing behavior of cascaded modulators, at first sight one could assume, that it is only determined by the first stage, because here sampling occurs for the first time. Thus, the CT SOFO modulator should show at least the same implicit AAF as the above discussed second-order modulator. In order to quantify the antialiasing performance of a CT cascaded ΣΔ modulator, a detailed analysis regarding the aliasing suppression is exemplary shown in the following on a CT SOFO ΣΔ modulator (Fig. 3.7b).

The first stage in a cascaded CT ΣΔ modulator provides an inherent antialiasing filter, in the same way as in a single-loop ΣΔ modulator. However, the error signal $u_2(t)$ that is fed into the second filter stage equals the sum of $u_2(t) = k_1\, x_1(t) + k_5\, x_2(t) + k_6\, y_1(t)$. Accordingly, $u_2(t)$ contains beside the residual quantization noise also aliased signal components. The input signal $u_2(t)$ is again filtered by the second stage and the digital cancellation filter DF_2. Next, the output signal of the second stage $y_2(n)$ is added to the filtered version (by DF_1) of the first stage $y_1(n)$. As a result, the the error signal $u_2(t)$ is ideally canceled out, reducing both the quantization noise and the aliasing signal components. Thus, a cascaded CT ΣΔ modulator features an improved *AAF* behavior beyond the first filter stage [20, 29, 97]. A quantitative analysis of single and multiloop ΣΔ modulators is shown in detail in [147]. The analysis of the CT SOFO modulator yields:

$$F_{\mathrm{AAF}}(s) \approx \frac{k_1 k_{51}\left(\frac{f_\mathrm{S}}{s}\right)^2 + k_1 k_2 k_{61}\left(\frac{f_\mathrm{S}}{s}\right)^3}{a_1 a_2 a_3 c_1 g_1\left(\frac{1}{(z-1)^3}\right)}, \tag{3.40}$$

where beside the expected CT third-order filter also a second-order filter emerges. This is due to the additional feed-forward path k_{51} from the output of the first integrator in the first stage to the input of the third integrator in the second stage, which arose during the CT–DT conversion (see Sect. 3.2.4). The overall antialiasing performance is given by

$$|F_{\mathrm{AAF}}(\omega)| \approx \frac{k_1}{a_1 a_2 a_3 c_1 g_1}\,\mathrm{sinc}^3\left(\frac{\pi f}{f_\mathrm{S}}\right) \times \sqrt{(k_{51} T w)^2 + (k_2 k_{61})^2}\,, \tag{3.41}$$

which illustrates as desired a third-order sinc-behavior. Nonetheless, the square root term caused by the second-order filter worsens the ideal antialiasing

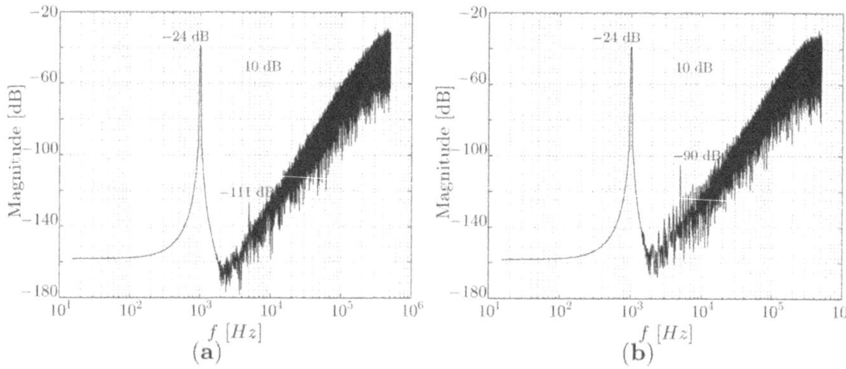

Fig. 3.18. Simulated PSD of a CT SOFO ΣΔ modulator concerning the AAF. $f_S = 1\,\text{MHz}$, $P_{\text{sig}} = -24\,\text{dB}$ at $f_{\text{sig}} = 1\,\text{kHz}$, $P_{\text{sig}} = +10\,\text{dB}$ at $f_{\text{sig}} = f_S - 5\,\text{kHz}$. (**a**) CT SOFO modulator; (**b**) CT SOFO modulator with mismatch

performance by roughly 10 dB. Furthermore, the above mentioned principle assumes an ideal loop filter cancelation. The complete cancelation of signals entering the system at the first-quantizer obviously depends on the perfect matching of the modulator coefficients k_i in Fig. 3.7b and the digital cancelation filters DF_i, as will be discussed in detail in Sect. 5.3.3. Nevertheless, in practical implementations any coefficient mismatch causes leakage such that both quantization noise and aliasing components of the first stage are partially seen at the modulator output again.

In order to confirm the theoretical considerations of the implicit AAF in CT cascaded ΣΔ modulator, simulations on a CT SOFO modulator are performed. The simulation result is shown in Fig. 3.18a, where the modulator is simulated under the same conditions as the second-order and third-order modulator in Sect. 3.5.1. The improved attenuation in comparison with the second-order single-loop ΣΔ modulator regarding the aliasing components can be clearly seen [29]. The impact of coefficient mismatch on the AAF performance is shown in Fig. 3.18b, where the SOFO modulator is simulated as above, but the coefficients $\{k_5,\ k_6\}$ have been afflicted with a 20% mismatch. The large increase of the aliased component is obvious. Nevertheless, at least the same AAF as for the second-order CT modulator will be ever achieved.

Concluding, the tremendous attenuation of aliased high-frequency signals by the implicit antialiasing filter can be seen as one of the key advantages of CT ΣΔ modulators. It can make the implementation of an additional front-end filter unnecessary. Especially in modern high-speed applications, where the OSR is kept as low as possible, this leads to a significant decrease of area and power consumption, when regarding the overall system.

3.6 Calculations with the CT Loop Filters

Above, the synthesis of single-stage and multistage, continuous-time ΣΔ modulator loop filters has been introduced by adoption of the impulse-invariant transformation. Nevertheless, before further discussing CT architectures and their nonideal behavior, the analytical treatment of these modulators has to be considered. We will need this in order to be able to include typical CT nonidealities into the calculations. Especially in the cascaded case compared to single-stage CT modulators this requires special attention, as the multiplication of the analog part by the $\mathcal{Z}$-domain digital cancellation filters cannot be done with the CT modulator representation. In contrast, the CT part has to be somehow converted to a $\mathcal{Z}$-domain equivalent first.

In order to perform the calculations on cascaded CT modulators the impulse invariant transformation can be used again; equivalently to the calculation of $\mathcal{S}$-domain equivalents for basic $\mathcal{Z}$-domain loop filter transfer functions from (3.12), which were given in Table 3.2, from the same expression also $\mathcal{Z}$-domain equivalents for basic $\mathcal{S}$-domain filters can be found. This has been done in [3, 132] for filters up to the third order, and is reviewed and extended in Table 3.6 up to the fourth order, according to the fourth-order 2–1–1 modulator in Fig. 3.8b. The calculations were done using the program MAPLE.

3.7 Alternatives for CT Filter Implementation

In the following, the most common continuous-time integrator implementation schemes are presented.

3.7.1 *gmC*-Integrator

gmC-integrators are based on a transconductance amplifier and a capacitor as shown in Fig. 3.19. The input voltage is fed through a transconductor gm, which produces a current $i = gm\,(V_{\text{in}+} - V_{\text{in}-}) = gmV_{\text{in}}$ to drive the capacitance C. Thus, the ideal transfer function of a *gmC*-integrator yields:

$$I(s) = k_i \frac{f_\text{S}}{s} = \frac{\text{gm}}{sC}\,. \tag{3.42}$$

This kind of integrator exhibits some considerable advantages over other techniques. Filters implemented with *gmC*-integrators are easily tunable, apply low current consumption and generate only small excess phase shift. The transconductance amplifier is required to have full signal swing at the output to maintain the dynamic range. Furthermore, in order to achieve high dc gains, usually cascading techniques are employed in the output stages. This in combination with the large signal swing requires large output current source

Table 3.6. DT equivalents for rectangular feedback DAC pulses for first-order to fourth-order CT low-pass loop filter poles, extended from [3, 132]

$\mathcal{S}$-domain	$\mathcal{Z}$-domain equivalents with f_S (Hz) $= 1/T_S$
$\frac{f_S}{s}$	$\frac{w_0}{z-1}, \quad w_0 = \beta - \alpha$
$\frac{f_S^2}{s^2}$	$\frac{w_1 z + w_0}{(z-1)^2}, \quad w_0 = \frac{(\beta^2-\alpha^2)}{2}, \quad w_1 = \frac{[\beta(1-\beta) - \alpha(1-\alpha)]}{2}$
$\frac{f_S^3}{s^3}$	$\frac{w_2 z^2 + w_1 z + w_0}{(z-1)^3}, \quad w_0 = \frac{(\beta^3-\alpha^3)}{6}, \quad w_1 = -\frac{(\beta^3-\alpha^3)}{3} + \frac{(\beta^2-\alpha^2)}{2} + \frac{(\beta-\alpha)}{2}$
	$w_2 = -\frac{(\beta^3-\alpha^3)}{6} - \frac{(\beta^2-\alpha^2)}{2} + \frac{(\beta-\alpha)}{2}$
$\frac{f_S^4}{s^4}$	$\frac{w_3 z^3 + w_2 z^2 + w_1 z + w_0}{(z-1)^4}, \quad w_0 = \frac{(\beta^4-\alpha^4)}{24}$
	$w_1 = -\frac{(\beta^4-\alpha^4)}{8} + \frac{(\beta^3-\alpha^3)}{6} + \frac{(\beta^2-\alpha^2)}{4} + \frac{(\beta-\alpha)}{6}$
	$w_2 = \frac{(\beta^4-\alpha^4)}{8} - \frac{(\beta^3-\alpha^3)}{3} + \frac{2(\beta-\alpha)}{3}$
	$w_3 = -\frac{(\beta^4-\alpha^4)}{24} + \frac{(\beta^3-\alpha^3)}{6} - \frac{(\beta^2-\alpha^2)}{4} + \frac{(\beta-\alpha)}{6}$

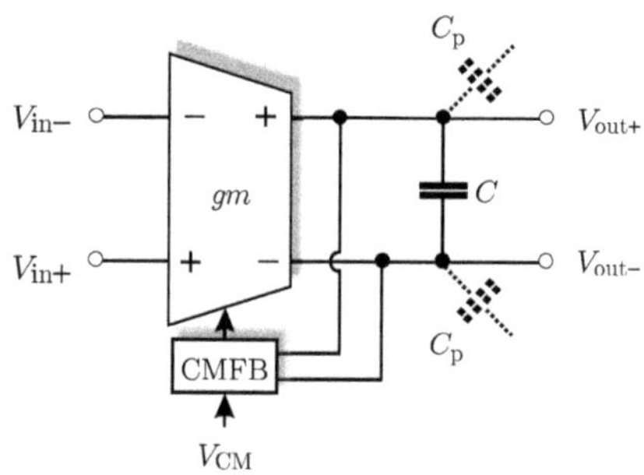

Fig. 3.19. Simplified schematic of a fully differential gmC-integrator with CMFB circuit and parasitic capacitances C_p

dimensions which leads to an increased noise budget [148]. Therefore, the entire SNR is almost limited by output noise sources. As a result this structure is only conditionally suitable for low-voltage design.

Another drawback is the sensitivity to parasitic capacitances C_p, which directly alters the integrator time constant. Besides, nonlinearities in the voltage-to-current conversion limit the entire THD of the integrator. Fully differential implementations reduce significantly even-order harmonics [149].

But, however, this increases the overall power drain due to the required CMFB. Otherwise a minimization of the odd-order harmonics is feasible by using linearization techniques such as source degeneration [127, 150, 151]. The adverse effect is the reduced power efficiency due to the fact that parts of the branch currents remain idle. In general, the transconductor employs MOS transistors operated in linear region as a resistor equivalent e.g., [9, 127, 149, 152, 153].An improvement of the *gmC* integrator can be found through addition of an active differential amplifier, which realizes a virtual ground node, see Sect. 3.7.3.

Note that *gmC* filters are also frequently used to build resonators with transfer functions similar to (3.43). Bandpass ΣΔ modulators employing this technique were presented in [105, 154].

3.7.2 *LC*-Resonator

CT ΣΔ bandpass converters are commonly used for digitizing signals at an intermediate frequency in radio receivers [155, 156]. After the antenna, the LNA and the mixer the entire IF band is digitized and mixed down to baseband. Since a modulator's sample rate f_S is typically four times the IF, the mix to baseband becomes particular simple in the digital domain. *LC* tank resonator structures (Fig. 3.20) are commonly used in the loop filter of bandpass ΣΔ modulators [155–159]. Here, the input voltage is fed through a transconductance amplifier, which drives the *LC* tank. This results in a bandpass shape transfer characteristic:

$$I(s) = k_i \frac{\omega_0 s}{s^2 + \omega_0^2} = \frac{\frac{gm}{C} s}{s^2 + \frac{1}{L_{\text{ind}} C}} . \tag{3.43}$$

The resonator design can be implemented with monolithic inductors on silicon [155, 157] or with off-chip inductors [156, 159]. Furthermore, the resonator structures can be implemented in a simple passive way [156] or by using an additional active transconductance stage [157], as shown in Fig. 3.20. Since the former implementation does not use active components, the resonator design is noiseless, highly linear and needs no power. Thus the architecture gains advantages in terms of noise, distortion, and power simultaneously [156]. The latter design contains an integrated inductor and a linear *Q*-enhanced *LC* filter. Monolithic inductors are lossy with a typical quality factor (Q) of 3–8 [157] while external devices feature a Q of better than 20. Having the noise shaping performance of a bandpass modulator in mind, the *LC* resonator should have an infinite Q to put the loop filter poles exactly on the jω axis. As a result tuning of the filter is typically required in order to limit the performance loss [105, 156].

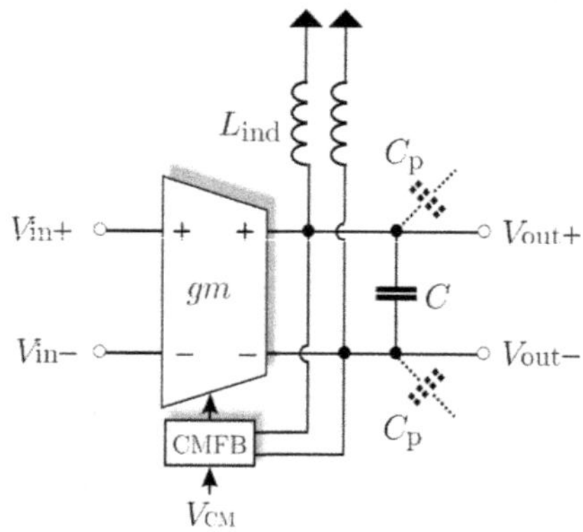

Fig. 3.20. Fully differential *LC*-resonator

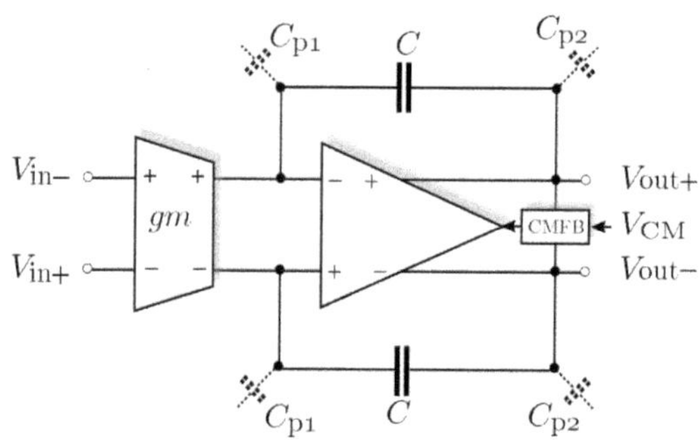

Fig. 3.21. Fully differential active *gmC*-integrator

3.7.3 Active *gmC*-Integrator

To overcome the drawbacks of *gmC*-filters, an additional amplifier is commonly used to shift its errors to second-order effects; an example is shown in Fig. 3.21. Here, the ideal integrator transfer function yields again

$$I(s) = k_i \frac{f_S}{s} = \frac{gm}{sC} . \tag{3.44}$$

The most important benefits are [160]:

- The influence of any parasitic capacitance C_{p1} connected at the amplifier input shows only small voltage fluctuations because of the virtual ground node. Other parasitic capacitances C_{p2} are driven by the amplifier.
- The transconductance amplifier does not need to provide large voltage swing and therefore no high output impedances are required again due to the virtual ground approach. This is especially important for low-voltage designs.

- The transconductance and voltage amplifier combination pushes the dominant pole to very low frequencies avoiding phase lag, which would be generated otherwise. Nevertheless, signal feedthrough via the integration capacitors feature additional excess phase. Compensation can be achieved by inserting a small impedance in series with the capacitance which can be a small resistor or a MOS transistor. The latter approach provides the additional opportunity of tuning its resistive value to cancel out precisely the parasitic pole by the created zero.
- The transconductance amplifier of the next filter stage represents only a capacitive load for the preceding voltage amplifier. Thus, a simple wideband amplifier can be used to drive the integrating capacitor which is a nice benefit over the traditional RC or MOSFET-C-technique.

The drawback of this scheme is still the limited linearity of the transconductance amplifier which determines the upper bound for the resolution of the entire filter or $\Sigma\Delta$ modulator. Furthermore, as mentioned above the additional amplifier requires beside the standard transconductance amplifier extra power which reduces the overall power efficiency of this structure. Architectures which feature a lower quiescent power consumption are the classical active-RC or MOSFET-C-filters, but at the expense of a resistive load for the previous filter stage.

3.7.4 Current-Mode Integrator

The current-mode counterpart to the gmC-filter are used, e.g., in [135, 161, 162]. In [161] a low-voltage bandpass filter using a digital CMOS technology was presented. Therefore, the capacitors are realized by a MOSFET gate capacitance. Furthermore, a low-power 100 kHz low-pass current-mode $\Sigma\Delta$ modulator was presented in [135]. Extended investigations of the current-mode approach have been made in [163]; here it was stated that in contrast to the formerly accepted advantage, current-mode filters perform neither better at high frequencies nor at low supply voltage levels. Additionally, this comparison showed that the performance is essentially identical to gmC-filters. A special case of current-mode integrators are log-domain integrators which provide a better power-efficiency as mentioned below.

3.7.5 Log-Domain Integrator

The trend toward lower power consumption and lower supply voltages has motivated the research of new solutions for continuous-time filters. The log-domain approach offers both low-voltage and low-power operation due to the voltage companding principle and the use of exponential current vs. voltage characteristics offering a high transconductance to current ratio [164, 165].

The first log-domain filters were implemented in BiCMOS processes. Log-domain filters can also be implemented by using MOS transistors biased in weak inversion [166, 167]. However, biasing a MOS device in weak inversion usually requires large W/L ratios which cause bandwidth limitations of typically a few MHz [166].

A power efficiency comparison with both log-domain and gmC-filters regarding the effect of lowering the supply voltage was presented in [168]. As a result, for low supply voltages of approximately $V_{\mathrm{dd}} \leq 2\,\mathrm{V}$ the log-domain technique shows a higher power efficiency.

3.7.6 Active *RC*-Integrator

Another commonly used integrator structure in CT ΣΔ modulators [17, 19, 34, 37, 101, 169] due to simplicity, linearity (provided highly linear resistors are available), parasitic insensitivity as well as the overall power consumption are active RC-integrators as shown in Fig. 3.22. The ideal integrator transfer function of the active RC-integrator is given by

$$I(s) = k_i \frac{f_{\mathrm{S}}}{s} = \frac{1}{sRC} \, . \tag{3.45}$$

Assuming the amplifier gain is high, the input nodes closely meet virtual ground conditions, so that the input resistor R performs a linear V/I conversion. Performance limitations regarding linearity result from the limited linearity of the V/I conversion (here the resistor R) itself as well as from the finite gain of the amplifier which leads to small residual voltage fluctuations at virtual ground. Further sources of distortion are caused by the non-linear amplifier transfer function as well as any voltage dependence of the integration capacitor.

Performance enhancements can be achieved by increasing the resistor area [170, 171] or by increasing the dc gain. Additionally, this refines also the matching of both input resistors (in the fully differential case), which cause otherwise even-order harmonics. A THD of 90 dB is commonly achieved with

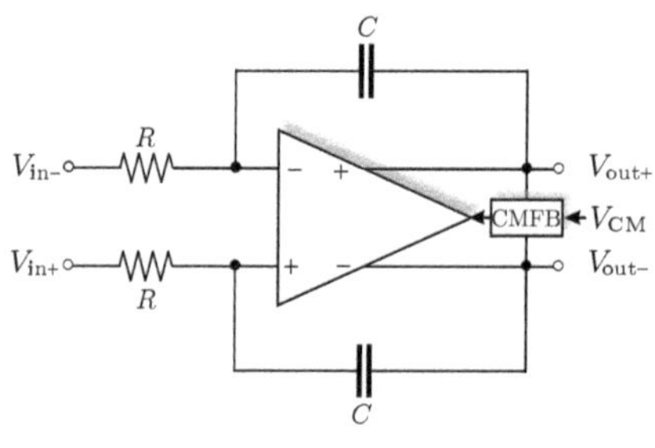

Fig. 3.22. Simple fully differential active RC-integrator, after [37]

integrated polysilicon resistors. The linearity of the integration capacitor is in general not critical, because first the inherent linearity of an integrated capacitor is better than that of the resistor (but still depends on the capacitor type) and second the integration capacitor builds a negative feedback loop around the amplifier, which further reduces distortion [172].

The nonlinearities in a fully differential amplifier principally arise from two sources: compressive voltage-to-current conversion of the input differential pair and voltage dependance of the output impedance [171]. As proposed in [128], the input resistors should be as large as possible to limit the amplifier nonlinearity. The upper limit for R is set by the allowed thermal noise level, which itself is fixed by the overall dynamic range requirements [35]. Moreover, the static current through the differential pair determines the linearity of the input stage, as calculated in [128]. Thus, a higher linearity in an active RC-integrator can be achieved by consuming more power. Accordingly, the amplifier has to provide high open loop gain, low noise contribution as well as large output signal swing.

A frequently used enhancement of the simple active RC-integrator is the insertion of a resistor R_Z in series to the integration capacitor C. This usually small resistor moves the parasitic right halfplane zero, which results from the amplifier input transconductance gm and the integration capacitance C, toward infinity [148].

3.7.7 Active MOSFET-C-Integrator

Active MOSFET-C-integrators are only a modification of the active RC-integrator. The resistor in Fig. 3.22 is replaced by a MOS transistor, operated in triode region; to guarantee this, the input voltage must be kept small. The ideal integrator transfer function is given by

$$I(s) = k_i \frac{f_S}{s} = \frac{1}{s r_{\text{out}} C}, \tag{3.46}$$

where r_{out} is the output resistance of the MOS device. A benefit of this structure is the simple ability of tuning by modulating V_{bias}.

Again this approach shows the sensitivity to parasitics, nonidealities and nonlinearities in a similar way as active gmC-filters. The even-order harmonics are canceled out by the balanced operation, whereas the odd-order nonlinearities still remain. The attainable linearity is 40–60 dB depending on the supply voltage and maximum signal swing. An impressive linearity improvement can be achieved by the CZARNUL-SONG circuit [173] or by a series connection of a resistor and MOSFET [174].

3.7.8 Conclusion on the Commonly Used CT Integrators

It is evident from the previously mentioned remarks that each of the specified structures features some important benefits regarding operable frequency range, tunability, mismatch insensitivity (frequency accuracy),

Table 3.7. Overview of various benefits and drawbacks of various CT filter approaches

	gmC-int.	CM-int.	Log.-int.	Active *gmC*	Active MOS-*C*	Active *RC*
Frequency range	+ + ++	++	+	+ + +	++	+ + +
Tunability	+ + +	++	+ + ++	+ + +	+ + +	+
Mismatch insensitivity	+ + +	++	+	++	+	+
Linearity	+	+	++	+ + +	++	+ + ++
Dynamic range	++	+	++	+	++	+ + +
Power requirements	+ + +	+	+ + ++	+	++	++
Low-voltage capability	+	+ + +	+ + ++	++	++	+ + +

Here, + denotes less suitable for the corresponding integrator characteristic whereas + + ++ mean very suitable

linearity, dynamic range, power requirements, signal swing, and low-voltage capability. Nevertheless, all these topologies also exhibit their own drawbacks.

A qualitative overview is given in Table 3.7: The log-domain approach shows advantages, if low-voltage capability and power consumption are of high interest while the desired frequency range is limited to a few MHz. If a higher frequency range is additionally demanded in connection with a high linearity, the active *RC*-filters are the preferred structure. For linearity requirements limited to about THD $= -60\,\text{dB}$, *gmC*-filters are favorable, if low power is a major interest. An overview of recently implemented CT ΣΔ modulators is presented in Tables 3.8 and 3.9. Obviously, many designs rely on a mixture of the different CT filter implementations, rather than realizing every integrator with the same architecture. Consequently, optimized features can be used where they are needed in the modulator loop.

3.8 Classification of Nonidealities in CT ΣΔ Modulators

As pointed out in Sect. 3.1, a CT ΣΔ modulator consists of three major building blocks: a continuous-time loop filter $H(s)$, a clocked internal quantizer (ADC) and a continuous-time feedback DAC. Due to variations during the IC

Table 3.8. Performance summary and resulting FOM according to (7.1) of most recently published low-pass ΣΔ modulators

Reference	Modulator architecture	Process	B (bit)	f_N (kHz)	P (mW)	V_{dd} (V)	FOM (pJ)
[19]	Fifth-order SL-FF, active-*RC*, *gmC*	0.18 µm SPFM CMOS	12	7,680	4.5	1.8	0.1
[36]	Third-order SL, active-*RC*	0.5 µm DPTM CMOS	13	4,000	4	1	0.11
[175]	Third-order SL, SC, single OpAmp	0.35 µm DPQM CMOS	12.2	0.18	2E-04	20	0.2
[37]	Third-order SL, active-*RC*	0.5 µm DPHM CMOS	13	50	0.135	1.5	0.3
[176]	Fifth-order SL, active-*RC*, *gmC*	0.18 µm SPFM CMOS	11	4,000	3.3	1.8	0.4
[17]	Fourth-order SL, active-*RC*, *gmC*	0.35 µm CMOS	14	200	1.8	1.8	0.5
[72]	Third-order SL, SC	0.5 µm DPTM CMOS	11	32	0.04	0.9	0.6
[169]	Fourth-order SL-FF, active-*RC*, *gmC*	0.8 µm CMOS	16.5	40	2.3	3.3	0.6
[34]	Third-order SL, active-*RC*	0.5 µm DPTM CMOS	13	50	0.25	1.5	0.6
[177]	Second-order SL, SC	0.18 µm DPQM CMOS	13.8	600	9	1.8	1
[55]	Second-order SL, SC	2 µm CMOS	15.5	41	2	5	1
[23]	Fourth-order SL-FF, active-*RC*, *gmC*	0.18 µm SPFM CMOS	11	30,000	70	1.8	1.1
[178]	Second-order SL, SC, SO	0.18 µm DPTM CMOS	12	16	0.08	0.7	1.2
[179]	Second-order SL, SC	0.5 µm CMOS	14.3	20	0.55	1.5	1.3
[126]	Third-order SL-FF, active-*RC*, *gmC*	0.5 µm SPFM CMOS	14.2	2,200	62	3.3	1.5
[180]	Second-order SL, SC, MOS-Caps	0.18 µm CMOS	11.8	32	0.19	0.9	1.6
[20]	Fourth-order 2-2, active-*RC*, *gmC*	0.18 µm SPFM CMOS	11	2,0000	68	1.8	1.66
[181]	Third-order SL, CM	0.18 µm SPHM CMOS	13.7	200	5	1.8	1.8

Table 3.9. Performance summary and resulting FOM according to (7.1) of most recently published low-pass ΣΔ modulators

Reference	Modulator architecture	Process	B (bit)	f_N (kHz)	P (mW)	V_{dd} (V)	FOM (pJ)
[63]	Third-order SL, SC	0.65 µm DPTM CMOS	15.8	2,500	295	5	2
[58]	Fifth-order ML	0.18 µm DPFM CMOS	14	4,000	150	1.8	2.2
[182]	Second-order, SC	0.13 µm CMOS	16.7	88	22	3	2.3
[183]	Second-order SL	0.5 µm TM CMOS	16	40	6.5	1.2	2.4
[71]	Second-order SL, SC	0.7 µm SP CMOS	12.5	6	0.1	1.5	2.8
[79]	Fifth-order SL-FF, active-RC, gmC	0.35 µm SPFM CMOS	13	400	10	3.3	3
[57]	Third-order SL, SC	0.6 µm DPDM CMOS	13	200	5.2	2.7	3.1
[184]	Third-order 2–1, SC	0.6 µm DPDM CMOS	15	50	5.4	1.8	3.2
[64]	Fourth-order 2–1–1, SC	0.5 µm DPTM CMOS	14.7	2,500	270	5	4
[107]	Fourth-order SL, SC	1.25 µm DP CMOS	13.7	1,000	58	5	4.3
[104]	Second-order SL, active-RC, SWS	0.5 µm MT CMOS	9.4	384	1.56	1	6
[185]	Third-order SL, SC	2 µm CMOS	19	0.8	2.7	5	6.4
[186]	Second-order SL, SC	0.35 µm DPTM CMOS	13.2	200	16.2	3	8.6
[112]	Second-order SL, active-RC	2 µm DP CMOS	12	240	15	5	15.2
[187]	Second-order SL, SC	1.2 µm DPDM CMOS	15.7	40	67.5	5	31.7
[188]	Second-order SL, SC, SO	0.5 µm CMOS	7.2	40	0.24	1	40.8
[74]	Fifth-order SL-FF, SC	0.6 µm TM CMOS	15.3	80	210	5	65
[189]	Third-order SL, SC	3 µm DPSM CMOS	20	1	120	20	114.4

manufacturing process and through circuit imperfections, which come from design (e.g., OpAmp gain) or are intrinsic (e.g., mismatch), each of these components deviates from its ideal behavior. Thus, the first classification of errors is their originating building block. Furthermore, the deteriorating influence of the different errors on the ideal modulator behavior can be used to specify two categories of nonidealities: those, which alter the ideal signal-transfer and noise-transfer function by altering their poles and zeros, and aside the nonidealities which introduce noise or distortion into the system [5].

Finally, a classification of nonidealities can be done concerning the error domain. First, there are the above mentioned mismatch or tolerance induced nonidealities altering the designed component magnitudes; second, there exist temporal nonidealities, which alter the signal flow either constantly or statistically in time; and last, the modulator is affected by designed imperfections, which result from trade-offs in the design. While the influence of these last nonidealities is minimized to achieve a certain modulator performance, the former are process dependent and mostly of statistical nature and countermeasures can only be taken on the architectural level.

In order to present a clearly arranged summary of the influence of nonidealities on continuous-time ΣΔ modulators, in the following chapters the first classification is used, i.e., identifying the errors by their originating building block. To give an overview regarding the mismatch type and the occurrence in the modulator loop, a detailed partitioning of several major error mechanisms concerning the three buildings blocks is shown in Table 3.10. Nonetheless, there are plenty of other nonidealities, where not each of them can be presented in a table!

3.8.1 Different Locations of Error Occurrence or Input Referred Nonidealities

Before dealing with the different nonidealities in detail, one common characteristic is discussed at this point, i.e., the above mentioned entrance point of the respective nonideal error signals, which introduce noise or distortion into the system. This is illustrated in Fig. 3.23. Here an Nth-order, single-loop, single-bit CT ΣΔ modulator is given with scaling coefficients k_i, which can be calculated from a DT-to-CT conversion as exemplarily shown in Sect. 3.2. Additionally to the input signal $u(t)$, modeled error signals have been introduced, where $E_i(s)|_{i=1,\ldots,N}$ are the input referred noise and distortion components of the respective integrator, $E_{\mathrm{q}}(s)$ are errors entering the system at the internal quantizer while $E_{\mathrm{d}}(s)$ finally are errors of the feedback DAC (Table 3.10).

Here, several important issues of ΣΔ modulator design can be seen clearly: First, it is obvious that both single-stage and cascaded modulators (which actually consist of single-stage modulators), are most sensitive to any nonideality, which affects the signal input. Beside any input referred noise and distortion of the first integrator $E_1(s)$, also the errors at the output of the feedback DAC $E_{\mathrm{d}}(s)$ directly add to the input node. Thus, any in-band

Table 3.10. Impact and point of interference of the major error mechanisms in a CT ΣΔ modulator

Basic block		Nonideality	Impact
Integrator	OpAmp	Finite and nonlinear gain	Increased noise floor, harmonic distortion
	OpAmp	Finite unity gain bandwidth	Increased noise floor, stability properties
	OpAmp	Finite slew rate	Quantization noise increase, harmonic distortion
	OpAmp	OpAmp gain nonlinearity	Harmonic distortion
	OpAmp	Thermal and $1/f$ noise	Increased noise floor
	VI conv.	Nonlinearity	Harmonic distortion
	VI conv.	Thermal and $1/f$ noise	Increased noise floor
	Gain	Time constant mismatch	Less aggressive noise shaping or stability issues
ADC		Metastability	Reduced max. stable amplitude, increased noise floor, harmonic distortion
ADC		Hysteresis	Increased noise floor
DAC		Delay	Stability, max. stable amplitude, noise, harmonic istortion
DAC		Unequal rise and fall times, intersymbol interference	Increased noise floor, harmonic distortion
DAC		Nonlinearity	Noise, harmonic distortion
Clock		Random clock jitter	Noise
Clock		Accumulated clock jitter	Skirt around signal

component of these error sources appears without suppression in the digital output of the modulator. This is why the input stage as well as the output of the DAC are commonly required to have a precision or resolution, which is equal to that of the overall modulator [123]. On the other hand, errors entering the system at other points are less critical. This is clearly understood when considering the quantization noise as an error source $E_\mathrm{q}(s)$ entering the system

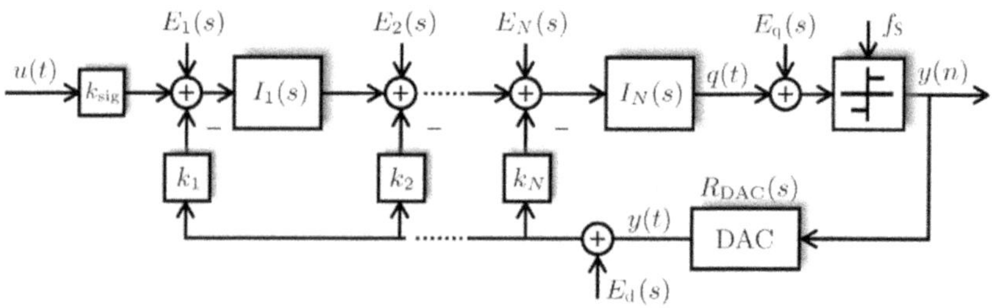

Fig. 3.23. Different inputs of nonideality induced errors illustrated for an Nth order, single-stage ΣΔ modulator

at the quantizer, i.e., behind the loop filter. As intended by the ΣΔ system and as discussed in Sect. 2.5.1, the quantization noise as well as any other error component entering the system at this point are heavily suppressed by the high, in-band loop-gain of the preceding loop filter $H(s)$. Equivalently, also all errors $E_i(s)|_{i=2,\ldots,N}$ in Fig. 3.23 are suppressed, each by its preceding part of the loop filter. Consequently, the second dominant error $E_2(s)$ is already suppressed by the first-order filter $k_1 I_1(s)$. Thus, since the design of the first integrator is essential for the overall modulator performance, it is often possible to reduce the size and power consumption of later filters [3, 34, 101, 113] and [4]. To emphasize the difference between noise and distortion, it should be mentioned at this point that a first-order noise suppression is proportional to OSR^{-3}, as shown in Sect. 2.5.1. This is first due to the spreading of the noise because of the oversampling and second due to the noise shaping by the filter function. Distortion components on the other hand are not affected by oversampling; thus only the transfer function of the preceding filters will suppress the distortion, and the attenuation will only be proportional to OSR^{-2}.

In cascaded structures, only the first integrator of the first stage is of special importance, because errors entering the cascaded systems in higher stages are suppressed by a filter of the order of the preceding stages [4] and [91]. This makes them even less stringent to design than later filter stages in the first cascade. Additionally, this is why errors from the feedback DAC of higher stages are high order attenuated; thus, their DAC linearity is of minor importance [95, 190].

3.8.2 Organization of the Following Chapters

The question in which order to present the design requirements and nonideal behavior of the different parts in the continuous-time ΣΔ modulator, the common way would be to start with the filter or integrator blocks, since it is used to impose the most difficult design constraints [4, 5, 110] in SC designs, and to end with the ADC, which is the most uncritical block of the ΣΔ modualtor. In contrast, going to the continuous-time implementation, the nonideal behavior and the design requirements of the feedback DAC become at least as important, as the filter itself. Additionally, as will be seen in the following, a compensation procedure derived for excess loop delay in the feedback DAC can be simultaneously adapted to counteract some nonideal behavior in the integrators. Consequently, we start with the description of the DAC nonidealities and countermeasures against it, subsequently we focus on the filter constraints, and finally, the ADC behavior is discussed in a glance.

When deciding, which performance measure should be taken for illustrating the impact of nonideal behavior, obviously the power spectrum, the IBN and the SNR can be taken. While the shape of the spectrum gives most information on the qualitative effects, the decision to chose the IBN in most of the following sections rather than the SNR is argued as follows: the SNR is only meaningful by simultaneously giving the used input amplitude, which complicates the understanding of simulation figures. Giving the maximum SNR, beside ideal and nonideal quantization noise, also distortion is seen, even without nonideal effects. In contrast, by giving the integrated IBN and choosing a small input amplitude, only the influence of a certain nonideality on the ideal quantization noise is illustrated, while by giving the IBN together with a high signal amplitude, also signal and nonideality dependent distortion appears. Still, we think this choice is a question of personal gusto, and typically circuit designers like more the illustration of the SNR, while engineers focusing on the system behavior prefer the IBN as a measure.

4

DAC Nonidealities in Continuous-Time $\Sigma\Delta$ Modulators

This section covers the influence of the nonideal behavior of the feedback digital-to-analog converter of continuous-time $\Sigma\Delta$ modulators. This is of special importance, since the CT implementation features an integration over time of a pulse-shaped feedback waveform. Thus, it is sensitive to nonidealities, which are not commonly seen in their DT counterparts. The calculation examples throughout this chapter focus on single-loop and multi-loop modulators with low-pass loop filter characteristic. Discussions on other implementations are given if useful and feasible. Additionally, possibilities to reduce or cancel the influence of the different nonidealities are reviewed and presented.

4.1 Feedback DAC Error Classification

Beyond the nonidealities of the filters or integrators within a $\Sigma\Delta$ modulator loop, the second major error contributor is the feedback digital-to-analog converter. This is easily understood from Fig. 3.23 and Sect. 3.8.1, where all DAC induced errors directly add to the input of the overall modulator through the first, outermost feedback path. Thus, equivalently to errors from the first integrator like noise and distortion, similar performance limiting behavior can be attributed to the feedback DAC.

If a typical rectangular feedback waveform is assumed, from Fig. 3.4 most different feedback DAC errors can be derived: first, obviously the feedback pulse is affected by timing errors, which varies its position and length. Here, usually two different effects are considered: the constant delay of the pulse by τ_d, known as *excess loop delay* [132], and the statistical variation of any edge or even the duration of the sampletime T_S in Fig. 3.4 caused by *clock jitter* [11, 12]. Also the edges can be nonideally rising and falling due to finite slew rate [3, 191]. Finally, the DAC levels can be affected by mismatch, which shows its influence mainly if multibit implementations are used, and which is therefore only briefly considered.

4.2 Excess Loop Delay in Continuous-Time $\Sigma\Delta$ Modulators

In [14, 105, 132] a timing nonideality has been considered, namely the excess loop delay, which is supposed to be a constant delay t_d between the ideal and the implemented feedback DAC pulse:

$$t_\mathrm{d} = \tau_\mathrm{d} T_\mathrm{S}\,, \tag{4.1}$$

which can arise from two different effects: first, due to a finite respond time of the DAC-output to the clock edges and its input, both of which are nonwanted, but are intrinsic characteristics of an integrated circuit [3]; additionally, there can be designed delays between the quantizer clocking edge, determining the sampling operation, and the subsequent latch feeding the DAC, in order to give the quantizer more decision time [192].

Principally, excess loop-delay causes two different nonideal effects in CT ΣΔ modulators, which are separately discussed in the following:

- Loop delay that shifts the DAC pulse but retains the entire pulse in the actual sampling instant. This is achieved if an RZ-DAC pulse is delayed no further than to the end of the current sampling period, see Fig. 4.1.
- Loop delay that shifts the DAC pulse in such a way that a part of it extends into the next sampling instant. This is obtained in either case for an NRZ-DAC with delay as shown in Fig. 4.2. Especially, in high speed ΣΔ modulators excess loop delay becomes so significant that even RZ pulses are shifted into the next clock cycle. Consequently, this shift is the more important case!

4.2.1 Coefficient Mismatch Through Excess Loop Delay

The scaling of a CT ΣΔ modulator, e.g., through DT–CT conversion, is done for a special feedback DAC pulse form and position, see Sect. 3.2. Consequently, a constant loop delay inside the desired sampling period as shown in Fig. 4.1, leads to coefficient mismatch: calculating and implementing the scaling coefficients for one special set of $\{\alpha T_\mathrm{S}, \beta T_\mathrm{S}\}$, but facing a shift of the feedback pulse from these designed values $\{\alpha T_\mathrm{S} + t_\mathrm{d}, \beta T_\mathrm{S} + t_\mathrm{d}\}$, results a mismatch of designed and required modulator scaling. Therefore, excess loop delay yields increased quantization in-band noise and possibly reduced stability (maximum stable amplitude) due to this scaling mismatch. Nonetheless, the order of the loop filter and the noise-transfer function remain unchanged.

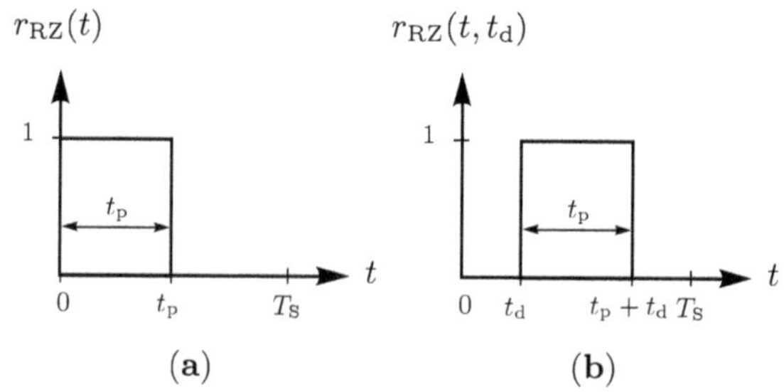

Fig. 4.1. RZ-DAC pulse. (**a**) Without loop delay t_d; (**b**) with loop delay t_d

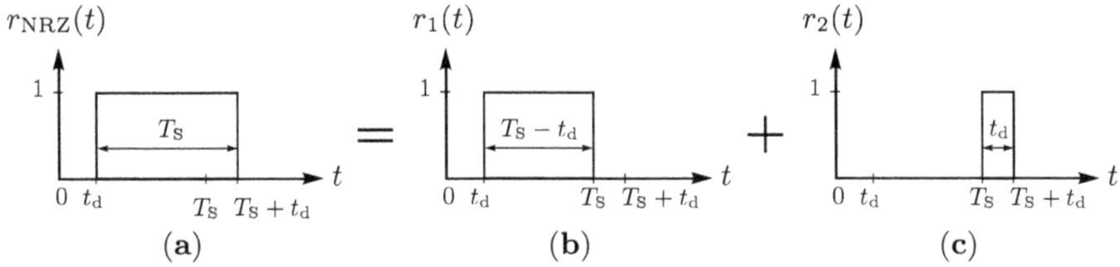

Fig. 4.2. Linear combination of (**a**) an NRZ-DAC pulse with loop delay t_d, established by (**b**) a DAC pulse in the present and (**c**) in the subsequent sampling instant

4.2.2 Increased Modulator Order Through Excess Loop Delay

In contrast, by using an NRZ pulse form every delay shifts a part of the feedback pulse into the next clock cycle; this also happens in the case of an RZ pulse, if the delay t_d exceeds the time slot between the end of the pulse and the end of the sample: $t_d > (1 - \beta)T_S$. For example [132] stated that a shift of the pulse into the next clock cycle increases the actual modulator order, which decreases the maximum stable amplitude and decreases the noise shaping performance.

Exemplarily, a second-order modulator as in Fig. 3.6b is considered incorporating a feedback DAC with ideal, NRZ pulse position $\{\alpha = 0, \beta = 1\}$, with an excess loop delay τ_d. Thus, the delayed pulse position becomes $\{\alpha = \tau_d, \beta = 1 + \tau_d\}$, illustrated in Fig. 4.2a. For further calculation, this pulse is split into two separate feedback pulses, one in the actual sample with a position given by $\{\alpha_1 = \tau_d, \beta_1 = 1\}$ and one part shifted into the next sample interval, i.e., delayed by exactly one sampletime, with a position given by $\{\alpha_2 = 0, \beta_2 = \tau_d\}$, shown in Fig. 4.2b, c [132]. With these two different feedback pulses, one obtains with Table 3.6 the equivalent DT loop filter of the first and second branch in Fig. 3.6b (with $f_S = 1$):

$$\begin{aligned}\frac{-k_1}{s^2} \to \mathrm{LF}_{\mathrm{CT-DT}}|_1 &= \frac{-k_1(\beta_1(2-\beta_1) - \alpha_1(2-\alpha_1))z - k_1(\beta_1^2 - \alpha_1^2)}{2(z-1)^2} \\ &\quad + \frac{-k_1(\beta_2(2-\beta_2) - \alpha_2(2-\alpha_2))z - k_1(\beta_2^2 - \alpha_2^2)}{2(z-1)^2} z^{-1}, \end{aligned} \tag{4.2}$$

$$\frac{-k_2}{s} \to \mathrm{LF}_{\mathrm{CT-DT}}|_2 = \frac{-k_2(\beta_1 - \alpha_1)}{z-1} + \frac{-k_2(\beta_2 - \alpha_2)}{z-1} z^{-1},$$

where the z^{-1} in the second terms indicates the shift by one sample. In combination, both $\mathrm{LF}_{\mathrm{CT-DT}}|_i$ form together the DT representation of the CT modulator with excess loop delay τ_d:

$$\begin{aligned}\mathrm{LF}_{\mathrm{2nd}}(z, \tau_d) = ((&-k_1\tau_d^2 + (2k_2 + 2k_1)\tau_d - 2k_2 - k_1)z^2 \\ &+ (2k_1\tau_d^2 + (-2k_1 - 4k_2)\tau_d + 2k_2 - k_1)z \\ &+ 2k_2\tau_d - k_1\tau_d^2)/(2z(z-1)^2)\,. \end{aligned} \tag{4.3}$$

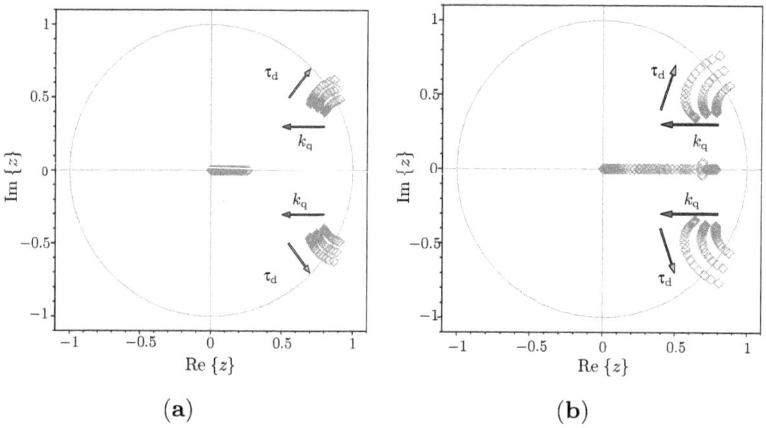

Fig. 4.3. Root loci of the NTF poles of (**a**) a second-order ($k_1 = 0.25,\ k_2 = 0.375$) and (**b**) a third-order ($k_1 = 0.05,\ k_2 = 0.3,\ k_3 = 0.64$) low-pass ΣΔ modulator with respect to excess loop delay τ_d and quantizer gain k_q

Note that for $\tau_\mathrm{d} = 0$, (4.3) becomes the ideal DT loop filter of a second-order modulator as in Fig. 3.6a. Expressions for different modulator orders or architectures are similarly found. Obviously, the loop filter in (4.3) contains terms to the power z^{-3}, thus indicating that the second-order, CT modulator with excess loop delay is now a third-order system.

The effect of excess loop delay τ_d on the NTF poles is illustrated in Fig. 4.3a, b for the considered second-order and also for a third-order modulator. Therefore, the NTF and its poles were calculated as in (2.19) with loop filters as the one in (4.3). τ_d was increased from 0.01 to 100% while a quantizer gain k_q was varied from 0.8 to 1.2. The poles move toward the unit cycle and exceed the stability boundary for a certain loop delay. Thus, the existence of excess loop delay may lead to an unstable modulator. Nevertheless, the NTF zeros remain on the unit circle, with an additional zero at $z = 0$. This is in accordance to the usually seen low influence of excess loop delay on the noise shaping behavior up to a certain amount of excess loop delay.

4.2.3 Alternative Approach to the Effect of Excess Loop Delay

Another approach to understand the effect of excess loop delay in continuous-time ΣΔ modulators was presented in [14], which can also be derived from the expression for the impulse-invariant transformation in (3.11). Here, the equivalence of the DT and the CT loop filter are achieved by mapping their impulse responses. Therewith, the poles of the loop filter and the order of the denominator are conserved. By introducing a delay in the DAC transfer function $R_\mathrm{DAC}(s)$, this relation is modified:

- In case of an RZ-DAC with the pulse not being shifted into the next clock cycle, the numerator order of the equivalent transfer function is the same as initial. For example for a second-order modulator, the numerator is of second-order; thus, with the two scaling coefficients and consequently two degrees of freedom, (3.11) can be resolved. This corresponds to adapted scaling coefficients.
- In case of a feedback pulse shifted into the next clock cycle, the numerator order of the equivalent transfer function increases by one; thus, with the original number of scaling coefficients, the equivalence in (3.11) cannot be fulfilled. Consequently, one has to introduce one more degree of freedom, which is outlined in Sect. 4.2.4.

4.2.4 Compensation for Excess Loop Delay in CT ΣΔ Modulators

Since excess loop delay cannot be completely avoided in a CT ΣΔ modulator circuit implementation, especially for high speed circuits it is necessary either to chose an architecture which is mostly insensitive to it, or to compensate for its influence.

Coefficient Mismatch

For RZ feedback pulses, rectangular or pulse-shaped, excess loop delay only results a gain error up to a certain amount of delay, see Sect. 4.2.1. But as obvious as the coefficient mismatch itself is the possibility for its compensation by coefficient tuning, as has been proposed in [132, 140]. Therefore, the scaling coefficients are not calculated for the ideal pulse position, but for its delayed version $\{\alpha T_\mathrm{S} + \tau_\mathrm{d}, \beta T_\mathrm{S} + \tau_\mathrm{d}\}$. Note that this method only works for any RZ pulse form, which is shifted inside one sample and not into the next clock cycle.

Increased Modulator Order

For compensating increased modulator order due to excess loop, a technique proposed in [14, 132] can be adopted, which is illustrated in Fig. 4.4: The pulse shape of the auxiliary feedback DAC, denoted with $\mathrm{DAC_{HRZ}}$, can be arbitrarily chosen, but it has to be delayed into the next clock cycle just as the feedback pulse of the main DAC, while concurrently to show a part of the pulse in the actual clock cycle. A feasible implementation is a half delay, RZ pulse, i.e., $\{\alpha_\mathrm{h} = 1/2,\ \beta_\mathrm{h} = 1\}$. Additionally, it can be fed back to any integrator input node in the loop.

The method to calculate the new scaling coefficients k_i^* which achieve compensation for excess loop delay in Fig. 4.4 is exemplarily presented for the second-order modulator: The idea is to determine the equivalent DT loop filter of the nonideal modulator in Fig. 4.4 with excess loop delay. Subsequently, this

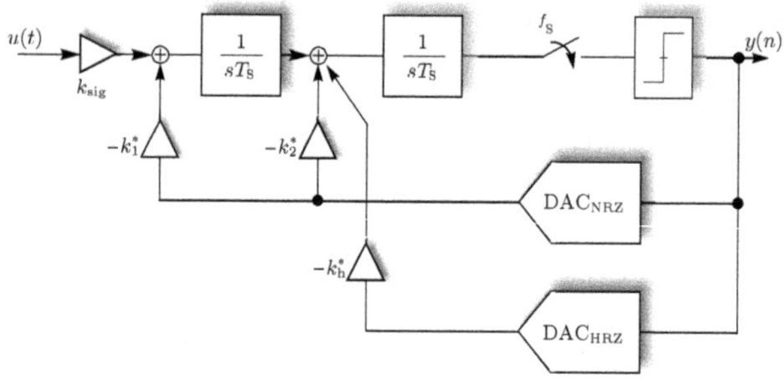

Fig. 4.4. Second-order CT ΣΔ modulator structure with excess loop delay compensation by using an auxiliary feedback DAC

DT transfer function is matched to the original, ideal loop filter. By using Table 3.6 the loop filter of the first branch in Fig. 4.4 (with NRZ feedback DAC and $f_S = 1$) yields:

$$\begin{aligned}\mathrm{LF_{CT}}|_1 = \frac{-k_1}{s^2} &\rightarrow \mathrm{LF_{CTDT}}|_1 \\ &= \frac{-k_1(\beta_{11}(2-\beta_{11}) - \alpha_{11}(2-\alpha_{11}))z - k_1(\beta_{11}^2 - \alpha_{11}^2)}{2(z-1)^2} \\ &+ \frac{-k_1(\beta_{21}(2-\beta_{21}) - \alpha_{21}(2-\alpha_{21}))z - k_1(\beta_{21}^2 - \alpha_{21}^2)}{2(z-1)^2} z^{-1}, \quad (4.4)\end{aligned}$$

where $\{\alpha_{11} = \tau_d,\ \beta_{11} = 1\}$ and $\{\alpha_{21} = 0,\ \beta_{21} = \tau_d\}$. For the loop filter of the second branch applies:

$$\mathrm{LF_{CT}}|_2 = \frac{-k_2}{s} \rightarrow \mathrm{LF_{CTDT}}|_2 = \frac{-k_2(\beta_{12} - \alpha_{12})}{z-1} + \frac{-k_2(\beta_{22} - \alpha_{22})}{z-1} z^{-1}, \quad (4.5)$$

where $\{\alpha_{12} = \tau_d,\ \beta_{12} = 1\}$ and $\{\alpha_{22} = 0,\ \beta_{22} = \tau_d\}$. For the loop filter of the compensation branch with the HRZ feedback DAC one obtains:

$$\mathrm{LF_{CT}}|_h = \frac{k_h}{s} \rightarrow \mathrm{LF_{CTDT}}|_h = \frac{k_h(\beta_{1h} - \alpha_{1h})}{z-1} + \frac{k_h(\beta_{2h} - \alpha_{2h})}{z-1} z^{-1}, \quad (4.6)$$

where $\{\alpha_{1h} = 1/2 + \tau_d,\ \beta_{1h} = 1\}$ and $\{\alpha_{2h} = 0,\ \beta_{2h} = \tau_d\}$. All three $\mathrm{LF_{CTDT}}|_i$ form together the DT representation of the CT modulator in Fig. 4.4 with excess loop delay τ_d in the feedback path. Optimally, this DT loop filter equals the original DT modulator loop filter in (4.7):

$$\mathrm{LF_{2nd}}|_{\mathrm{ideal}}(z) = -\frac{a_1 a_2}{(z-1)^2} - \frac{a_2}{z-1} \quad (4.7)$$

with optimally DT–CT converted scaling coefficients in (3.21), this becomes:

$$\mathrm{LF_{2nd}}|_{\mathrm{ideal}}(z) = -\frac{1}{2}\frac{(k_1 + 2k_2)z}{(z-1)^2} - \frac{1}{2}\frac{k_1 - 2k_2}{(z-1)^2}. \quad (4.8)$$

Note, this ideal loop filter is also found by setting $\tau_\mathrm{d} = 0$ in (4.3). By solving the combined equation from (4.8) and (4.4)–(4.6)

$$\mathrm{LF}_{\mathrm{2nd}}|_{\mathrm{Fig.\,4.4}}(z) = \sum_i \mathrm{LF}_{\mathrm{CTDT}}|_i \stackrel{!}{=} \mathrm{LF}_{\mathrm{2nd}}|_{\mathrm{ideal}}(z)\,, \tag{4.9}$$

one obtains the *new* scaling coefficients of the compensated modulator in Fig. 4.4:

$$k_1^* = k_1, \quad k_2^* = \frac{3k_1\tau_\mathrm{d}}{2} + 2k_2, \quad k_\mathrm{h}^* = k_1\tau_\mathrm{d} + 2k_2\,, \tag{4.10}$$

where k_i^* are the scaling coefficients of the compensated modulator, while k_i are the coefficients of the original, ideal modulator. Similarly, coefficients for other modulator orders or architectures can be found. The advantage of this technique is that it ideally cancels out the influence of feedback excess loop delay, while it requires only an easy additional DAC implementation as the HRZ one.

Alternatively, a technique proposed in [193] can be used. It applies an auxiliary feedback path after the last integrator (e.g., on a second-order modulator, Fig. 4.5). The compensation is exemplarily illustrated for the second-order modulator in the following: Again, the idea is to calculate the equivalent DT loop filter of the modulator in Fig. 4.5 and subsequently to find the coefficients k_i^* thus as to yield an ideal DT modulator loop filter.

The equivalent DT loop filter of a second-order, NRZ feedback modulator with excess loop delay τ_d was given in (4.3). The additional feedback path in Fig. 4.5, adds a new term into that loop filter, which yields:

$$\mathrm{LF}_{\mathrm{2nd}}|_{\mathrm{Fig.\,4.5}}(z) = \mathrm{LF}_{\mathrm{2nd}}(z, \tau_\mathrm{d}) - \frac{k_\mathrm{h}^*}{z}\,. \tag{4.11}$$

By comparing this DT equivalent of the modulator in Fig. 4.5 with excess loop delay to the ideal loop filter of a second-order modulator, either (4.8) or (4.3) with $\tau_\mathrm{d} = 0$, the scaling coefficients for the compensated modulator are found:

$$k_1^* = k_1, \quad k_2^* = k_1\tau_\mathrm{d} + k_2, \quad k_\mathrm{h}^* = \frac{k_1\tau_\mathrm{d}^2}{2} + k_2\tau_\mathrm{d}\,, \tag{4.12}$$

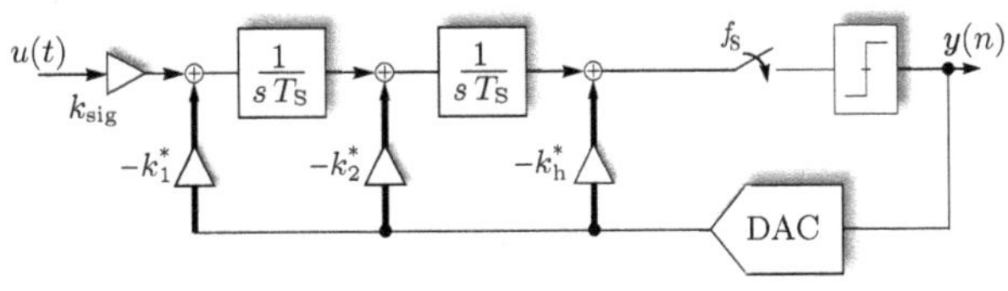

Fig. 4.5. Second-order CT ΣΔ modulator with excess loop delay cancellation by using an additional feedback path

where k_i^* are the scaling coefficients of the compensated modulator in Fig. 4.5, while k_i are the coefficients of the original, ideal modulator. Similarly, coefficients for other modulator orders or architectures can be found. The disadvantage of this concept is that it requires significant hardware: most important in CT modulator, which employ current mode feedback is the auxiliary current-to-voltage converter. Alternatively, a current-mode quantizer has to be chosen [21, 126].

Robustness of the Compensation Approach

The above introduced compensation procedures are feasible structures to counteract excess loop delay in continuous-time ΣΔ modulators. But another interesting point arises from the robustness of the compensation approach and the corresponding architectures, especially in the presence of integrator gain variations, which can be very high in CT filters, see Sect. 5.3.

Here, it turns out that the compensation with either the additional feedback path or DAC even improves the modulator robustness [3]. A feasible measure for the sensitivity and its improvement was proposed in [16], where a filter gain margin is used. This gain margin is the ratio of the minimal to maximal value by which the filter transfer function may be varied until the modulator performance drops below a given limit. Interestingly, for both bandpass and low-pass continuous-time ΣΔ modulators employing one of the compensation architectures shown in Figs. 4.4 and 4.5, the optimal robustness in terms of maximum gain margin occurs for nonzero excess loop delay, i.e., it seems feasible to explicitly introduce an excess loop delay and to use one of the compensation schemes to achieve better reliability [16].

4.2.5 Simulation Results on Excess Loop Delay

Simulation results of a second-order (Fig. 3.6b) and third-order (Fig. 3.12a) CT modulators as well as of the CT 2–1–1 cascade (Fig. 3.8b) are presented in Fig. 4.6a, c. Additionally, both compensation approaches are exemplarily simulated for the second-order modulator in Fig. 4.6d.

In order to show the reduced maximum stable amplitude as well as the better behavior with RZ feedback for the uncompensated modulators, the simulations were performed at one low and one high input signal magnitude, and additionally for NRZ and RZ feedback DAC. Also the compensated second-order modulators were simulated for the technique in Figs. 4.5 and 4.4 for one low and one high input amplitude, but only for NRZ feedback!

The expected behavior is clearly confirmed: first, the tolerable loop delay decreases with higher amplitude and NRZ feedback. Furthermore, the higher third-order single-loop modulator is more sensitive due to the higher order modulator in the first stage, which is prone to instability (with even increased order due to excess loop delay). Note that the loss in performance of the single-loop second-order and cascaded CT 2–1–1 modulator is approximately

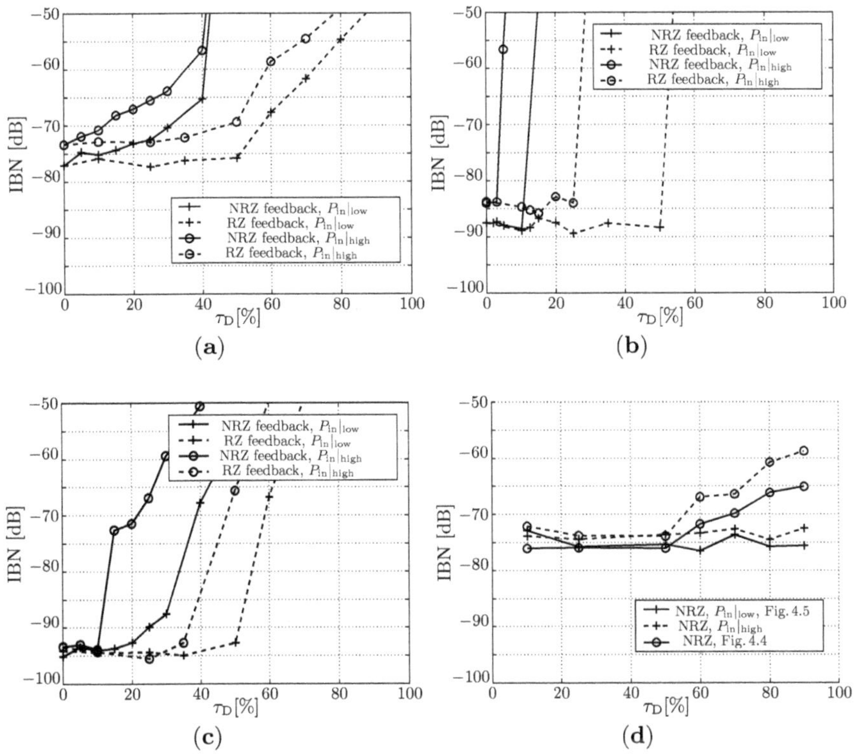

Fig. 4.6. Simulated CT modulators, NRZ and RZ implementation, under the influence of excess loop delay τ_D. $f_{sig} = f_{\overline{B}}/3$ and $P_{in}|_{high} = -9$ dB or $P_{in}|_{low} = -23$ dB. (**a**) CT second-order, OSR = 48; (**b**) CT third-order, OSR = 48; (**c**) CT 2–1–1, OSR = 24, after [32]; (**d**) compensated CT second-order NRZ

comparable: this gives raise to the conclusion that the influence of the induced gain errors is not critical and the first stage dominates the nonideal behavior.

In contrast, by choosing a RZ feedback pulse form $\{\alpha_{ideal} = 0,\ \beta_{ideal} = 1/2\}$, without any countermeasures a loop delay up to almost $\tau_d \approx 40\%$ can be tolerated for the second-order single-loop and 2–1–1 modulator, while still a $\tau_d \approx 25\%$ is acceptable for the third-order single-loop modulator.

By using the presented compensation approach, an enhanced nonsensitive behavior is also achieved for NRZ feedback DAC. Note that the additional HRZ feedback DAC only works as intended, as long as it partially remains in the original clock cycle. Thus, for $\tau_d > 50\%$, the performance starts to decrease. Alternatively, a more complicated delayed RZ pulse can be used, e.g., $\{\alpha_h = 1/4,\ \beta_h = 1\}$ [132], which would increase the *nonsensitive* range to $\tau_d = 75\%$.

In contrast, the compensation through the additional compensation path to the input of the quantizer shows almost ideal performance until the simulation limit $\tau_d = 100\%$, which is its clear advantage over the other technique.

4.2.6 Extension to Other Architectures

The effect of excess loop delay, i.e., coefficient mismatch, increased in-band noise, and reduced stability (maximum stable amplitude), hold true for all architectures (feedforward, feedback, etc.) and feedback waveforms (rectangular, pulse-shaped). Only the severity of the influence alters, as could be seen for the RZ and the NRZ modulators. Nonetheless, also the compensation approaches, which were discussed in Sect. 4.2.4 can be adopted: for any feedback pulse waveform, which is not shifted into the next clock cycle, the method of coefficient tuning is feasible.

For significantly large loop delay or NRZ feedback pulses, the additional feedback path, either with a second DAC waveform (different from the original one), Fig. 4.4, or with a simple additional feedback path, Fig. 4.5, can be used. For example, this was done for bandpass modulators in [14, 105, 132], or for feedforward structures in [21, 126].

4.3 Clock Jitter in Continuous-Time ΣΔ Modulators

Clock jitter, i.e., statistical variations of the sampling frequency, depends on the purity of the clock source. Typically, ΣΔ modulators were found to be rather tolerant to timing jitter; [110] noted that compared to the Nyquist rate converter, the jitter sensitivity can be improved by the oversampling ratio of the modulator. Nonetheless, this advantage holds only true, as long as discrete-time implementations are considered.

In contrast, continuous-time modulators are affected more severely by clock jitter than their discrete-time counterparts, as shown manifold in the published literature [8, 9, 11, 12, 24, 122]; most commonly this is regarded as the major disadvantage of CT ΣΔ implementations. The intention of the following discussions is to review this dependency and to give general, while easily understandable, expressions, which serve as a basis for the jitter-tolerant CT ΣΔ modulator architectures reviewed in Sect. 4.3.4.

4.3.1 Jitter Effects in CT ΣΔ Modulators

The difference of clock jitter influence on DT and CT ΣΔ modulators appears since both feature different sources of clock jitter errors in the modulator loop [11]. Figure 4.7 shows typical block diagrams of a DT and a CT ΣΔ modulator with respect to the influence of clock jitter [11, 132].

Suppose nonuniform sampling takes place, the decision of the quantizer in DT modulators remains correct in first-order approximation, because the integrators are designed to settle to a given accuracy within half the sampling period. Therefore, the performance degradation due to clock jitter is primarily caused by errors in the front-end S/H circuit, where jitter caused sampling misalignment produces an equivalent amplitude error that degrades the SNR.

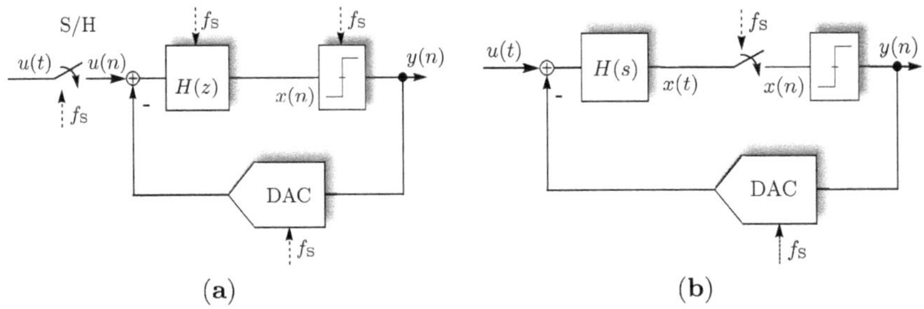

Fig. 4.7. Jitter error sources in typical DT and CT ΣΔ modulators, after [31]. (**a**) DT modulator; (**b**) CT modulator

Nonetheless, the introduced noise is rather uncritical for most applications [9, 11, 110, 194], and the increase of the IBN due to clock jitter is proportional to OSR^{-3}.

A CT implementation faces two error sources due to clock jitter: first, similar to the sampling errors in the DT modulator, the sampled internal quantizer is prone to jitter affected sampling errors. But these errors enter the system at the point of maximum error suppression, i.e., at the quantizer, and hence may be neglected in practice. The dominant influence of clock jitter in CT implementations appears through errors resulting from the feedback DAC: this is because a CT ΣΔ modulator integrates the feedback waveform over time. Thus, a statistical variation of the feedback waveform results in a statistical integration error and consequently in increased noise!

Effect of Pulse Delay Jitter in CT Modulators

Pulse delay jitter is not the dominant effect in CT ΣΔ modulators [24], which is shortly explained as follows: When considering a CT loop filter scaling, as exemplarily shown for a second-order modulator in Sect. 3.2.1, it turns out that the CT feedback integrator scaling coefficient k_1 of the first branch is linearly varying with pulse length, but independent from its absolute position, see (3.21). Consequently, the influence of the feedback signal of the first branch on the system behavior is not changing, when the feedback pulse is moved inside its own sample.

In contrast the feedback scaling coefficients of later feedback branches are indeed dependent on the absolute pulse position $\{\alpha,\ \beta\}$, as can be seen, e.g., in (3.21). Thus, a jitter induced pulse delay will cause an error which enters the system here, but which will be at least first-order noise-shaped and has therefore a negligible effect, which is in accordance to the simulations in [24].

Effect of Pulse Width Jitter in CT Modulators

A variation of the length of a feedback pulse directly modulates the amount of feedback charge, which appears during the corresponding clock cycle. Due to the commonly used rectangular pulse form and the corresponding constant value, this error is directly proportional to the clock jitter variance σ_t^2, while it adds directly to the modulator input through the outermost feedback branch.

The resulting jitter noise depends on the amount of timing error in each clock cycle σ_t^2, the magnitude of the feedback pulse at the time of jitter affected switching, and the feedback path with its scaling coefficients, through which the jitter noise enters the modulator.

4.3.2 Calculation of the Jitter Influence for Rectangular Feedback

The noise performance of CT ΣΔ modulators incorporating a rectangular RZ-DAC under clock jitter influence is estimated. For calculation purposes, it is usually assumed that the variations of the sampling instances follow a uniformly distributed random process, which makes clock jitter to cause a white noise, flat spectral increase of the modulator output [11, 24, 34]. Beside the uniformly distributed random clock jitter, another important type of clock jitter is the accumulated one [3, 195, 196]. The difference will be addressed further in Sect. 4.3.8.

For reasons of simplicity, first only single-bit implementations are regarded, while multibit quantization is discussed in Sect. 4.3.3. A typical RZ pulse stream with amplitude $\hat{I}_{\mathrm{DAC}}$ is shown in Fig. 4.8a. Here, in every clock cycle a RZ pulse is stimulated with appropriate polarity and an amplitude of $\hat{I}_{\mathrm{DAC}}$. The RZ pulse can be described with:

$$I_{\mathrm{DAC}} = \begin{cases} \hat{I}_{\mathrm{DAC}}, & \alpha T_{\mathrm{S}} \leq t < \beta T_{\mathrm{S}} \\ 0, & \text{otherwise}, \end{cases} \tag{4.13}$$

where $\{\alpha,\ \beta\}$ are the start and end point of the rectangular RZ-DAC pulse. An integration of I_{DAC} over one period, gives the amount of feedback charge in one cycle. Including a jitter timing error t_{j} to the integration limits, yields

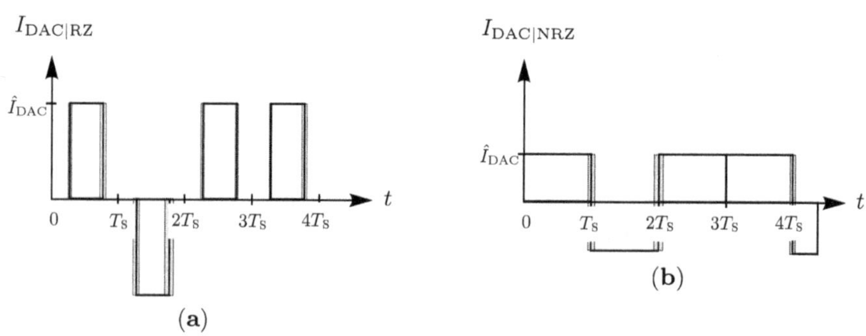

Fig. 4.8. RZ- and NRZ-DAC pulse sequence of $\{1, -1, 1, 1\}$ under clock jitter influence. The gray shaded areas depict the sampling uncertainty. (**a**) RZ-DAC pulse; (**b**) NRZ-DAC pulse

the feedback charge under the influence of clock jitter. To ease the subsequent expressions, first only single-sided jitter is assumed:

$$Q_{\mathrm{DAC}} = \int_{\alpha T_{\mathrm{S}}}^{\beta T_{\mathrm{S}}+t_{\mathrm{j}}} I_{\mathrm{DAC}}\,\mathrm{d}t = \hat{I}_{\mathrm{DAC}}\left(t_{\mathrm{j}} + (\beta-\alpha)\,T_{\mathrm{S}}\right). \tag{4.14}$$

To obtain the jitter induced feedback charge error $\mathfrak{E}_{Q_{\mathrm{DAC}}}$ within one period, (4.14) is subtracted from the ideal case (jitter free, $t_{\mathrm{j}} = 0$), which yields:

$$\mathfrak{E}_{Q_{\mathrm{DAC}}} = \hat{I}_{\mathrm{DAC}}\,t_{\mathrm{j}}\,. \tag{4.15}$$

The variance of a function $g(t)$ of a statistical variable t, where the variable t is distributed with the probability density function $f(t)$, is calculated using:

$$\sigma_{g(t)}^2 = \int_{t=-\infty}^{+\infty} \left(g(t) - \langle G\rangle\right)^2 f(t)\mathrm{d}t\,, \tag{4.16}$$

where $\langle G\rangle$ is the expected mean value of $g(t)$, defined as:

$$\langle G\rangle = \mathcal{E}[g(t)] = \int_{t=-\infty}^{+\infty} g(t)f(t)\mathrm{d}t\,. \tag{4.17}$$

By assuming the value of clock jitter t_{j} at the different sampling times being distributed as dc-free Gaussian, white noise with a standard deviation of σ_t, the probability density function $f(t)$ becomes:

$$f(t)|_{\mathrm{Gauss}} = \frac{1}{\sqrt{2\pi}}\,\mathrm{e}^{-t^2/2\sigma_t^2}\,. \tag{4.18}$$

By choosing (4.15) as $g(t)$ in (4.17), the resulting mean value of the jitter noise becomes:

$$\mathcal{E}[\mathfrak{E}_{Q_{\mathrm{DAC}}}] = \int_{t=-\infty}^{+\infty} \left(\hat{I}_{\mathrm{DAC}}\,t_{\mathrm{j}}\,\frac{1}{\sqrt{2\pi}}\,\mathrm{e}^{-t_{\mathrm{j}}^2/2\sigma_t^2}\,\mathrm{d}t_{\mathrm{j}}\right) = 0\,. \tag{4.19}$$

By choosing again (4.15) as $g(t)$, (4.19) as $\langle G\rangle$ and the Gaussian distribution in (4.18) as $f(t)$ in (4.16), the variance σ_e^2 of the jitter induced charge error $\mathfrak{E}_{Q_{\mathrm{DAC}}}$ for an RZ feedback DAC is found:

$$\sigma_e^2|_{\mathrm{RZ}} = \int_{t=-\infty}^{+\infty} \left[\left((\hat{I}_{\mathrm{DAC}}\,t_{\mathrm{j}}) - 0\right)^2 \frac{1}{\sqrt{2\pi}}\,\mathrm{e}^{-t_{\mathrm{j}}^2/2\sigma_t^2}\right]\mathrm{d}t_{\mathrm{j}} = \hat{I}_{\mathrm{DAC}}^2\sigma_t^2\,. \tag{4.20}$$

If the calculation is done with a two sided jittered RZ pulse, i.e., in (4.14) both integration limits are jittered as also shown in Fig. 4.8a, this results in an additional factor of 2 in (4.20), under the assumption that both jitter instants are statistically independent! This can be included by an activity factor, which is typically $A_{\mathrm{RZ}} = 2$ for an RZ-DAC implementation, see Fig. 4.8a.

The expression in (4.20) is the variance of the charge error due to jitter in the RZ feedback DAC pulse. Translating this charge into a signal or noise variance, it has to be related to time, i.e., to one clock period:

$$E_{\mathrm{j}}^2|_{\mathrm{RZ}} = \frac{\sigma_e^2|_{\mathrm{RZ}}}{T_{\mathrm{S}}^2} = \hat{I}_{\mathrm{DAC}}^2 \left(\frac{\sigma_t}{T_{\mathrm{S}}}\right)^2 A_{\mathrm{RZ}}\,. \tag{4.21}$$

The amplitude of the DAC pulse is determined by the feedback scaling coefficient and the DAC step width $\varDelta$. If a single-bit modulator is assumed:

$$\hat{I}_{\mathrm{DAC}} = k_i|_{\mathrm{RZ}} \frac{\varDelta}{2}\,. \tag{4.22}$$

The jitter noise in (4.21) enters the modulator through each feedback path, with different amplitudes determined by $k_i|_{\mathrm{RZ}}$, which is the RZ feedback scaling coefficient of the ith feedback branch . In order to obtain the resulting in-band noise, these noise components are related to the modulator input as shown in Sect. 3.8.1. The dominant error through the outermost first branch yields:

$$\mathrm{IBN}_{\sigma_t}|_{\mathrm{RZ}} \approx \frac{E_1^2|_{\mathrm{RZ}}}{k_1^2|_{\mathrm{NRZ}}\mathrm{OSR}} = \left(\frac{\varDelta}{2}\right)^2 \left(\frac{k_1|_{\mathrm{RZ}}}{k_1|_{\mathrm{NRZ}}}\right)^2 \left(\frac{\sigma_t}{T_{\mathrm{S}}}\right)^2 \frac{A_{\mathrm{RZ}}}{\mathrm{OSR}}\,, \tag{4.23}$$

where $\varDelta$ is the DAC full scale range in the case of single-bit internal quantization. This calculation procedure is adoptable to every feedback waveform, which will be subsequently shown in a glance. But first the expression in (4.23) has to be adapted to the second rectangular DAC implementation, i.e., the NRZ pulse in Fig. 4.8b.

In the case of NRZ feedback pulses, two effects are important to note: first, jitter affects the transferred feedback charge only, if the feedback signal varies its state. In other words, since the jitter noise only depends on the difference between two adjacent feedback pulses, in a NRZ implementation the activity of the pulse stream determines the sensitivity to jitter. Consequently, the above introduced activity factor, which was $A_{\mathrm{RZ}} = 2$ since always two edges per clock cycle are jittered in a RZ pulse, see Fig. 4.8a, has to be reduced *below one* for the NRZ pulse, since a state transition does not occur in every cycle: several authors found the NRZ activity factor for single-bit quantization to be $A_{\mathrm{NRZ}} \approx 0.7$ for large input signals [3, 8, 11].

Additionally, if a transition occurs in a NRZ feedback DAC, the step is always a complete step-width $\varDelta$, while this was only $\varDelta/2$ in the case of the RZ feedback, see Fig. 4.8. Thus, in contrast to (4.22) for the NRZ modulator the jittered edge transition has an amplitude of:

$$\hat{I}_{\mathrm{DAC}|\mathrm{NRZ}} = k_i|_{\mathrm{NRZ}} \varDelta\,. \tag{4.24}$$

Consequently, the jitter induced in-band noise for a CT ΣΔ modulator with rectangular, NRZ feedback DAC yields:

$$\mathrm{IBN}_{\sigma_t}|_{\mathrm{NRZ}} \approx \Delta^2 \left(\frac{k_1|_{\mathrm{NRZ}}}{k_1|_{\mathrm{NRZ}}}\right)^2 \left(\frac{\sigma_t}{T_{\mathrm{S}}}\right)^2 A_{\mathrm{NRZ}} = \Delta^2 \left(\frac{\sigma_t}{T_{\mathrm{S}}}\right)^2 \frac{A_{\mathrm{NRZ}}}{\mathrm{OSR}}, \quad (4.25)$$

where Δ is the DAC full scale range in the case of single-bit internal quantization.

In order to estimate the ratio between clock jitter noise in single-bit CT ΣΔ modulators employing RZ or NRZ feedback DAC, note that the RZ scaling coefficient in (4.23) is typically twice the value than the NRZ scaling coefficient, see (3.21) with $\{\beta - \alpha = 0.5\}$. Thus,

$$\frac{\mathrm{IBN}_{\sigma_t}|_{\mathrm{RZ}}}{\mathrm{IBN}_{\sigma_t}|_{\mathrm{NRZ}}} = \frac{A_{\mathrm{RZ}}}{A_{\mathrm{NRZ}}} \approx \frac{2}{0.7} \approx 4.5\,\mathrm{dB}, \quad (4.26)$$

which is in direct agreement with [12]. Consequently, for reduced jitter sensitivity a NRZ feedback DAC is more feasible than its RZ counterpart!

4.3.3 Reduction of Clock Jitter Influence Using Multibit DACs

Commonly, the only known method for the reduction of the clock jitter sensitivity in CT ΣΔ modulators has been to use multibit internal quantization: Fig. 4.8 illustrates that a single-bit feedback is the worst case concerning the jitter sensitivity. When using multibit feedback and NRZ pulses, the difference of two adjacent feedback pulses will differ mostly by only one LSB, thus reducing the jitter influence by orders of magnitude [8, 126, 191], as illustratively shown in Fig. 4.9a. Simultaneously, simulations show that in the case of multibit quantization, the activity factor of the NRZ feedback DAC increases from 0.7 (see (4.25)) to approximately $A_{\mathrm{NRZ,MB}} \approx 1$, since the sequences with no transitions decrease. Thus, the clock jitter improvement is approximately 6 dB for each additional bit (just like the quantization noise improvement) [191]:

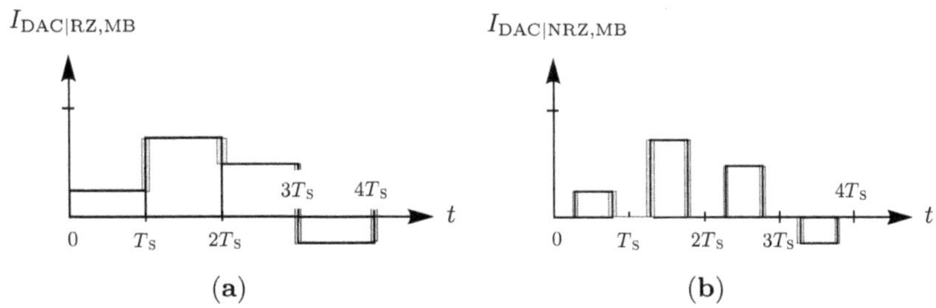

Fig. 4.9. Illustrative multibit NRZ and RZ pulse sequence under clock jitter influence. The gray shaded areas depict the clock edge uncertainties. (**a**) Multibit NRZ-DAC pulse; (**b**) Multibit RZ-DAC pulse

$$
\begin{aligned}
\mathrm{IBN}_{\sigma_t}|_{\mathrm{NRZ,MB}} &\approx \Delta^2 \left(\frac{\sigma_t}{T_\mathrm{S}}\right)^2 \frac{A_{\mathrm{NRZ,MB}}}{\mathrm{OSR}} \\
&= \frac{\mathrm{FS}^2}{(2^{B_{\mathrm{int}}}-1)^2} \left(\frac{\sigma_t}{T_S}\right)^2 \frac{A_{\mathrm{NRZ,MB}}}{\mathrm{OSR}} .
\end{aligned} \tag{4.27}
$$

When using multibit quantization combined with RZ feedback, it can be imagined from Fig. 4.9b that the just discussed advantage alleviates. This is because the pulse height and thus the jitter affected transition edge depends on the quantizer decision, and the signal has to change by significantly more than one LSB at every sample instant. Thus, even if the medium height of the pulses slightly decreases compared to the single-bit approach, depending on the input signal to the modulator the resulting jitter noise decrease is only minor. This has usually only been included by adopting the activity factor, found through simulations [197]. Nonetheless, an analytical approach can be found, which is presented slightly different to [198] in the following:

The feedback signal in a multibit ΣΔ modulator tracks the input signal almost perfectly with an LSB deviation of $\pm\Delta/2$. By assuming a sine-wave input signal with amplitude $\hat{u}$, the quadratic mean value of the RZ feedback DAC yields:

$$
I^2_{\mathrm{DAC|RZ,MB}} = k_i|_{\mathrm{RZ,MB}} \sqrt{\frac{\left(\frac{\hat{u}}{\sqrt{2}}+\frac{\Delta}{2}\right)^2 + \left(\frac{\hat{u}}{\sqrt{2}}-\frac{\Delta}{2}\right)^2}{2}} , \tag{4.28}
$$

which obviously is lower than (4.22), where $\Delta = \mathrm{FS}$ of the single-bit DAC is higher than the maximum input amplitude or the LSB in (4.28).

By inserting (4.28) into (4.21) and relating it to the modulator input, similar to (4.23) the expression for multibit, RZ feedback is found:

$$
\mathrm{IBN}_{\sigma_t}|_{\mathrm{RZ,MB}} \approx \left(\frac{\hat{u}^2}{2} + \frac{\Delta^2_{\mathrm{MB}}}{4}\right) \left(\frac{k_1|_{\mathrm{RZ,MB}}}{k_1|_{\mathrm{NRZ,MB}}}\right)^2 \left(\frac{\sigma_t}{T_\mathrm{S}}\right)^2 \frac{A_{\mathrm{RZ,MB}}}{\mathrm{OSR}} , \tag{4.29}
$$

where $k_{1|\mathrm{MB}}$ is the scaling coefficient of the multibit modulator in its outermost feedback branch, $\Delta = \mathrm{FS}/(2^{B_{\mathrm{int}}}-1)$ is one least significant bit and $A_{\mathrm{RZ,MB}}$ is the activity factor for the multibit RZ feedback pulse. Obviously, from Fig. 4.9b follows that $A_{\mathrm{RZ,MB}} \approx 2$. Note that (4.29) approximates an optimal curve for a large number of DAC bits, i.e., $\Delta \ll \hat{u}$. Then, the jitter performance is only determined by the input amplitude and not by the DAC bit width!

4.3.4 Reduction of Clock Jitter Influence Using Shaped Feedback Waveform DACs

Beyond the usage of multibit quantization, there have been only few publications presented in the past, which dealt with possibilities for a reduction of the clock jitter sensitivity in CT ΣΔ modulators: both, [81, 143] discussed the general possibility to use peaking or shaped feedback pulses, which show a reduced amplitude at the switching instant and thus a reduced charge variation due to a jittered clock.

Nonetheless, it took several years then to see promising techniques in the literature: the first technique was published in [136] and extensively reviewed in [31, 33, 199–201], while it was previously patented in [202]. Here, an exponentially decaying feedback pulse is realized with a switched-capacitor–resistor (SCR) arrangement, which intends to imitate the sloping feedback pulse-form of switched-capacitor implemented, DT modulators in the feedback path of continuous-time architectures. Consequently, the jitter noise is exponentially suppressed.

A second promising technique has been published in [130, 203], which uses a cosine-shaped feedback waveform. Therefore, from an on-chip PLL circuit a cosine function is generated, which is locked to the modulator system clock. Here, the amplitude and slope of the feedback waveform are zero around the switching instants, since the cosine waveform can be approximated quadratically around the switching point. Thus, the cosine feedback yields a second-order jitter suppression.

Finally, [38] proposed to use linearly or quadratically sloping feedback waveforms, which present simple to realize techniques decreasing the jitter sensitivity by first-order or second-order.

Clock Jitter in Linearly Sloping DACs

Considering a shaped, decaying feedback pulse, the most obvious is a linearly sloping pulseform as shown in Fig. 4.10. This triangular DAC pulse is described by:

$$I_{\mathrm{DAC}} = \begin{cases} \hat{I}_{\mathrm{DAC}} \left(1 - \frac{t-\alpha T_{\mathrm{S}}}{(\beta-\alpha)T_{\mathrm{S}}}\right) , & \alpha T_{\mathrm{S}} \le t < \beta T_{\mathrm{S}} \\ 0 , \text{ otherwise} , \end{cases} \tag{4.30}$$

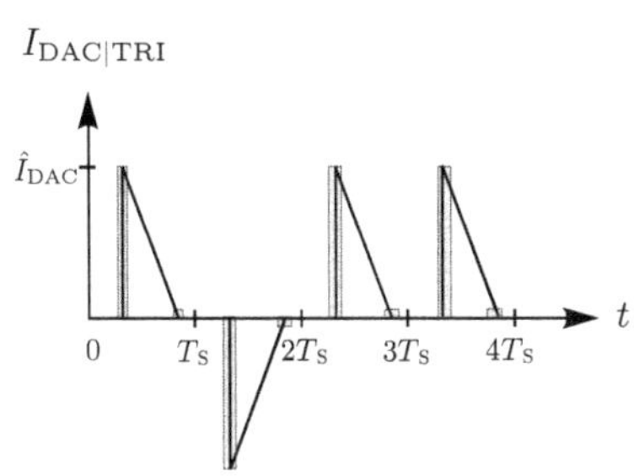

Fig. 4.10. Triangular-characteristic DAC pulse sequence $\{1, -1, 1, 1\}$. The gray shaded areas depict the sampling uncertainty

where $\{\alpha,\ \beta\}T_\mathrm{S}$ are the start and end point of the pulse. In order to obtain the amount of error charge $\mathfrak{E}_{Q_\mathrm{DAC}}$ that is fed back by the DAC in one clock cycle, I_DAC is integrated over one period including one jittered edge and subsequently subtracted from the ideal jitter free charge ($t_\mathrm{j} = 0$):

$$\mathfrak{E}_{Q_\mathrm{DAC}} = \int_{\alpha T_\mathrm{S}}^{\beta T_\mathrm{S}+t_\mathrm{j}} I_\mathrm{DAC}\,\mathrm{d}t - \int_{\alpha T_\mathrm{S}}^{\beta T_\mathrm{S}} I_\mathrm{DAC}\,\mathrm{d}t = \frac{\hat{I}_\mathrm{DAC}\,t_\mathrm{j}^2}{2\,(\alpha-\beta)\,T_\mathrm{S}}\,, \tag{4.31}$$

which is now a quadratic function of t_j.

By assuming again zero-mean, Gaussian jitter noise (σ_t) and by inserting (4.31) into (4.17) and (4.16), the variance $\sigma_e^2|_\mathrm{TRI}$ of the charge error $\mathfrak{E}_{Q_\mathrm{DAC}}$ for the linearly sloping feedback is obtained. Then, a translation of the charge error variance into a noise variance through relation to the sampling time yields:

$$E_\mathrm{j}^2|_\mathrm{TRI} = \frac{\sigma_e^2|_\mathrm{TRI}}{T_\mathrm{S}^2} = \frac{1}{2(\beta-\alpha)^2}\hat{I}_\mathrm{DAC}^2\left(\frac{\sigma_t}{T_\mathrm{S}}\right)^4 A_\mathrm{TRI}\,. \tag{4.32}$$

Here, A_TRI is again an activity factor, which takes into account the number of jittered edges of a feedback pulse, which was assumed to be one in (4.31) for simplicity. Similar to the RZ pulse, obviously $A_\mathrm{TRI} \approx 2$, if two edges are affected by zero-mean uncorrelated Gaussian jitter with variance σ_t^2.

The jitter IBN is again calculated by relating the jitter noise in (4.32) to the modulator input, which yields for the dominant, first branch:

$$\mathrm{IBN}_{\sigma_t}|_\mathrm{TRI} = \frac{1}{2(\beta-\alpha)^2}\left(\frac{\Delta}{2}\right)^2\left(\frac{k_1|_\mathrm{TRI}}{k_1|_\mathrm{NRZ}}\right)^2\left(\frac{\sigma_t}{T_\mathrm{S}}\right)^4\frac{A_\mathrm{TRI}}{\mathrm{OSR}}\,. \tag{4.33}$$

As (4.33) indicates, the clock jitter induced in-band noise is proportional to σ_t^4. This first-order clock jitter immunity results from the fact that the DAC amplitude is zero at the sampling instances. Obviously, this is a considerable improvement in comparison with (4.21).

Clock Jitter in Quadratically Sloping DAC

The former presented triangular-pulse shape DAC uses a linear decaying characteristic. Consider now the same transfer function but with a square-law behavior. The normalized quadratic-characteristic DAC pulse ideally located at $\{\alpha,\ \beta\}T_\mathrm{S}$ is then given by:

$$I_\mathrm{DAC} = \begin{cases} \hat{I}_\mathrm{DAC}\left(1-\frac{t-\alpha T_\mathrm{S}}{(\beta-\alpha)T_\mathrm{S}}\right)^2, & \alpha T_\mathrm{S} \le t < \beta T_\mathrm{S} \\ 0, & \text{otherwise}\,. \end{cases} \tag{4.34}$$

In order to obtain the amount of error charge $\mathfrak{E}_{Q_\mathrm{DAC}}$ that is fed back by the DAC in one clock cycle, I_DAC is integrated over one period including one

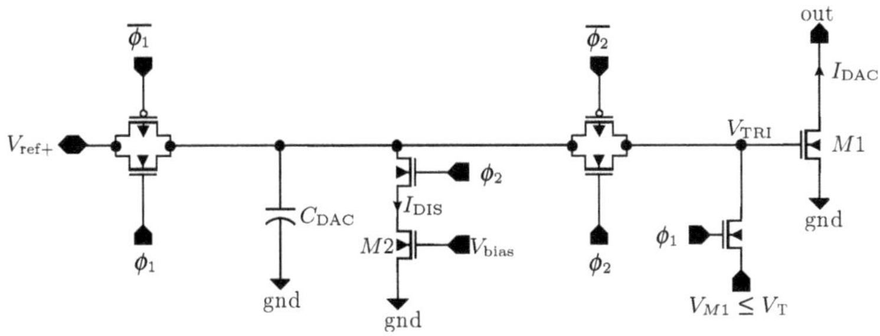

Fig. 4.11. Illustrative implementation of a DAC cell being able to generate a triangular as well as a quadratic-characteristic current

jittered edge and subsequently subtracted from the ideal jitter free charge ($t_{\text{j}} = 0$):

$$\mathfrak{E}_{Q_{\text{DAC}}} = \int_{\alpha T_{\text{S}}}^{\beta T_{\text{S}}+t_{\text{j}}} I_{\text{DAC}}\,\text{d}t - \int_{\alpha T_{\text{S}}}^{\beta T_{\text{S}}} I_{\text{DAC}}\,\text{d}t = \frac{\hat{I}_{\text{DAC}}\,t_{\text{j}}^3}{3\,(\alpha-\beta)^2\,T_{\text{S}}^2}\,. \tag{4.35}$$

By assuming again zero-mean, Gaussian jitter noise (σ_t) and by inserting (4.35) into (4.17) and (4.16), the variance $\sigma_e^2|_{\text{QUAD}}$ of the charge error $\mathfrak{E}_{Q_{\text{DAC}}}$ for the quadratically sloping feedback is obtained. Then, a translation of the charge error variance into a noise variance through relation to the sampling time T_{S} yields:

$$E_{\text{j}}^2|_{\text{QUAD}} = \frac{\sigma_e^2|_{\text{QUAD}}}{T_{\text{S}}^2} = \frac{5}{3(\beta-\alpha)^4}\hat{I}_{\text{DAC}}^2\left(\frac{\sigma_t}{T_{\text{S}}}\right)^6 A_{\text{QUAD}}\,. \tag{4.36}$$

Here, A_{QUAD} is again an activity factor, which takes into account the number of jittered edges of a feedback pulse, which was assumed to be one in (4.35) for simplicity of the calculation. Similar to the RZ pulse, obviously $A_{\text{QUAD}} \approx 2$, if two edges are affected by zero-mean uncorrelated Gaussian jitter with variance σ_t^2 [11, 204].

The jitter IBN is again calculated by relating the jitter noise in (4.36) to the modulator input, which yields for the dominant, first branch:

$$\text{IBN}_{\sigma_t}|_{\text{QUAD}} = \frac{5}{3(\beta-\alpha)^4}\left(\frac{\Delta}{2}\right)^2\left(\frac{k_1|_{\text{QUAD}}}{k_1|_{\text{NRZ}}}\right)^2\left(\frac{\sigma_t}{T_{\text{S}}}\right)^6\frac{A_{\text{QUAD}}}{\text{OSR}}\,. \tag{4.37}$$

The clock jitter induced in-band noise is proportional to σ_t^6. Figure 4.11 shows an exemplary DAC architecture that is able to generate a triangular as well as a quadratic-pulse shape output current. The mode of operation is as follows:

- During the first part of the clock phase, the capacitor C_{DAC} is charged to $V_{\text{ref+}}$ by closing the switches ϕ_1 and $\overline{\phi_1}$ (switches ϕ_2 and $\overline{\phi_2}$ are both open). Additionally, the output transistor $M1$ is biased to a voltage that

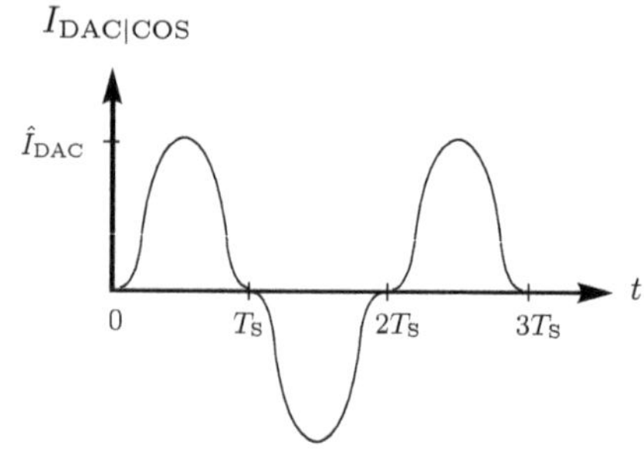

Fig. 4.12. Cosine DAC pulse shape

is near the threshold voltage of $M1$, $V_{M1} \leq V_{\mathrm{T}}$. As a result, the switch on sequence is speed up, while souring only a small constant subthreshold current. This current offset can be eliminated on demand by an additional fixed current source.

- In the second phase, the switches ϕ_1 and $\overline{\phi_1}$ are opened again and the capacitor C_{DAC} is discharged by closing ϕ_2 and $\overline{\phi_2}$. The discharging is accomplished by a fixed current source ($M2$), which sinks the current I_{DIS} to the virtual ground node in order to entail a linear decaying voltage slope V_{TRI}. This voltage can be used to generate a linear decaying output current I_{DAC} if $M1$ is biased in triode region or a quadratic-characteristic output current if $M1$ operates in saturation region.

Besides, instead of using a transistor to accomplish the voltage-to-current conversion [200], it is also possible to use an amplifier or transconductance stage for that purpose [191]. Nonetheless, the shown DAC was successfully simulated in a 1.8 V, 0.18 µm standard CMOS process [38].

The same jitter immunity as with the quadratic feedback DAC can be achieved by using a cosine-pulse shape DAC as shown in Fig. 4.12. Suppose the cosine wave is ideally locked to the clock, the DAC pulse has the same frequency as the sampling frequency or integer multiples of it. Thus, jitter influences the feedback pulse when it has both zero value and zero slope. Accordingly, a first-order and second-order insensitivity to clock jitter is obtained [130]. Note that the resulting IBN expression is slightly different to (4.37), but can be calculated the same way by Taylor approximation of the cosine waveform around the switching point!

Details on various error sources, expansion features as well as implementation issues for the cosine DAC are presented in [130]. Recently, even a circuit level implementation of this cosine DAC has been presented [203].

Clock Jitter in Exponentially Sloping DACs

Finally, an exponential feedback DAC is considered [19, 31, 136], whose waveform is illustrated in Fig. 4.13. This exponential DAC pulse is described by:

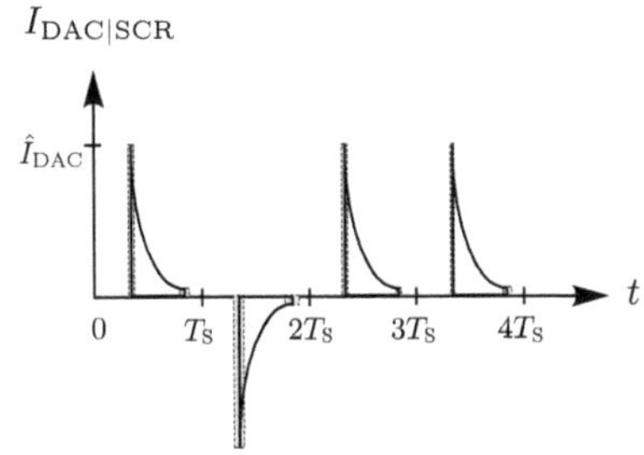

Fig. 4.13. SCR-DAC pulse sequence $\{1, -1, 1, 1\}$. The gray shaded areas depict the sampling uncertainty

$$I_{\mathrm{DAC}} = \begin{cases} \hat{I}_{\mathrm{DAC}}\, \mathrm{e}^{-(t-\alpha T_{\mathrm{S}}/\tau_{\mathrm{DAC}})}\,, & \alpha T_{\mathrm{S}} < t < \beta T_{\mathrm{S}} \\ 0\,, \text{ otherwise}\,, \end{cases} \tag{4.38}$$

where $\{\alpha,\ \beta\}T_{\mathrm{S}}$ are the start and end point of the SCR-DAC pulse and τ_{DAC} is its time constant. To obtain the feedback error charge due to clock jitter within one period, the nonideal charge is subtracted from the ideal case (jitter free, $t_{\mathrm{j}} = 0$):

$$\begin{aligned} \mathfrak{E}_{Q_{\mathrm{DAC}}} &= \int_{\alpha T_{\mathrm{S}}}^{\beta T_{\mathrm{S}}+t_{\mathrm{j}}} I_{\mathrm{DAC}}\,\mathrm{d}t - \int_{\alpha T_{\mathrm{S}}}^{\beta T_{\mathrm{S}}} I_{\mathrm{DAC}}\,\mathrm{d}t \\ &= \hat{I}_{\mathrm{DAC}}\,\tau_{\mathrm{DAC}}\,\mathrm{e}^{-T_{\mathrm{S}}(\beta-\alpha)/\tau_{\mathrm{DAC}}}\left(1-\mathrm{e}^{-\frac{t_{\mathrm{j}}}{\tau_{\mathrm{DAC}}}}\right). \end{aligned} \tag{4.39}$$

Applying a Taylor expansion on (4.39) yields:

$$\mathfrak{E}_{Q_{\mathrm{DAC}}} = \hat{I}_{\mathrm{DAC}}\,\mathrm{e}^{-T_{\mathrm{S}}(\beta-\alpha)/\tau_{\mathrm{DAC}}}\,t_{\mathrm{j}}\,, \tag{4.40}$$

which is again a linear function of t_{j} as in (4.15), but this time drastically reduced through the exp-function. By assuming again zero-mean, Gaussian jitter noise (σ_t) and by inserting (4.40) into (4.17) and (4.16), the variance σ_e^2 of the charge error $\mathfrak{E}_{Q_{\mathrm{DAC}}}$ for the exponential feedback is obtained. Then, a translation of the charge error variance into a noise variance through relation to the sampling time yields:

$$E_{\mathrm{j}}^2|_{\mathrm{SCR}} = \frac{\sigma_e^2|_{\mathrm{SCR}}}{T_{\mathrm{S}}^2} = \hat{I}_{\mathrm{DAC}}^2\left(\mathrm{e}^{-T_{\mathrm{S}}(\beta-\alpha)/\tau_{\mathrm{DAC}}}\right)^2\left(\frac{\sigma_t}{T_{\mathrm{S}}}\right)^2 A_{\mathrm{SCR}}\,. \tag{4.41}$$

A_{SCR} is again the activity factor, which takes into account the number of jittered edges of a feedback pulse, which was assumed to be one in (4.39) for simplicity. Similar to the RZ pulse, obviously $A_{\mathrm{SCR}} \approx 2$, if two edges are affected by zero-mean uncorrelated Gaussian jitter with variance σ_t^2.

The jitter IBN is again calculated by relating the jitter noise in (4.41) to the modulator input, which yields for the dominant, first branch:

$$\mathrm{IBN}_{\sigma_t}|_{\mathrm{SCR}} = \left(\frac{\Delta}{2}\right)^2\left(\frac{k_1|_{\mathrm{SCR}}}{k_1|_{\mathrm{NRZ}}}\right)^2\left(\mathrm{e}^{-T_{\mathrm{S}}(\beta-\alpha)/\tau_{\mathrm{DAC}}}\right)^2\left(\frac{\sigma_t}{T_{\mathrm{S}}}\right)^2\frac{A_{\mathrm{SCR}}}{\mathrm{OSR}}\,. \tag{4.42}$$

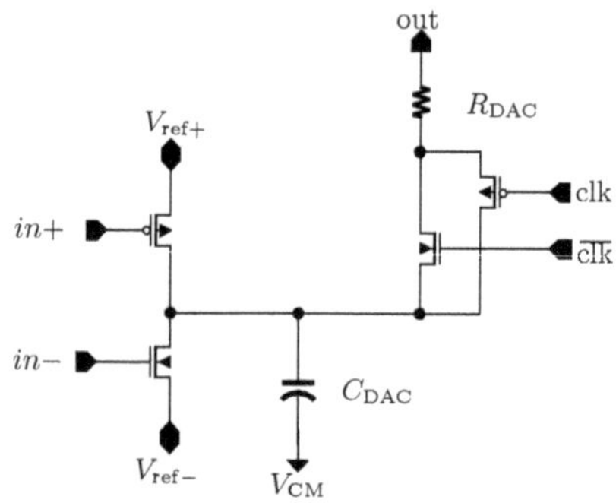

Fig. 4.14. Simplified schematic of the unit SCR-DAC cell [33]

Note the exponential suppression compared to (4.23) and that the clock jitter influence can be lowered by scaling down the slope τ_{DAC} of the feedback pulse.

A circuit level implementation of an exponentially sloping DAC is presented in Fig. 4.14 [33, 199]. The implementation is based on a switched capacitor DAC, in which the capacitor C_{DAC} is charged to the reference voltage $\pm V_{\text{ref}}$, depending on the quantizer decision in $\pm$. The discharging is accomplished by means of an auxiliary feedback resistance R_{DAC} (accordingly called SCR-DAC) [31, 136]. The feedback time-constant is determined by:

$$\tau_{\text{DAC}} = R_{\text{DAC}} C_{\text{DAC}} . \tag{4.43}$$

In contrast to the not well-defined on-resistance of pure SC circuits, the resistor in the CT SCR feedback is used twofold: first, it allows to regain full controllability on the feedback waveform, since CT modulators are strongly dependent on the exact waveform, which is continuously processed. Second, the resistor limits the maximum feedback signal and thus relaxes the speed requirements of the active circuitry by some amount. Nonetheless, also the switch-on resistance could be used for this purpose.

4.3.5 Further Possibilities for CT ΣΔ Modulators with Reduced Clock Jitter Sensitivity

Another possibility to achieve reduced clock jitter sensitivity was introduced in [26]. As already proposed in [73], additional filters are added into the feedback paths of a typical ΣΔ modulator. The resulting architecture is feasible to counteract tones in the output spectrum, as well as obligatory needed to stabilize multirate modulators [100]. Oliaei [26] proposes the additional feedback filters in order to counteract analog imperfections, since the feedback DAC signals are spectrally filtered. Thus, in the case of SC, DT ΣΔ modulators, the main benefit was found to be either reduced power consumption or increased linearity. In the case of CT modulators instead, the additional FIR feedback filters could drastically decrease the jitter sensitivity, since now the actual feedback pulse is transiently build from more than one DAC signal, or equivalently, is spectrally shaped in the band of interest! Simultaneously, the technique leaves enough degrees of freedom to ensure stability of the overall ΣΔ loop. Nonetheless, up to now no implementation of the technique has been presented.

Further on, Hernandez et al. [28] showed the feasibility of tuned delay line integrators to implement continuous-time ΣΔ modulators with an insensitivity to clock jitter as well as excess loop delay. Therefore, a combination of an OTA and an external delay line were combined in order to build a CT version of a DT integrator transfer function. Due to the delay and accumulation nature of the delay lines, the feedback DAC yields staircase shaped pulses, which is the reason for the lower jitter sensitivity. The main disadvantage of the presented technique is indeed the use of the transmission delay lines themselves, which were externally implemented and trimmed as coaxial cables in [28].

Finally, in [27] an idea was proposed to counteract the jitter noise of a ΣΔ modulator on the architectural level: therefore, the intrinsic modulator jitter behavior, independent from the input signal and its activity, has been taken as a quality measure of the jitter performance. Then, by changing the overall NTF at high frequencies, the area of the feedback pulses due to the signal-*independent*, intrinsic white quantization noise of the modulator is minimized. This is done by moving some of the NTF zeros closer to half of the sampling frequency $f_S/2$. This way, a ΣΔ modulator is obtained with significantly reduced jitter noise in the absence of a strong input signal! Nonetheless, the main drawback arises exactly in the *null-input* assumption. The argument in [27] for the signal free, intrinsic modulator jitter performance relies on communication systems, whose input signal power is well below the dynamic range of the modulator and the signal features a wide noise-like spectrum. If this is not the case, the optimization procedure should not work as ideal.

4.3.6 CT Loop Filters Employing Shaped Feedback Waveforms

As presented in Sect. 3.2, the loop filter scaling coefficients for CT ΣΔ modulators with feedback pulses other than rectangular can be exemplarily derived by means of the modified $\mathcal{Z}$-variant or the impulse-invariant transformation. Therefore, the same scheme as for NRZ or RZ feedbacks can be employed and subsequently a CT loop filter is obtained from a given DT one [25, 31]. This was exemplarily presented for a second-order modulator by using the modified $\mathcal{Z}$-transformation in Sect. 3.2.2. Alternatively, the CT loop filter can be directly synthesized as shown in Sect. 3.3 for NRZ feedback, and then a CT–DT transformation with rectangular feedback and subsequently a DT–CT conversion with the desired pulse-shaped DAC is performed.

Table 4.1. Equivalent CT loop filter coefficients for the triangular and quadratic-pulse shape DAC with $\{\alpha,\ \beta\} = \{1/2, 1\}$

	Triangular-characteristic pulse shape DAC	Quadratic-characteristic pulse shape DAC
Mod2	$k1_{\mathrm{TRI}} = 4a_1a_2$ $k2_{\mathrm{TRI}} = 4a_2 - \frac{4}{3}a_1a_2$	$k1_{\mathrm{QUAD}} = 6a_1a_2$ $k2_{\mathrm{QUAD}} = 6a_2 - \frac{9}{4}a_1a_2$
Mod3	$k1_{\mathrm{TRI}} = 4a_1a_2a_3$ $k2_{\mathrm{TRI}} = 4a_2a_3 - \frac{10}{3}a_1a_2a_3$ $k3_{\mathrm{TRI}} = 4a_3 - \frac{4}{3}a_2a_3 + \frac{31}{36}a_1a_2a_3$	$k1_{\mathrm{QUAD}} = 6a_1a_2a_3$ $k2_{\mathrm{QUAD}} = 6a_2a_3 - \frac{9}{2}a_1a_2a_3$ $k3_{\mathrm{QUAD}} = 6a_3 - \frac{9}{4}a_2a_3 + \frac{99}{80}a_1a_2a_3$

Table 4.2. Equivalent CT loop filter coefficients for the exponentially decaying, SCR-DAC with$\{\alpha,\ \beta\} = \{1/2, 1\}$

	Exponential SCR-DAC $\tau_{\mathrm{DAC}} = 0.2T_{\mathrm{S}}$	Exponential SCR-DAC $\tau_{\mathrm{DAC}} = 0.07T_{\mathrm{S}}$
Mod2	$k1_{\mathrm{SCR}} \approx 1.36,\ k2_{\mathrm{SCR}} \approx 2.25$	$k1_{\mathrm{SCR}} \approx 3.57,\ k2_{\mathrm{SCR}} \approx 5.6$
Mod3	$k1_{\mathrm{SCR}} \approx 0.27,\ k2_{\mathrm{SCR}} \approx 1.13$ $k3_{\mathrm{SCR}} \approx 2.31$	$k1_{\mathrm{SCR}} \approx 0.71,\ k2_{\mathrm{SCR}} \approx 2.9$ $k3_{\mathrm{SCR}} \approx 5.8$

For simple second-order and third-order ΣΔ architectures with distributed feedback, the scaling coefficients are exemplarily presented in Tables 4.1 and 4.2. These coefficients were derived with the time invariant transformation for the optimally scaled second-order and third-order DT modulators [6], exemplarily shown for the second-order modulator in Fig. 3.6a. For the linear and quadratically sloping feedback DACs, the CT feedback coefficients are given in Table 4.1. Since the general expressions become too complex, the feedback is assumed to be active during the second part of the clock cycle $\{\alpha,\ \beta\} = \{1/2, 1\}$. For the exponentially sloping DAC, the expressions are also too complex to be shown. Therefore, only two illustrative sets of scaling coefficients are given in Table 4.2, which were used in the subsequent simulations. For the general derivation of the scaling coefficients, the reader is forwarded to the published literature, e.g., [31].

4.3.7 Trade-off for Reduced Clock Jitter Sensitivity

The different possibilities for reduced sensitivity to clock jitter all come along with their own disadvantages:

- Using multibit internal quantization only shows its full advantage, if concurrently NRZ feedback is used and thus the step width is optimally reduced [126]. On the other hand, RZ feedback is much more feasible if high sampling rates are necessary, which has been seen for example in the case of excess loop delay. Simultaneously, multibit quantization needs linearization in the feedback DAC, which has to be implemented without causing another excessive excess loop delay, and which is even harder to achieve with NRZ feedback [21, 126].
- All pulse shaped feedback waveforms feature higher scaling coefficients and thus higher feedback signal peaks than their rectangular counterparts. Consequently, they obviously impose increased requirements on the transient behavior and speed of the integrators! This will be shown in a glance in Chap. 5, where also the influence of different feedback implementations on the integrator requirements is discussed. Nonetheless, note that there is still work and research to do to figure out the relation of different feedback realizations and integrator speed requirements.
- Beyond the increased speed requirements of the integrators, circuit nonidealities also show impact on the jitter reduction of the pulse-shaped feedback DACs: as discussed in [130], nonideal locking of the DAC pulses with the switching clock reduces the jitter suppression, since the feedback signal is not switched in its minimum any more. The same is true for the linearly and quadratically sloping DAC.
- The SCR feedback with its passive capacitor discharge shows an additional nonideal behavior in the presence of finite amplifier speed, which arises from the moving virtual ground node: if the latter varies during the discharge process, the discharge time constant is virtually increased [31, 124]. Thus, finite amplifier bandwidth causes increased jitter sensitivity. This will be shown in a glance in Sect. 5.4.4.

4.3.8 Discussion on the White Clock Jitter Model

Above, for the simplicity and illustration of the calculation procedure, the variations of the sampling instances were assumed to follow a uniformly distributed random process, which makes clock jitter to cause a white noise causing a flat spectral increase of the modulator output [11, 24, 34]. This jitter, known as independent jitter, translates to sampling instants which are statistically distributed around the ideal, jitter free sampling instants!

Beside, there exists accumulated or long term jitter, where the sampling intervals are independent random variables, but the jitter of the sampling instants is accumulated and depends on the sum of the variations before. Thus, the deviation of the sampling instants from ideally periodic instants will be subject to a statistical distribution with a much higher variance [3, 195, 196]. Accumulated or long-term jitter appears in a VCO clock with finite stability [205], and it adds noise skirts around the input signal frequency.

In [12] the influence of accumulated jitter noise was approximated to be used also in the expressions derived for the independent, white jitter. In [206], a general expression was derived, estimating the achievable SNR under the influence of a sampling clock with long term jitter σ_{tj}^2 and a signal $x(t)$, which is directly able to include the effect of accumulated jitter. Nonetheless, the same author found this expression to be rather pessimistic for real applications [206]: This is clear when understanding the free running oscillator, which produces the accumulated jitter, can be statistically modeled as a Wiener process and accumulates indefinitely over time. Thus, accumulated jitter for an infinite number of samples yields a zero SNR.

Consequently, [206] proposed to take only a limited number of samples into account, which then results a finite value for the effective jitter and also a nonzero value for the SNR. The argument in [206] is that most "final" users of the sampled information provide only a short memory, or even feature a low frequency jitter cancellation. As a result, in the derived expressions the effective jitter equals the white jitter, if white jitter noise is assumed to be dominant, while for accumulated jitter, the effective jitter variance increases linearly with the number of samples. Therefore, the expressions throughout Sects. 4.3.2–4.3.4 can be used with the adopted jitter definition in [206].

The influence of both different jitter forms on ΣΔ modulators has been discussed in [12] for CT modulators, in [207] for DT modulators as well as reviewed through simulations in [195] for DT and CT modulators. There it could be shown that in the presence of accumulated clock jitter the signal-dependent noise-skirts dominate the total in-band noise, alleviating the difference between various modulator implementations concerning clock jitter.

4.3.9 Simulation Results on Clock Jitter

Simulation results concerning the jitter sensitivity of various CT ΣΔ implementations are presented in Fig. 4.15. Here, a third-order CT distributed feedback modulator with NRZ and RZ $\{\alpha = 0,\ \beta = 1/2\}$ feedback employing single-bit and multi-bit internal quantization are simulated. Additionally, the linearly and sloping as well as with exponentially decaying feedback are presented. For comparison sake, the original third-order DT, single-bit modulator [6] is also simulated under the influence of clock jitter. The modulator scaling coefficients were obtained by means of a DT–CT conversion of the original third-order single-bit modulator in [6] $\{a_1 = 0.2,\ a_2 = 0.5,\ a_3 = 0.3\}$ and the third-order multibit modulator in [8] $\{a_1 = 0.3,\ a_2 = 0.6,\ a_3 = 2\}$.

In the simulations, the input signal was chosen with $P_{\text{sig}} = -9$ or $-6\,\text{dB}$ for the single-bit or multi-bit modulators, respectively, while $f_{\text{sig}} = f_{\text{B}}$ and $\text{OSR} = 48$. To simplify the simulation, only one edge of the feedback pulses was affected by jitter, thus all activity factors are $A_{\text{X}} = 1$, while the activity factor for the rectangular, NRZ single-bit modulator is still $A_{\text{NRZ}} \approx 0.7$, see (4.25). The calculation for the DT modulator was done as proposed in [9, 11]. First, note the drastically increased clock jitter sensitivity from the DT to the

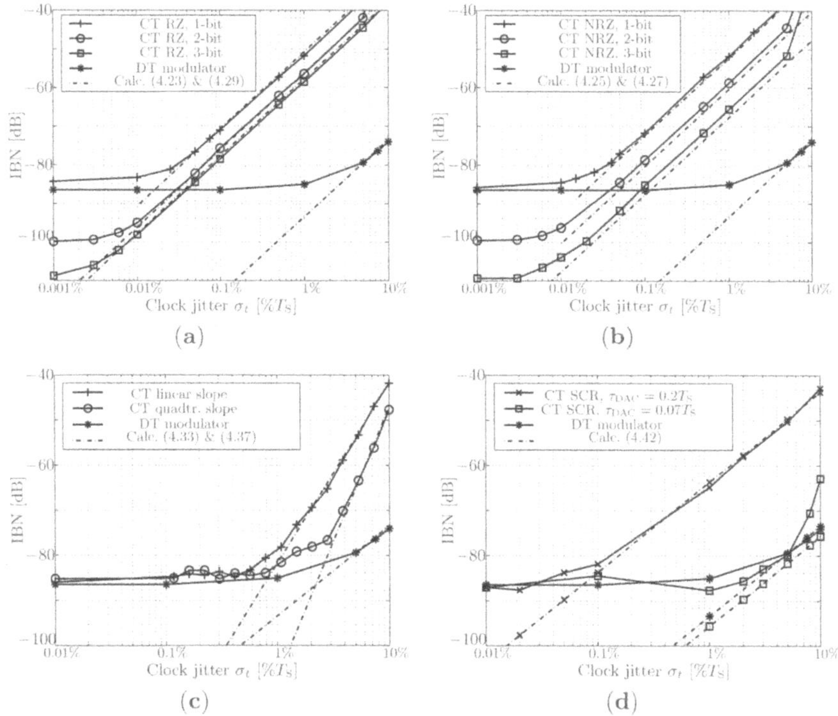

Fig. 4.15. Simulated IBN of third-order modulators (DT and CT) with different CT implementations under clock jitter influence. $P_{\text{sig}} = -9$ or -6 dB, $f_{\text{sig}} = f_B$ and OSR = 48. $A_X = 1$, $A_{\text{NRZ,SB}} = 0.7$. Note the different axis scaling. (**a**) CT RZ, 1,2 and 3-bit; (**b**) CT NRZ, 1,2 and 3 Bit; (**c**) CT linear and quadratic sloping; (**d**) CT SCR [31]

CT modulator with rectangular single-bit feedback DAC in Fig. 4.15a, which is in accordance to (4.23). If both edges of the RZ pulse were simulated being jittered, the nonideal IBN for the RZ modulator would increase by another 3 dB, yielding the approximately 4.5 dB shift compared to the NRZ, single-bit simulation in Fig. 4.15b, which was expected in (4.26).

By using multibit internal quantization, the difference for NRZ and RZ rectangular feedback appears in Fig. 4.15a, b. With a 2-bit DAC, both the RZ and NRZ feature better jitter performance, caused by the decreased feedback step width. But while the NRZ modulator keeps on to improve (by 6 dB) for a 3-bit DAC, the RZ modulator only improves very little. For a large number of internal bits, the RZ feedback modulator approximates a performance, which is only determined by the signal amplitude, see (4.29). Thus, as expected, multibit internal quantization is only significantly helpful when NRZ feedback is employed!

Finally, the proposed shaped feedback waveforms, linearly, quadratically or exponentially decaying, are simulated. Figure 4.15c, d show that these architectures are able to reduce the clock jitter sensitivity by orders of magnitudes: the linearly and the quadratic feedback result in second-order and third-order dependencies of the nonideal noise to clock jitter, i.e., the slopes decrease with 40/60 dB per decade, in contrast to the 20 dB per decade in the case of the rectangular feedback. This is in agreement with (4.23) compared to (4.33) and (4.37). A sinusoidal feedback would perform similar to the quadratically sloping feedback [130].

The exponentially decaying feedback additionally shows an adjustable jitter dependency: by adjusting the time constant of the slope to $\tau_{\mathrm{DAC}} = 0.2T_{\mathrm{S}}$, the jitter behavior is improved by one order of magnitude compared to the RZ modulator. For $\tau_{\mathrm{DAC}} = 0.07T_{\mathrm{S}}$, almost the DT insensitivity is obtained!

Different Approaches for Clock Jitter Simulations

In publications on the topic of CT ΣΔ modulators, very regularly simulations concerning the behavior due to clock jitter are presented. Nonetheless, there exist only few which explicitly deal with the way how the clock jitter is generated and its influence is determined. Therefore, the following several approaches are outlined in a glance:

- The straightforward method is a spice level simulation of the CT ΣΔ modulator with a jittered clock generator. Here, exemplarily, an AHDL model can be used for this purpose [208]. The advantage of this method is its direct relation to the real circuit and it includes all influences of clock jitter in the modulator. On the other hand, the obvious disadvantage is that these simulations are extremely time consuming and they are not feasible during the start of a design phase.
- A higher level simulation can be performed using MATLAB. Also here, a clock pulse with jittered edges can be implemented and used for the determination of the sampling instants and/or for the generation of the feedback waveform. This is advantageous concerning the simulation time compared to the circuit level simulator, and at the same time the effect of jitter is still realistically modeled. Nonetheless, the simulation accuracy has to be set very high in order to cover the extremely low timing jitter variations.
- Somewhat more imprecise, but with the advantage of drastically increased simulation time, is the complete modeling of the error caused by clock jitter. This has been exemplarily shown in [27, 209]; here, the dominant feedback charge error due to clock jitter is modeled as an additive error source with statistical properties and a dedicated transfer function, thus as to closely track the realistically seen behavior. The major disadvantage is here that only the dominant clock jitter influence is considered, but which could clearly be enhanced by a more sophisticated approach.

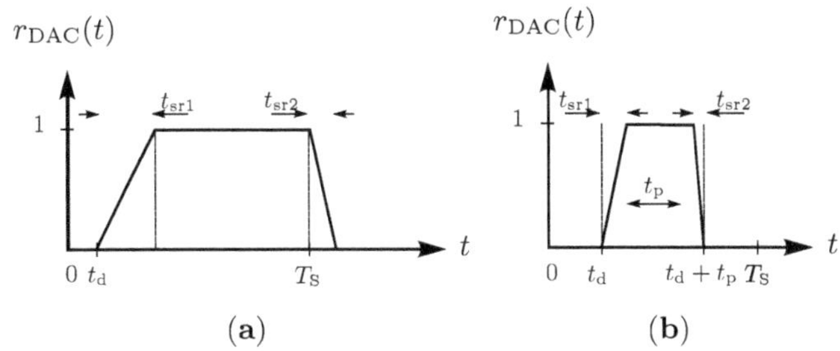

Fig. 4.16. NRZ and RZ feedback DAC pulse shapes with finite rise and fall times. (**a**) Slewing NRZ-DAC pulse; (**b**) slewing RZ-DAC pulse

- Finally, another feasible approach for fast simulation is to use a statistically varying feedback amplitude instead of a statistically distributed clock cycle. Thus, the feedback charge variance is the same, while the simulation is much easier!

4.4 DAC Slew Rate Limitation

In practical realizations of CT $\Sigma\Delta$ modulators the implemented feedback DAC usually exhibits beside a certain delay also slew rate limitations as well as different rise and fall times. Both nonidealities tend to degrade the performance, unless these errors or their effects on the overall modulator are made small enough. Especially in CT $\Sigma\Delta$ modulators employing an NRZ-DAC (Fig. 4.16a), the impact of both errors cause intersymbol interference (ISI) due to the fact that the resulting errors depend on the output sequence of the modulator.

If the rise and fall times of the 1-bit feedback signal are matched, the area error under a sequence of bits does not depend on the order of bits. However, if the rise and fall times are mismatched, the resulting error depends on the sequence of bits, which causes additional noise and tones within the spectrum that fold into the baseband [191]. Furthermore, a declaration with regard to the maximal tolerable DAC asymmetry maintaining the overall modulator performance, is given in [3, 191]. An improvement regarding ISI can be achieved by using a fully differential DAC so that the same error occurs in both signal paths: differentially the error is canceled out in first-order [113, 126]. Nonetheless, different switching errors alleviate the advantages again [210]. Another technique employs two RZ-DAC; the first switches on and off in the first part of T_{S} while the second operates in the same way in the second part. Since the remaining error is independent of the sequence of bits, this technique prevents harmonic distortion. Fortunately the resulting mismatch can be merely considered as a coefficient mismatch (gain error) [191].

The third technique replaces the NRZ-DAC by a single RZ-DAC (Fig. 4.16b), so that in spite of different rise and fall times, the error remains constant in every sample instant. Therefore, this explains that a compensation of the resulting error can be made in terms of tuning the scaling coefficients. Similarly to the compensation of pure excess loop delay, a given CT modulator loop filter is transformed to its DT equivalent using the more realistic feedback waveform and transformation in Fig. 3.3d. Then, the scaling coefficients are tuned as to match this nonideal loop filter with the ideal second-order DT filter. Resulting scaling coefficients k_i^* for a second-order CT modulator with RZ feedback (ideal scalings k_i) are presented in (4.44):

$$k_1^* = \frac{k_1}{(\beta - \alpha)}, \quad k_2^* = \frac{k_1(\beta + \alpha + 2(1 + \tau_d) + \tau_{sr})}{2(\beta - \alpha)} + \frac{k_2}{(\beta - \alpha)}. \tag{4.44}$$

Adopting these modified scaling eliminates the impact of excess loop delay τ_d as well as finite rise $t_{\mathrm{sr1}} = \tau_{\mathrm{sr1}} T_\mathrm{S}$ and fall times $t_{\mathrm{sr2}} = \tau_{\mathrm{sr2}} T_\mathrm{S}$ for a modulator with RZ feedback (and delays, which do not shift the pulse into the next clock cycle). For the sake of simplicity, the rise and fall times are supposed to be equal $\tau_{\mathrm{sr1}} = \tau_{\mathrm{sr2}} = \tau_{\mathrm{sr}}$. Nevertheless, this assumption can be simply removed to get universally valid coefficients. Similar expression are found for other loop filter orders or architectures.

4.5 DAC Nonlinearity

Beyond the transient errors in the feedback DAC, which were discussed above, also the DAC levels themselves can be affected by mismatch, which is briefly considered subsequently:

If the different DAC-levels are affected by mismatch, the positive and negative level of the feedback pulse are slightly different, resulting in a DAC gain and offset error, as illustrated already in Fig. 2.3 for the single-bit quantizer. Typically, ΣΔ modulators are very tolerant to both of these errors [5], and usually only little performance degradation results in addition to an offset.

In the case of multibit internal quantization and therewith multibit DAC, the variation of the feedback levels yields a signal-dependent feedback charge error, which is directly fed to the modulator input; thus, they cause unsuppressed distortion! Consequently, the feedback DAC, even if it is low-resolution, requires a linearity better than the overall modulator [5]. Note that this requirement is always fulfilled for a single-bit internal quantizer, because a two level DAC is intrinsically linear.

Actually, there have been many publications dealing with techniques to achieve the linearity requirements, based on design consideration [67] as well as digital or analog correction and linearization techniques [4, 8, 55, 57, 63, 68, 70, 107, 185, 211–213]. However, all of these possibilities make use of rather complex circuitry, causing an increase of area and power consumption as well as switching activity.

Concerning the functionality of the linearization in CT ΣΔ implementations, note that most of the techniques are afflicted with additional time delay, which is a critical issue in CT modulators. Therefore, for CT multibit implementations circuits were proposed to combine both linearization and CT feedback: Instead of implementing the linearization in the feedback DAC, Doerrer et al. [21] used a dynamic element matching technique in the ADC, where more time is available to perform the switching. In contrast, Yan and Sanchez-Sinencio [126] designed the modulator loop filter with an explicit latched loop delay and a compensation for it as shown in Sect. 4.2.4, thus giving an explicit time slot to perform internal A/D and D/A conversion!

Recently, more publications have been presented, which deal with first and second-order nonidealities in current-steering DACs in CT $\Sigma\Delta$ modulators [23, 210]. Nonetheless, it is beyond the scope of this book to go into the details of linearization techniques, but employing them in CT $\Sigma\Delta$ modulators will surely be of much interest in future publications!

5

Filter Nonidealities in Continuous-Time $\Sigma\Delta$ Modulators

This section is intended to give an insight into typical sources of nonideal behavior in the loop filter of continuous-time $\Sigma\Delta$ modulators. It will be shown that certain nonidealities, which adversely affect the performance of DT $\Sigma\Delta$ modulators, have similar effects on continuous-time $\Sigma\Delta$ modulators. Consequently, they are not considered in detail due to the existing broad range of literature on these topics. Since little information is available on cascaded CT $\Sigma\Delta$ modulators, also major importance is attached to the influence of typical continuous-time nonidealities on these architectures. Thus, throughout this chapter single-loop and multiloop modulators are considered with a focus on single-bit quantizer realizations as well as low-pass loop filters. Where techniques are known to counteract or at least compensate for the nonideal behavior, these are presented in a glance.

5.1 Analytical Description

The loop filter transfer function is the main performance determining part in a $\Sigma\Delta$ modulator, because it defines the noise-transfer function and therewith the quantization noise-shaping behavior. Principally, the complete loop filter usually consists of several first-order filters, which are commonly arranged in a feedback or forward architecture, shown as DT implementations in Figs. 2.14 and 2.16. Taking CT $\Sigma\Delta$ modulators into consideration, the single filters are realized using either RC-integrator or gmC-integrator or even LC-resonators for bandpass noise shaping. Without loss of generality, in the following active RC-integrators are considered.

5.1.1 Analytical Description of the Nonideal CT Filter Behavior

Figure 5.1 shows a typical schematic of an RC-integrator with n_{A} inputs incorporating an amplifier, whose transfer function is given by $A(s)$. When considering a single input, the scaled integrator corner frequency ω_{I} in (3.18) is adjusted by the RC-product. If different input scaling coefficients are required, this can only be realized by different resistors. Corresponding to Sect. 3.7.6, the following ideal relation can be found for the integrator corner frequency in CT modulators:

$$\omega_{\mathrm{I}s}|_i = \frac{1}{R_i C} \stackrel{!}{=} |k_i f_{\mathrm{S}}| \ . \tag{5.1}$$

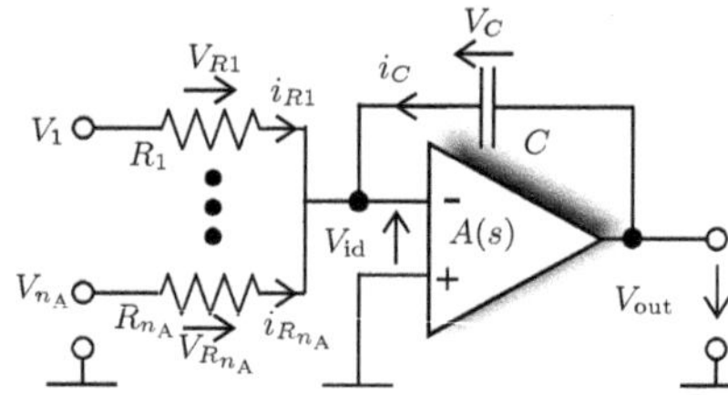

Fig. 5.1. Active *RC* integrator with single pole amplifier and n_A input paths

Here, the CT scaling coefficients k_i, which can be found, e.g., from a DT–CT conversion, are mapped into different resistors. For an illustrative explanation of the dependencies of f_S, ω_I and the actual implementation with R and C, please see also Appendix C.

The calculation of the ITF of the ith input path to the integrator output in Fig. 5.1 yields:

$$\mathrm{ITF}_i(s) = \frac{k_i f_S}{s\,(1 + \frac{1}{A(s)}) + \frac{1}{A(s)} \sum_{l=1}^{N} k_l f_S} \overset{A(s)\to\infty}{\approx} \frac{k_i f_S}{s}\,, \tag{5.2}$$

which obviously equals the ideal integrator-transfer function in (3.18) for an infinite, constant amplifier gain $A(s)$. For any nonideal OpAmp transfer function $A(s)$, the ITF of the different input paths will be modified, and depend also on the scaling coefficients of all other integrator inputs. This is usually not emphasized in publications, but was shown, e.g., in [30]. Illustratively, this is understood with the moving virtual ground node of the nonideal differential amplifier in Fig. 5.1, i.e., $V_{id} \neq 0$: if this node varies, there will be some current flowing through every resistor connected to it. Thus, the amplifier output due to an input signal applied to R_1 will be altered by other input resistors R_i.

Note that the k_i in (5.2) include also unity gain input paths to an integrator. All nonidealities of the loop filter integrators, which linearly alter the noise-transfer function, can be derived from (5.1) and (5.2).

5.1.2 Quantitative Impact of Nonideal CT Filter Behavior

Analyzing the impact of linear nonideal behavior of the loop filter on ΣΔ modulators is done by calculating the dependency of the integrated in-band noise on a special nonideality. Therefore, first the loop filter transfer function is described in terms of the nonideal variable v. This can be done either using a DT or its equivalent CT loop filter:

$$H(z, v) \text{ or } H(s, v)\,. \tag{5.3}$$

If the white noise model for the internal quantizer is adopted with its effective gain (2.25), together with its loop filters $H(z)$ or $H(s)$ the output of any single-loop architecture (see Fig. 2.10) yields:

$$\begin{aligned} Y(z,v) &= U(z)\frac{k_\mathrm{q}H(z,v)}{1+k_\mathrm{q}H(z,v)} + E(z)\frac{1}{1+k_\mathrm{q}H(z,v)}\,, \\ \text{or } Y(s,v) &= U(s)\frac{k_\mathrm{q}H(s,v)}{1+k_\mathrm{q}H(s,v)} + E(s)\frac{1}{1+k_\mathrm{q}H(s,v)}\,, \end{aligned} \tag{5.4}$$

where the expression with the input signal U is the STF and with the quantization noise E is the NTF of the corresponding loop (2.11).

Note, if a cascaded modulator as in Fig. 2.20 is considered, more than one expression as in (5.4) exists, all of which are combined over digital recombination filters $\mathrm{DF}_i(z)$. Therefore, it is advantageous to start the calculation with the DT loop filter $H(z,v)$ in (5.3), since this $\mathcal{Z}$-domain loop filter can be directly multiplied with the DT cancellation filter. If a cascaded CT loop filter is directly synthesized (without DT–CT conversion [20]), it is recommended to perform a CT–DT conversion before doing the combination of the cancellation logic with the calculated loop filters. Finally, the overall resulting (cascaded) modulator output expression yields:

$$Y_\mathrm{res}(z,v) = U(z)\mathrm{STF}_1(z,v) + \sum_{i=1}^{M} E_i(z)\mathrm{NTF}_i(z,v)\,. \tag{5.5}$$

Thus, the IBN contains one ($M = 1$) or several ($M > 1$ for cascaded modulators) quantization noise components E_i, all of which add to the integrated in-band noise. In order to calculate this IBN, the various resulting NTF_i are extracted from (5.5) and z is approximated for low frequencies by $z = \mathrm{e}^{s/f_\mathrm{S}} \approx 1 + s/f_\mathrm{S}$ as in (2.15). Then, the IBN yields:

$$\mathrm{IBN}(v) \approx \sum_i \left(\frac{\sigma_{ei}^2}{f_\mathrm{S}} \int_{-f_\mathrm{B}}^{f_\mathrm{B}} |\mathrm{NTF}_i(f,v)|^2 \,\mathrm{d}f \right). \tag{5.6}$$

Note that the expression in (5.6) generally becomes not solvable in a closed formula, if the order of the loop filter $H(z)$ increases to 3 or more. If so, it can be approximated by a Taylor series, again for low frequencies ($f_\mathrm{B} \to 0$), which corresponds to an assumed OSR $\gg 1$. Additionally, only the Taylor coefficients up to the order $2N + 1$ are taken into account, where N is the modulator order. This way, only the intrinsic ideal modulator in-band noise and the noise components with less noise shaping than the ideal are considered. The presented way of calculation is used in the following for various nonidealities.

5.2 Finite OpAmp Gain

One of the most well-studied nonideal effects in $\Sigma\Delta$ modulators is that of finite dc gain of the operational amplifiers, among others in [4, 5, 66, 110, 214, 215]. From (5.2), for a frequency-independent, but finite amplifier gain A_dc, the transfer function of the ith input path of an RC-integrator as in Fig. 5.1 can be derived to:

$$\mathrm{ITF}_{A_{\mathrm{dc}}}(s)|_i = \frac{k_i f_{\mathrm{S}}}{s\,(1+\frac{1}{A_{\mathrm{dc}}}) + \frac{1}{A_{\mathrm{dc}}}\sum_{l=1}^{n_{\mathrm{A}}} k_l f_{\mathrm{S}}} \approx \frac{k_i f_{\mathrm{S}}}{s + \frac{1}{A_{\mathrm{dc}}}\sum_{l=1}^{n_{\mathrm{A}}} k_l f_{\mathrm{S}}}\,, \tag{5.7}$$

i.e., the ITF of the ith input becomes a first-order pole transfer function with a dc gain of $A_{\mathrm{dc}}k_i/\sum_l k_l$ and a pole at $f_{\mathrm{S}}\sum_l k_l/A_{\mathrm{dc}}$, which is obviously displaced away from dc. Following (2.11) the poles of the loop filter become the zeros of the NTF. Thus, *all* zeros of the NTF are pushed away from dc, which reduces the amount of attenuation of the quantization noise in the baseband, as can be imagined from the ideal NTF in Fig. 2.13. Usually, this nonideality is known as *leaky integration* and it affects continuous-time and discrete-time modulators in exactly the same way.

The expressions for the in-band noise due to leaky integrators are calculated following the way outlined in Sect. 5.1.2. The expression for an exemplary second-order modulator as presented in Fig. 3.6b in Sect. 3.2.1 as well as the expression for a general single-loop modulator with Nth-order distributed feedback loop filter are given in (5.8) [5]:

$$\begin{aligned}
\mathrm{IBN}_2(A_{\mathrm{dc}}) &\approx \frac{\Delta^2}{12}\frac{1}{k_{1|\mathrm{NRZ}}^2 k_{\mathrm{q}}^2}\left(\frac{\pi^4}{5}\frac{1}{\mathrm{OSR}^5} + \frac{2\pi^2}{3}\frac{1}{\mathrm{OSR}^3 A_{\mathrm{dc}}^2} + \frac{1}{\mathrm{OSR}A_{\mathrm{dc}}^4}\right),\\
\mathrm{IBN}_N(A_{\mathrm{dc}}) &\approx \frac{\Delta^2}{12k_{1|\mathrm{NRZ}}^2 k_{\mathrm{q}}^2}\left[\frac{1}{A_{\mathrm{dc}}^{2N}\mathrm{OSR}}\right.\\
&\quad\left.+\sum_{i=1}^{N}\left(\frac{\pi^{2i}N(N-1)\cdots(N-i+1)}{(2i+1)\,\mathrm{OSR}^{2i+1}A_{\mathrm{dc}}^{2(N-i)}i!}\right)\right].
\end{aligned} \tag{5.8}$$

Thus, in the case of single-loop ΣΔ modulation, the dc gain of the incorporated amplifiers should be in the range of the oversampling ratio $A_{\mathrm{dc}} \approx \mathrm{OSR}$ of the modulator [5, 66, 110]. This requirement keeps every part of the in-band noise proportional to $1/\mathrm{OSR}^{2N+1}$, i.e., the ideal noise shaping suppression.

In contrast to single-loop architectures, multiloop ΣΔ modulators suffer tremendously from integrator leakage. This is due to the principle of noise cancellation of the lower stages by the higher loops in the cascade. These are designed to eliminate the ideal quantization noise from the previous stages, but they are not able to also reject the nonideal noise components. Thus, these nonideal parts of the quantization noise leak into the overall output. This is illustrated by calculating the IBN of the 2–1–1 CT modulator (see Fig. 3.8b) under the influence of finite dc gain:

$$\begin{aligned}\mathrm{IBN}_{211}(A_{\mathrm{dc}}) \approx\ & \frac{\pi^8}{9}\frac{\Delta_3^2}{12}\frac{1}{c_1^2c_2^2\mathrm{OSR}^9} + \frac{\pi^6}{7}\frac{\Delta_3^2}{12}\frac{1}{c_1^2c_2^2\mathrm{OSR}^7A_{\mathrm{dc3}}^2} \\ & + \frac{\pi^4}{5}\frac{\Delta_2^2}{12}\frac{1}{c_1^2\mathrm{OSR}^5A_{\mathrm{dc2}}^2} + \frac{4\pi^2}{3}\frac{\Delta_1^2}{12}\frac{1}{k_{1|\mathrm{NRZ}}^2k_q^2\mathrm{OSR}^3A_{\mathrm{dc1}}^2},\end{aligned} \tag{5.9}$$

where $A_{\mathrm{dc}i}$ is the integrator dc gain in the ith stage. Obviously, in the 2–1–1 cascaded modulator the required amplifier dc gain $A_{\mathrm{dc}i}$ differs for the respective modulator stage: In the last stage, the same requirement as for single-loop modulators applies. But for earlier stages, the minimum dc gain drastically increases, dependent on the order of the respective and preceding stages and the order of the overall modulator. Thus, in the second stage of the 2–1–1 modulator roughly $A_{\mathrm{dc2}}|_{211} \approx \mathrm{OSR}^2$ is required. Moreover, the requirements increase in the first stage of the 2–1–1 modulator to $A_{\mathrm{dc1}}|_{211} \approx \mathrm{OSR}^3$. Note, if last stage multibit quantization is applied, these additionally increase proportional to $(\Delta_1/\Delta_M)^2$.

Taking into account that (5.10) and the other equations are only estimates which well correspond to simulation results, for a general cascaded modulator the influence of dc gain may be further approximated. The IBN of a cascaded modulator with finite dc gain in the amplifiers approximately yields:

$$\mathrm{IBN}_{\mathrm{casc}}(A_{\mathrm{dc}}) \approx \mathrm{IBN}_{\mathrm{casc|ideal}} + \sum_{i=1}^{M-1}\left(\mathrm{IBN}_i\left(A_{\mathrm{dc}}\right) - \mathrm{IBN}_{i|\mathrm{ideal}}\right), \tag{5.10}$$

i.e., the ideal IBN of the cascaded modulator (2.36) plus the *nonideal* parts due to integrator leakage of the IBN in all previous stages $(1,\ldots,M-1)$. Note, expressions (5.8–5.10) only take leakage into account. If also nonlinearity effects introduced by the amplifier are considered, the required dc gain maintaining the overall performance has to be higher (see Sect. 5.6.3) [128].

5.2.1 Simulation Results

Simulation results are given in Fig. 5.2, where the CT second-order and third-order single-loop, single-bit modulators are simulated, while exemplarily the 2–1–1 CT modulator is simulated with OSR = 24 and last stage single-bit and multibit quantization. The accordance to the above presented calculations can be seen. Furthermore, the drastic increase of dc gain requirements for the 2–1–1 modulator is obvious, especially when multibit, last stage quantization is applied.

5.3 Integrator Gain or Time-Constant Error

Variations of the integrator gain, determined by the scaling coefficients are a well known nonideal characteristic of single-loop and even more of cascaded

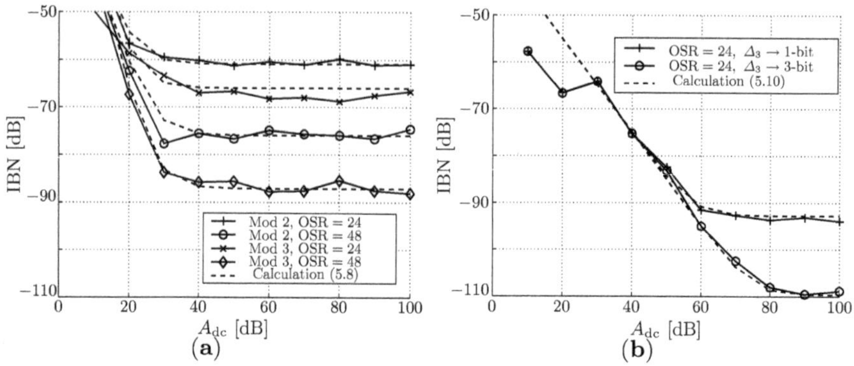

Fig. 5.2. Simulated CT modulators under the influence of finite OpAmp gain. $P_{\text{sig}} = -23\,\text{dB}$ and $f_{\text{sig}} = f_{\text{B}}/3$. (**a**) sb, CT second-order and third-order; (**b**) CT 2–1–1 [32]

ΣΔ modulators. In DT modulators implemented in SC technique, the integrator gains are mapped into capacitor ratios, which are intrinsically precise and variations lower than 0.1% are typical [5, 8, 216].

In contrast, in CT ΣΔ modulators, integrator gains are mapped into resistor–capacitor products [3, 9, 37], which are known to largely vary over process, temperature, etc.; process variations of the absolute component values of 10–20% are not unusual, which increases the possible variation of the RC-product to more than 30% [217]. Therefore this error is a major one to investigate. Equation (5.11) shows the chosen model for the RC-integrator transfer function ITF(s) subject to ITF(s) subject to a tolerance δ_{RC} of the k_i is the integrator scaling coefficient of the ith path and GE is the resulting equivalent gain-error of the integrator:

$$\text{ITF}_{RC}(s)|_i = \frac{1}{sR_iC\,(1+\delta_{RC})} \overset{(5.1)}{=} \frac{f_{\text{S}}}{s}\frac{k_i}{(1+\delta_{RC})} = \text{GE}_{RC}\frac{k_i f_{\text{S}}}{s}\,. \quad (5.11)$$

With this transfer function for the integrators, the influence of absolute integrator gain errors on CT ΣΔ modulators can be calculated.

5.3.1 Effective Quantizer Gain and Integrator Gain Errors

Before calculating the in-band noise due to integrator gain variations, first an additional remark on the effective quantizer gain is given: as outlined in Sect. 2.7.5 an ideal single-bit quantizer counteracts a scaling coefficient of the last integrator in the loop in front of the quantizer: ideally, regardless if a comparator input signal is scaled or not, the decision if it is positive or negative is made.

Consequently, if in addition to the integrator scaling coefficient an additional integrator gain error arises, the latter just multiplies to the original scaling coefficient. Thus, as long as the internal quantizer performs an ideal single-bit decision, its effective gain cancels the gain error of the last integrator and (2.25) becomes:

$$\begin{aligned} k_\mathrm{q} &= 2/a_N/\mathrm{GE}_N \text{ for } N\text{th-order loops} \\ \text{or } k_\mathrm{q} &= 1/a_1/\mathrm{GE}_1 \text{ for first-order loops}, \end{aligned} \tag{5.12}$$

where a_N is again the last integrator scaling coefficient of the original, equivalent DT modulator (see Sect. 2.7.5), while GE_N is the gain error of the last integrator, DT or CT. Note: this holds also true for *multibit internal quantization*, if the input signal is small enough that only the LSB is switched on/off. This observation is important in order to achieve accurate matching between the analytical and simulation results!

5.3.2 Single-Loop Modulators

It has been frequently reported [10, 37, 214] that integrator gain variations have only little influence on single-loop $\Sigma\Delta$ modulators, as long as these do not rely on optimally spread zeros (see Sect. 2.7.3). This is, because a variation of the integrator gains varies the poles of the noise-transfer function, while not affecting the zeros at dc. Thus, the slope of the ideal NTF in Fig. 2.13 is only slightly modified. (Note that this is quite different for architectures with optimally spread zeros or bandpass modulators, where also the placement of the NTF zeros depends on the passive components!)

More precisely, it is important to classify two different effects: first, a positive variation of the passive components, which yields reduced integrator scaling coefficients k_i in (5.1), and therefore less aggressive noise-shaping behavior. Following the calculations for the performance of a scaled modulator in (2.21), this will increase the in-band noise, but will not affect its stability. Second, a negative shift of the RC-product is equivalent to increased scaling coefficients k_i and thus more aggressive noise shaping. Consequently, after a short, slight increase in performance, this tolerance finally results in instability caused by an insufficient loop scaling (Sect. 2.7.4). Note that this shift into instability is not covered within the following linear calculations, since it strongly depends on the input amplitude and the ideal modulator loop filter with its out-of-band gain.

Again, the IBN due to integrator gain error GE_i is calculated as proposed in Sect. 5.1.2 and yields for an Nth-order single-loop modulator:

$$\mathrm{IBN}_N(\mathrm{GE}) \approx \frac{\pi^{2N}}{2N+1}\frac{\Delta^2}{12}\frac{1}{k^2_{1|\mathrm{NRZ}}k_\mathrm{q}^2\mathrm{OSR}^{2N+1}}\prod_{i=1}^{N}\frac{1}{\mathrm{GE}_i^2}, \tag{5.13}$$

where GE_i is the gain error of the ith integrator in a single-loop $\Sigma\Delta$ modulator (as in Figs. 3.6a and 3.12a), which may exemplarily arise from RC-mismatch, see (5.11). Clearly, gain errors decrease the modulator performance, but even the error afflicted terms are still proportional to $1/\mathrm{OSR}^{2N+1}$, thus the modulator still shows Nth-order noise shaping. Note that (5.13) slightly differs for multibit and single-bit internal quantization, since the effective quantizer gain differs in the presence of an integrator gain error (see Sect. 5.3.1).

5.3.3 Cascaded Modulators

The calculations for cascaded modulators become more sophisticated, and many simplifications have to be performed in order to get closed expression formulas. For the above discussed 2–1–1 modulator, the integrated in-band noise due to absolute integrator gain errors becomes:

$$\mathrm{IBN}_{211}(\mathrm{GE}) = \frac{\pi^8}{9}\frac{\Delta_3^2}{12}\frac{1}{c_1^2 c_2^2 \mathrm{OSR}^9} + \frac{\pi^6}{7}\frac{\Delta_2^2}{12}\frac{\left(\frac{1}{\mathrm{GE}_2}-1\right)^2}{c_1^2 \mathrm{OSR}^7} + \frac{\pi^4}{5}\frac{\Delta_1^2}{12}\frac{\left(\left(\frac{1}{\mathrm{GE}_{11}}-1\right)+\left(\frac{1}{\mathrm{GE}_{12}}-1\right)\right)^2}{k_{1|\mathrm{NRZ}}^2 k_\mathrm{q}^2 \mathrm{OSR}^5}, \tag{5.14}$$

where GE_{il} is the gain error of the lth integrator in the ith stage and k_q is the effective quantizer gain (2.25) as explained in Sect. 5.3.1.

As the case for leaky integration, again the modulator architecture as well as the OSR determine the tolerable gain error GE and thus RC-variation δ_{RC}: the first loop shows the most significant noise leakage, where the noise components due to integrator gain error are only second-order noise shaped. Thus, the allowable RC-tolerance δ_{RC} in (5.11) becomes proportional to OSR^{-2} [5, 91]. Note, if last stage multibit quantization is applied, the requirements on the RC-tolerance become more stringent since they are proportional to $(\Delta_1/\Delta_M)^2$. The general solution for a M stage cascaded modulator, which showed good matching for various cascaded architectures, yields:

$$\mathrm{IBN}_\mathrm{casc}(\mathrm{GE}) \approx \frac{1}{\prod_{ci=1}^{M-1} c_{ci}}\left(\frac{\pi^{2N_M}}{2N_M+1}\frac{\Delta_M^2}{12}\frac{1}{k_{1M}^2 k_{\mathrm{q}M}^2 \mathrm{OSR}^{2N_M+1}}\right) + \sum_{i=1}^{M-1}\frac{1}{\prod_{ci=1}^{i-1} c_{ci}}\left(\frac{\pi^{2N_{\Sigma i}}}{2N_i+1}\frac{\Delta_i^2}{12}\frac{1}{k_{1i}^2 k_{\mathrm{q}i}^2 \mathrm{OSR}^{2N_{\Sigma i}+1}}\sum_{l=1}^{N_{\Sigma i}}\left(\frac{1}{\mathrm{GE}_{il}}-1\right)^2\right), \tag{5.15}$$

where M is the number of loops in the cascade, N_i is the order of the ith loop, $N_{\Sigma i} = \sum_{i=1}^{M-1} N_i$ is the summed loop order, GE_{il} is the gain error of the lth integrator in the ith stage, c_{ci} are the interstage coupling coefficients, and $k_{1i|\mathrm{NRZ}}$ is the first integrator NRZ scaling coefficient and $k_{\mathrm{q}i}$ is the effective quantizer gain of the ith stage.

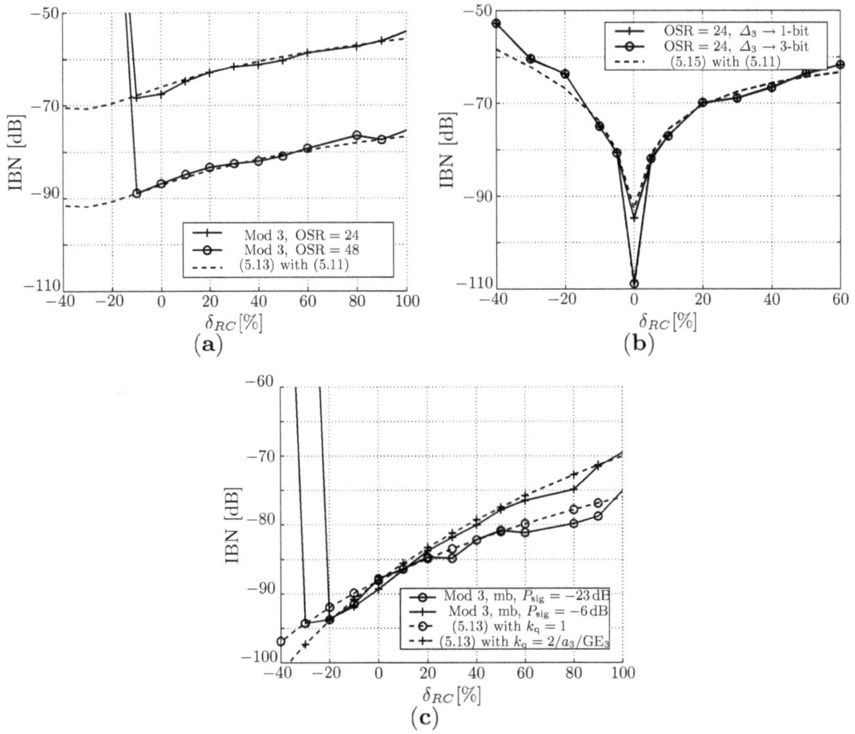

Fig. 5.3. Simulated CT modulators under the influence of absolute integrator gain error. $P_{\text{sig}} = -23\,\text{dB}$ and $f_{\text{sig}} = f_{\text{B}}/3$. (**a**) sb, CT third-order; (**b**) CT 2–1–1, after [32]; (**c**) mb, CT third-order

5.3.4 Simulation Results

Simulation results are shown in Fig. 5.3, where single-loop and multiloop modulators are simulated with integrator gain tolerance according to (5.11). An input signal of $-23\,\text{dB}$ has been applied to the third-order single-bit and the cascaded architecture, while the signal power was also chosen higher for the single-loop multibit modulator. Equations (5.13)–(5.15) and the observations above are clearly confirmed. The earlier degradation for negative tolerance ($\delta_{RC} < 0$) is due to the above mentioned overload and instability.

The simulation for the multibit modulator confirms the discussion on the effective quantizer gain in the presence of gain-errors in Sect. 5.3.1: For low-input signals the quantizer acts as single-bit one and ideally cancels the gain error of the last integrator, thus showing less gain error dependency. For large input signals, the quantizer approximates the ideal unity gain, and thus the gain error of the last integrator appears!

The presented simulations were done using NRZ feedbacks. Nonetheless, different feedback waveforms, rectangular RZ, pulse-shaped, realized with active *RC* or current source feedbacks do not have an impact on the behavior under integrator gain errors. The only result may be that the gain errors of forward and feedback paths vary (e.g., resistor in forward and current source in feedback path) and thus the equations above have to be adapted. This was done exemplarily for the exponentially shaped SCR-feedback in [31].

Concluding, single-loop CT ΣΔ modulators appear quite tolerant against integrator gain variations, and they can be built without severe performance degradation over a wide range [36]. In contrast, cascaded CT ΣΔ modulators cannot be implemented with an acceptable resolution *without any countermeasures* against the influence of gain errors [20, 125].

5.3.5 Compensation of Gain Errors in Single-Loop ΣΔ Modulators

Even if single-loop, low-pass ΣΔ modulators are not extremely sensitive to integrator gain errors, especially the design of high-order modulators still imposes design constraints, due to their tendency to instability. This is seen in Fig. 5.3a, where for integrator gain variation of $\delta_{RC} < -10\%$ the third-order modulator becomes unstable. Thus, without any countermeasures, the designer of higher order CT modulators has to leave a wide gap between the theoretically possible performance, i.e., aggressive noise shaping, and a set of scaling coefficients which is feasible to keep the modulator stable over all possible gain variations [23, 37].

Alternatively, a tuning of the CT integrator gain can be adapted. This has been used frequently in the past for achieving accurate CT filters [105, 149] or also to tune bandpass continuous-time ΣΔ modulators, whose sensitivity to integrator gain mismatch is more comparable to the sensitivity of cascaded architectures! Here, both gm [218] as well as integrator *C* tuning are commonly implemented [219]. Similar techniques can be used for tuning of single-loop modulators: Xia et al. [220] proposed integrator *C* tuning, since gm tuning relies more on accurate frequency tuning than on achieving good linearity. Additionally, the *C*-tuning approach in [220] requires only little circuit overhead and power consumption and has been successfully implemented in a multibit CT ΣΔ modulator [126].

Concluding, for achieving more optimal performance in higher order, single-loop low-pass ΣΔ modulators, an adoption of integrator gain tuning, either *R*, *gm*, or *C*, can be useful, but it has not been commonly used and needed in the recent years!

5.3.6 Compensation of Gain Errors in Cascaded ΣΔ Modulators

As seen, cascaded CT modulators cannot be implemented, unless a way to cancel or correct the intrinsically large errors due to integrator gain variations.

A straightforward way to do so is by tuning the passive components, e.g., the integrator capacitor as proposed for single-loop modulators in [221]; but this would require a large, accurate capacitor array for every integrator, thus increasing the core area of the circuit enormously. Additionally, the relative matching of the components in an array as well as components of different arrays had to be very accurate, which is doubtful to be achieved, because they would cover rather a large chip area. Thus, a pure *analog* correction is not feasible.

A possibility to digitally cancel or at least compensate the influence of gain-errors can be deduced from the general structure of a cascaded modulator in Fig. 2.20 and (2.32)–(2.34), where the digital cancellation logic of a cascaded modulator was found to be a digital representation of the analog signal or noise-transfer functions. Now, if a loop filter $H(z)$ is affected by a gain error GE and $H(z)$ is assumed to have high gain in the band of interest, the signal-transfer function and noise-transfer function in (2.11) become (without the assumption of any effective quantizer gain or loop filter scalings for simplicity reasons):

$$\mathrm{STF}_{\mathrm{GE}}(z) = \frac{\mathrm{GE}H(z)}{1+\mathrm{GE}H(z)} = \frac{1}{\frac{1}{\mathrm{GE}H(z)}+1} \overset{H(z)\gg}{\approx} 1\,, \tag{5.16}$$

$$\mathrm{NTF}_{\mathrm{GE}}(z) = \frac{1}{1+\mathrm{GE}H(z)} = \frac{\frac{1}{\mathrm{GE}H(z)}}{\frac{1}{\mathrm{GE}H(z)}+1} \overset{H(z)\gg}{\approx} \frac{1}{\mathrm{GE}H(z)}\,. \tag{5.17}$$

Thus, under the influence of a gain error, the analog STF remains ideal, while the analog NTF still shows the zeros, where the loop filter has its poles, but it is now scaled with the inverse filter gain error 1/GE. Thus, the conditions given in (2.33) are not fulfilled any more because the nonideally varied analog NTF should match to its fixed digital representation. Consequently, noise-leakage will appear. On the other hand, from these equations it is directly visible that a multiplication of a digital representation of the analog integrator gain error into the digital recombination logic could perfectly restore analog-to-digital matching.

This has been basically introduced in [98], where a calculation was performed for a two-stage, 2–2 modulator, which did not include scaling coefficients or interstage coupling coefficients. Thus, an extension had to be performed to generally counteract gain errors in scaled $\Sigma\Delta$ modulators, especially in the common 2–1 and 2–1–1 architectures [32, 125]: Generally, for a given architecture the integrator gain errors, the scaling coefficients as well as effective quantizer gains have to be taken into account. The approach is to multiply the outputs $y_i(n)$ of each modulator stage with its digital cancellation logic DF_i and additionally with correction-factors Corr_i, $i = 1, \ldots, M$. By combining the resulting expressions:

$$Y(z) = \sum_{i=1}^{M} Y_i(z)\,\mathrm{DF}_i(z)\,\mathrm{Corr}_i = f(U_1, E_i|_{i=1,\ldots,M})\,, \tag{5.18}$$

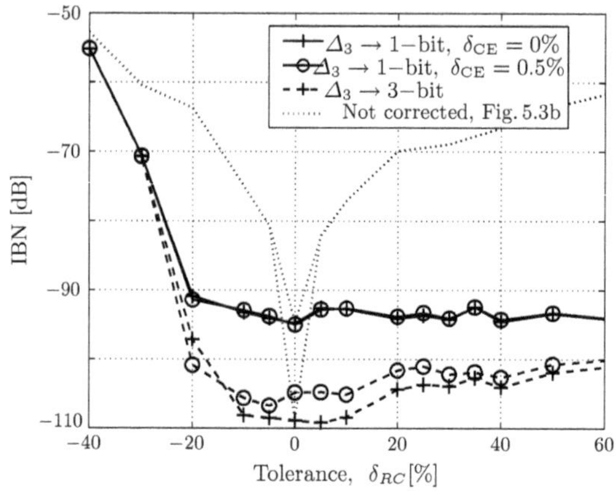

Fig. 5.4. Simulated CT 2–1–1 modulator with (non)ideal correction of an integrator gain error δ_{RC}. OSR $= 24$, $P_{\text{sig}} = -23\,\text{dB}$, $f_{\text{sig}} = f_{\text{B}}/5$. $\text{Corr}_i = \text{Corr}_i|_{\text{ideal}}/(1+\delta_{\text{CE}})$

a set of equations is obtained for achieving ideal cancelation of low-order quantization noise, which leaks to the output through all but the last stage: $E_i|_{i=1,\ldots,M-1} = 0$. Then, the correction factors Corr_i have to be multiplied to the output $Y_i(z)$ in stage i in Fig. 2.20.

For a two-stage modulator, (2.33) together with (5.16, 5.17) instructs to either multiply DF_2 by the same inverse gain error, or multiply DF_1 with the non-inverse gain-error of the first stage. Similar conditions can be found for a three-stage architecture as the 2–1–1 modulator shown in Fig. 3.8b, where digital corrections can either be performed in DF_2 and DF_3 or alternatively in DF_1 and DF_2:

$$\text{2–1–1:}\quad \text{Corr}_1 = \text{Corr}_2\,\text{GE}_{11}\text{GE}_{12},\ \ \text{Corr}_2 = \text{Corr}_3\,\text{GE}_{21}\,, \tag{5.19}$$

where GE_{il} are the integrator gain errors of the lth integrator of the ith stage. To show the functioning of this ideal gain error correction, similar to Fig. 5.3b simulation results for the 2–1–1 modulator are presented, this time with ideal correction as in (5.19). Additionally, in a second simulation a slight relative correction mismatch (δ_{CE}) has been introduced. Obviously in Fig. 5.4, digital correction is a feasible method for correcting integrator gain errors in CT cascaded ΣΔ modulators [20, 97, 125]. This will also be experimentally shown in Chap. 7.

5.4 Finite Amplifier Gain-Bandwidth Product

Another important cause of nonideal behavior are errors due to the integrator dynamics; here, the first influence to consider is a finite gain-bandwidth product (GBW) of the operational amplifiers in the integrators, which introduces non-dominant poles into the integrator transfer function.

Finite GBW is known from DT modulator implementations to cause distortion and increased in-band noise, e.g., as shown in [5]. The requirements can actually be very high: Boser, Wooley, Gregorian and Temes [110, 222] found that in many sampled-data analog filters, the unity-gain bandwidth of the OpAmps has to be at least an order of magnitude higher than the sampling rate; in DT $\Sigma\Delta$ modulators, this is somewhat relaxed. Baird and Fliz and Medeuro et al. [107, 223] have shown finite GBW to cause an error equivalent to an integrator gain error resulting in an increased in-band noise. Consequently, single-loop DT modulator implementations have been published, where the amplifier GBW could be reduced to 1–2f_S [72, 110]. Nevertheless, in cascaded DT implementations the unity-gain bandwidth still has to be chosen a factor of at least 5–10f_S due to their increased sensitivity to nonidealities [8, 60–62, 94].

In contrast, CT implementations are claimed throughout the published literature to work with much lower GBW of the OpAmps [9, 36, 121, 122]. This has been largely attributed to the lack of the high-current peaks of switched capacitor implemented DT circuits. Nevertheless, while Chan and Martin [121] noted that a finite GBW of around the sampling rate gives only negligible performance loss, in [34] for a third-order modulator a margin of around 1.5f_S was found; Wongkomel [122] finally claimed an integrator nondominant pole of 2–3 times the sampling frequency. Recently, a basic [34, 36, 125] and an advanced [30] model were introduced, which showed that finite amplifier GBW in CT $\Sigma\Delta$ modulators can even be chosen lower than the sampling frequency. This is subsequently outlined.

5.4.1 Basic Analytical Description of Finite GBW

Incorporating the influence of finite gain bandwidth into a continuous-time modulator can by done by replacing the amplifier in Fig. 5.1 by a single-pole amplifier [125], whose transfer function is given by (5.20):

$$A(s) = \frac{A_{\mathrm{dc}}}{\frac{s}{\omega_{\mathrm{A}}}+1}, \quad \mathrm{GBW} = A_{\mathrm{dc}}\,\omega_{\mathrm{A}} \quad [\mathrm{rad/s}], \tag{5.20}$$

where ω_{A} is the dominant pole of the amplifier inside the integrator, while GBW is its gain-bandwidth product. Incorporating (5.20) into (5.2) yields an integrator transfer function of the ith input path due to finite GBW of the OpAmp as follows:

$$\mathrm{ITF}_{\mathrm{GBW}}(s)|_i \overset{A_{\mathrm{dc}}\gg}{\approx} \frac{k_i f_{\mathrm{S}}}{s} \frac{\frac{\mathrm{GBW}}{\mathrm{GBW}+\sum_l |k_l f_{\mathrm{S}}|}}{\frac{s}{\mathrm{GBW}+\sum_l |k_l f_{\mathrm{S}}|}+1}. \tag{5.21}$$

This ITF consists of a scaled integrator in series with a gain error and an additional second integrator pole. Both depend on all realized integrator corner frequencies $\omega_{\mathrm{I}}|_i$ in (5.1) of the respective integrator.

Thus, an integrator gain error and a second pole due to the finite GBW can be defined as in (5.22) and the integrator transfer function of the ith input path in (5.21) simplifies to (5.23):

$$\mathrm{GE}_{\mathrm{GBW}} = \frac{\mathrm{GBW}}{\mathrm{GBW} + \sum_{l=1}^{n_{\mathrm{A}}} \omega_{\mathrm{I}}|_l} \,, \qquad \omega_{\mathrm{p}} = \mathrm{GBW} + \sum_{l=1}^{n_{\mathrm{A}}} \omega_{\mathrm{I}}|_l \,, \tag{5.22}$$

$$\mathrm{ITF}_{\mathrm{GBW}}(s)|_i \approx \frac{k_i f_{\mathrm{S}}}{s} \frac{\mathrm{GE}_{\mathrm{GBW}}}{\frac{s}{\omega_{\mathrm{p}}} + 1} \,, \tag{5.23}$$

where $\omega_{\mathrm{I}}|_l = k_l f_{\mathrm{S}}$ is the corner frequency of the scaled lth integrator, see (5.1). By neglecting the influence of the second integrator pole, according to this model, finite GBW in CT ΣΔ modulators has a rather similar influence as *RC*-variations, i.e., both cause integrator gain errors, which were already discussed in Sect. 5.3. Thus, the expressions for the integrated in-band noise due to finite GBW are approximated from the corresponding expressions for the integrator gain errors in (5.13–5.15) by replacing GE with the gain error expression in (5.22).

Simulation Results

Figure 5.5 presents simulation results for a second-order and third-order single-loop and for a 2–1–1 cascaded CT modulator, all with NRZ-feedback implementation. Note that in the figures the x-axis parameter c is intended as a comparison of the GBW of the OpAmps and the sampling frequency f_{S} of

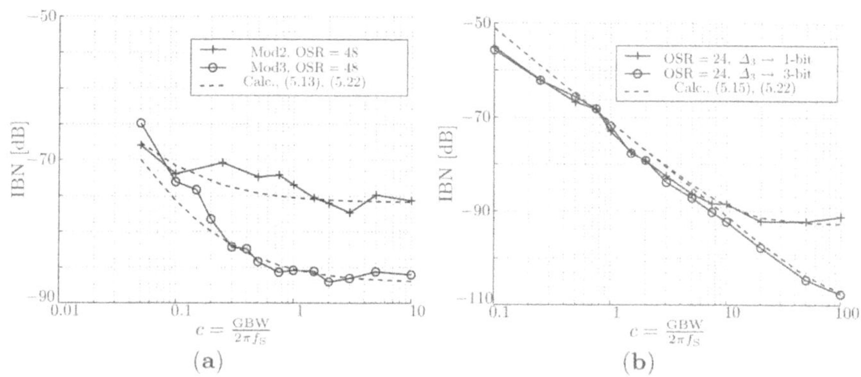

Fig. 5.5. Simulated CT modulators under the influence of finite gain bandwidth. NRZ-feedback implementation. $P_{\mathrm{sig}} = -23\,\mathrm{dB}$ and $f_{\mathrm{sig}} = f_{\mathrm{B}}/3$. Note the different axis scaling for single-loop and cascaded modulators. (**a**) sb, CT second- and third-order, NRZ; (**b**) CT 2–1–1, NRZ [32]

the modulator. Due to the unit of GBW = rad/s in (5.20) and f_S = Hz, the ratio is chosen as [34, 125]:

$$c = \frac{\text{GBW}}{2\pi f_S} . \tag{5.24}$$

Thus, c represents the ratio of the GBW to the sampling frequency. Concerning the nomenclature and the dependencies of the sampling frequency f_S, the CT integrator corner frequency ω_I as well as the GBW and the factor c, please see also the illustrative explanation in Appendix C.

For the single-loop architectures simulated here, $c = 1$ seems to be a reasonable limit without severe performance degradation, which is in accordance to other publications [34, 121, 122]. Only for very low values of the GBW the calculations predict better behavior, which is attributed to the neglected influence of the second integrator pole in the above considerations.

For the cascaded modulators, the simulation shows an unacceptable loss of ideal performance for decreasing GBW-magnitudes. The reason for this is self-evident when considering the GBW-induced gain error and the extremely high sensitivity of cascaded modulators to it, which is also confirmed with the calculation results from IBN(GE) in (5.15), which matches with the simulation results [32, 125].

5.4.2 Extended Model for Single-Loop Modulators

The above considerations and simulations showed that the influence of finite GBW can be mainly reduced to the corresponding gain error. But while this matches very well for the sensitive cascaded modulators, the single-loop architectures show significant deviation, especially for very low GBW values. Furthermore, simulations with exemplary RZ feedback waveform show slightly different (*better than* NRZ!) behavior [30], which is not predicted by the gain-error model and both are attributed to the neglected nondominant integrator pole in (5.22).

Therefore in [21, 30, 124] a more sophisticated finite GBW model was derived, which represents the nondominant pole in a way giving a direct insight into its influence on the modulator and allowing the definition of a compensation approach. The idea is based on modeling the nonideal poles as feedback delays and is mainly intended for modulators with rectangular feedback. The model procedure is shown in Fig. 5.6 by exemplarily modifying the block diagram of a second-order modulator.

By pulling the nonideal poles behind the summing points, the modified Fig. 5.6b is obtained. For the input transfer function the poles ω_{pi} do not have an impact as long as $\omega_{pi} \gg f_{sig}$, i.e., the input signal frequency. The feedback transfer functions are scaled single or double pole systems. Due to the considered rectangular feedback pulse, the combination of the feedback transfer functions with the integrators in the forward path is modeled as nonideal integrators with delayed output slopes.

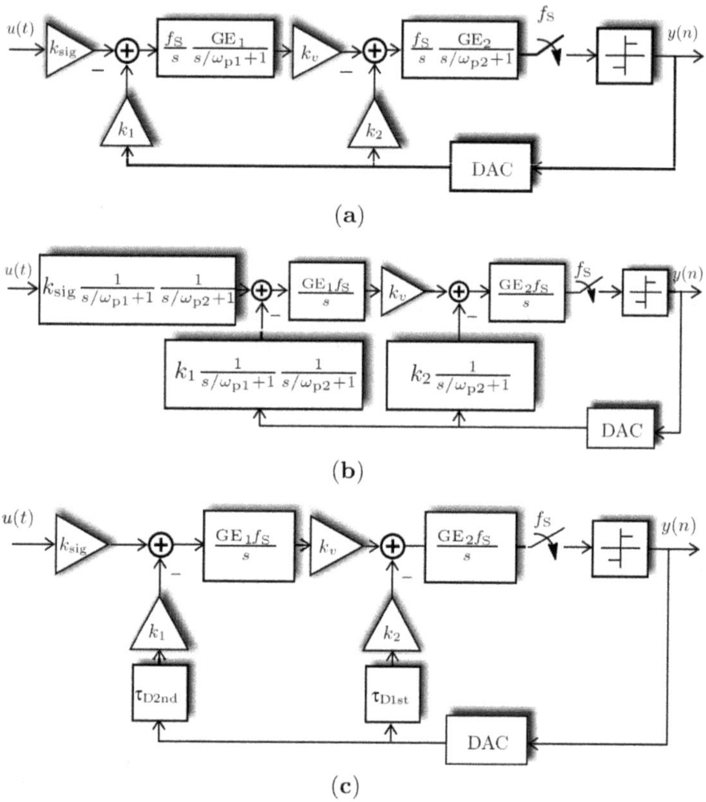

Fig. 5.6. Derivation of a finite GBW model for a second-order modulator, after [30]. (**a**) Integrators with gain error and OpAmp pole; (**b**) modified block diagram; (**c**) complete model for finite GBW modulator

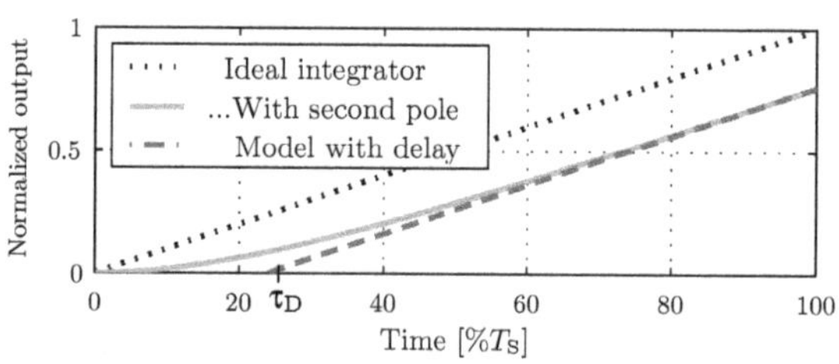

Fig. 5.7. Nonideal and ideal integrator step response, after [30]

The possible adoption of the delay model is illustratively simulated in Fig. 5.7 for the ideal integrator with single pole delay, where T_S is the sampling time and the GBW of the OpAmp is exemplary chosen: $c = 0.5$. The corresponding method, i.e., the replacement of a multipole system by a delayed first-order pole system is a well-known technique in the control systems theory [224, 225]. This approximation results the modeled second-order modulator where the influence of finite GBW is approximated as gain errors and feedback delay (Fig. 5.6c).

An extension to a third (or higher) order modulator is straightforward: the outermost feedback loop sees an additional third-order pole system, the second loop a second-order pole and finally the innermost loop a single-order pole system. Self-evident, a cascade of more poles in outer loops or even the inclusion of zeros will drastically alter the step response from the simple one in the example of Fig. 5.7 [30]. Nevertheless, simulations show that the simple model results quite accurate conformity in the simulations. Other architectures than the considered common feedback are feasible, since the model relies on the delayed feedback with integrators in the forward path. Feedback waveforms other than rectangular can also be modeled, but their feasibility for the upcoming compensation approach is doubtful. Both are still a matter for further research.

The feedback delays $t_{\mathrm{D}i}$ of the GBW model can be calculated from the expressions of a tangent to the step responses of the nonideal integrator-transfer functions; these consist of the respective feedback transfer function and its subsequent integrator in Fig. 5.6c. A calculation procedure was presented in [30], but also a graphical determination of the delays can be performed. The calculation results for a first-order to third-order feedback poles yield [30]:

$$
\begin{aligned}
t_{\mathrm{D1st}} &= \frac{1-\mathrm{e}^{-\omega_{\mathrm{p2}}/f_{\mathrm{S}}}}{\omega_{\mathrm{p2}}}, \\
t_{\mathrm{D2nd}} &= \frac{\omega_{\mathrm{p1}}^2(1-\mathrm{e}^{-\omega_{\mathrm{p2}}/f_{\mathrm{S}}})-\omega_{\mathrm{p2}}^2(1-\mathrm{e}^{-\omega_{\mathrm{p1}}/f_{\mathrm{S}}})}{\omega_{\mathrm{p1}}\omega_{\mathrm{p2}}(\omega_{\mathrm{p1}}-\omega_{\mathrm{p2}})}, \\
t_{\mathrm{D3rd}} &= \frac{\dfrac{1-\mathrm{e}^{-\omega_{\mathrm{p1}}/f_{\mathrm{S}}}}{\omega_{\mathrm{p1}}^3\omega_{\mathrm{p3}}}+\dfrac{1-\mathrm{e}^{-\omega_{\mathrm{p3}}/f_{\mathrm{S}}}}{\omega_{\mathrm{p3}}^3\omega_{\mathrm{p2}}}+\dfrac{1-\mathrm{e}^{-\omega_{\mathrm{p2}}/f_{\mathrm{S}}}}{\omega_{\mathrm{p2}}^3\omega_{\mathrm{p1}}}}{\dfrac{1}{\omega_{p3}^2\omega_{p2}}+\dfrac{1}{\omega_{\mathrm{p2}}^2\omega_{\mathrm{p1}}}+\dfrac{1}{\omega_{\mathrm{p1}}^2\omega_{\mathrm{p3}}}-\dfrac{1}{\omega_{\mathrm{p2}}^2\omega_{\mathrm{p3}}}-\dfrac{1}{\omega_{\mathrm{p3}}^2\omega_{\mathrm{p1}}}-\dfrac{1}{\omega_{\mathrm{p1}}^2\omega_{\mathrm{p2}}}} \\
&\quad+\frac{-\dfrac{1-\mathrm{e}^{-\omega_{\mathrm{p1}}/f_{\mathrm{S}}}}{\omega_{\mathrm{p1}}^3\omega_{\mathrm{p2}}}-\dfrac{1-\mathrm{e}^{-\omega_{\mathrm{p3}}/f_{\mathrm{S}}}}{\omega_{\mathrm{p3}}^3\omega_{\mathrm{p1}}}-\dfrac{1-\mathrm{e}^{-\omega_{\mathrm{p2}}/f_{\mathrm{S}}}}{\omega_{\mathrm{p2}}^3\omega_{\mathrm{p3}}}}{\dfrac{1}{\omega_{\mathrm{p3}}^2\omega_{\mathrm{p2}}}+\dfrac{1}{\omega_{\mathrm{p2}}^2\omega_{\mathrm{p1}}}+\dfrac{1}{\omega_{\mathrm{p1}}^2\omega_{\mathrm{p3}}}-\dfrac{1}{\omega_{\mathrm{p2}}^2\omega_{\mathrm{p3}}}-\dfrac{1}{\omega_{\mathrm{p3}}^2\omega_{\mathrm{p1}}}-\dfrac{1}{\omega_{\mathrm{p1}}^2\omega_{\mathrm{p2}}}}, \\
\tau_{\mathrm{Di}} &= \frac{t_{\mathrm{D}i}}{T_{\mathrm{S}}}, \qquad (5.25)
\end{aligned}
$$

where $\omega_{\mathrm{p}i}$ are the poles resulting from (5.22) for the different integrators.

Discussion and Simulation Results for the Extended Model

Before simulating the model, a short note is given what is expected: gain errors affect all single-loop modulators in a similar way, while feedback delays affect higher order loops more than lower ones, and NRZ-feedback modulators are degraded more than RZ ones. Both were discussed in Sect. 4.2, where excess loop delay was analyzed, and this is now also expected for the influence of finite GBW.

The model is confirmed by simulations. Therefore, a second-order and third-order, single-loop, distributed feedback CT ΣΔ modulator is simulated, both with RZ and NRZ-feedback waveform. In (5.20), a single-pole amplifier model was regarded. In practice, the amplifier will more be a two pole system, possibly also with a dominant zero. Also this is included in the simulations, where first the single-pole (1p) approximation, and partially the two pole, one zero (2p1z) amplifier model is assumed. Exemplarily, in the latter case a dominant second amplifier pole ($2 \times$ GBW) and amplifier zero ($5 \times$ GBW) were chosen for all simulations. This way, an amplifier with a phase margin of approximately 50° is taken into account. Note that for the 2p1z approximation, the delays in (5.25) have to be adopted [30].

All simulations were done in a circuit simulator (SPECTRE), where the modulators have been implemented using RC-integrators. To account for large signal induced distortion, all simulations were performed at $f_{\mathrm{sig}} = f_{\mathrm{B}}/5$ and $P_{\mathrm{sig}} = -9\,\mathrm{dB}$ for the second-order and to $P_{\mathrm{sig}} = -15\,\mathrm{dB}$ for the third-order modulators. Obviously in Fig. 5.8, for all simulations the model and the modulator with finite GBW match well, even for very low values of GBW. Additionally, the slightly different behavior of RZ and NRZ, and even the inclusion of additional poles and zeros is covered by the model [30].

5.4.3 Compensation for Finite GBW-Induced Errors in CT ΣΔ Modulators

Based on the above presented models for finite GBW, compensation approaches were presented, both for cascaded and for single-loop modulators.

Single-Loop CT ΣΔ Modulators

To compensate for finite GBW Induced errors in single-loop modulators, the more precise model with gain errors and loop delay is taken into account: gain errors, which are deterministic, i.e., more or less known before the implementation, can be roughly counteracted with adopted scaling coefficients. In contrast, as shown in Sect. 4.2, possibilities for compensating loop delays depends on the form and position of the feedback waveform: RZ pulses, which are not moved into the next clock cycle are much easier compensated (adoption of scaling coefficients) than pulses, which are moved into the next clock cycle. The latter require a change of the architecture, as e.g., the use of additional feedback paths. One of the feasible architectures was shown in Fig. 4.4 [14, 132].

Exemplarily, Fig. 5.9 gives this architecture of a second-order modulator with different (GBW induced) feedback delays, (5.25), in the feedback paths and a compensation path for loop delay. The HRZ pulse is a half sample delayed pulse with $\{\alpha = 1/2, \quad \beta = 1\}$. The main DAC can either be a NRZ DAC or an RZ DAC seeing a feedback delay which shifts the pulse into the next clock cycle.

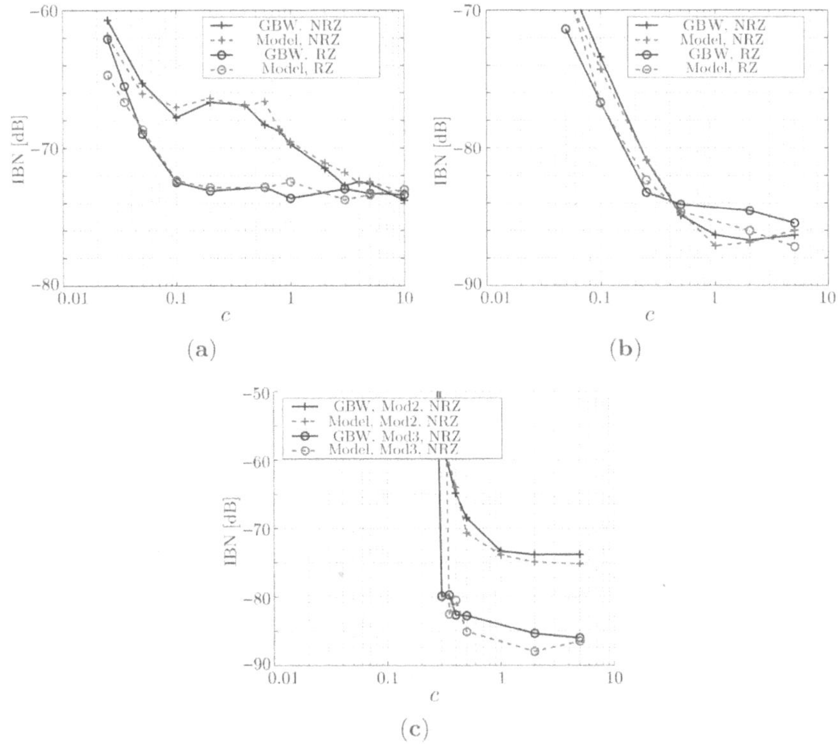

Fig. 5.8. IBN of various modulators with finite GBW OpAmps and the model according to Fig. 5.6c. $c = \mathrm{GBW}/2\pi f_S$, after [30]. (**a**) Second-order modulator, 1p model; (**b**) third-order modulator, 1p model; (**c**) second- and third-order modulator, 2p1z

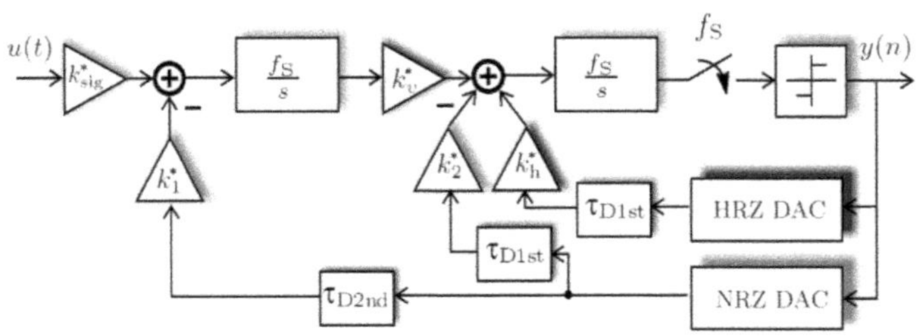

Fig. 5.9. Compensation for (GBW) feedback delay in a second-order modulator [30]

As for excess loop delay in Sect. 4.2, the objective is to transform the nonideal CT modulator in Fig. 5.9 into its DT equivalent. Subsequently, the scaling coefficients are chosen such that this DT equivalent matches the *ideal* DT modulator. Since an NRZ-feedback pulse is considered, as proposed in [132] and presented in Sect. 4.2 the overlap of the feedback signal into the next clock cycle has to be taken into account, which is done by splitting it into two fractions, one in the actual and one in the next clock cycle (z^{-1}). Using Table 3.6 for the loop filter of the first branch in Fig. 5.9 yields (with $f_S = 1$):

$$\begin{aligned} \mathrm{LF_{CT}}|_1 &= \frac{-k_1^*}{s^2} \to \mathrm{LF_{CTDT}}|_1 \\ &= \frac{-k_1^*(\beta_{11}(2-\beta_{11}) - \alpha_{11}(2-\alpha_{11}))z - k_1^*(\beta_{11}^2 - \alpha_{11}^2)}{2(z-1)^2} \\ &\quad + \frac{-k_1^*(\beta_{21}(2-\beta_{21}) - \alpha_{21}(2-\alpha_{21}))z - k_1^*(\beta_{21}^2 - \alpha_{21}^2)}{2(z-1)^2} z^{-1}, \end{aligned} \tag{5.26}$$

where $\{\alpha_{11} = \tau_{\mathrm{D2nd}},\ \beta_{11} = 1\}$ and $\{\alpha_{21} = 0,\ \beta_{21} = \tau_{\mathrm{D2nd}}\}$. For the loop filter of the second branch applies:

$$\mathrm{LF_{CT}}|_2 = \frac{-k_2^*}{s} \to \mathrm{LF_{CTDT}}|_2 = \frac{-k_2^*(\beta_{12} - \alpha_{12})}{z-1} + \frac{-k_2^*(\beta_{22} - \alpha_{22})}{z-1} z^{-1}, \tag{5.27}$$

where $\{\alpha_{12} = \tau_{\mathrm{D1st}},\ \beta_{12} = 1\}$ and $\{\alpha_{22} = 0,\ \beta_{22} = \tau_{\mathrm{D1st}}\}$. For the loop filter of the compensation branch with the HRZ feedback DAC one obtains:

$$\mathrm{LF_{CT}}|_{\mathrm{h}} = \frac{k_{\mathrm{h}}^*}{s} \to \mathrm{LF_{CTDT}}|_{\mathrm{h}} = \frac{k_{\mathrm{h}}^*(\beta_{1\mathrm{h}} - \alpha_{1\mathrm{h}})}{z-1} + \frac{k_{\mathrm{h}}^*(\beta_{2\mathrm{h}} - \alpha_{2\mathrm{h}})}{z-1} z^{-1}, \tag{5.28}$$

where $\{\alpha_{1\mathrm{h}} = 1/2 + \tau_{\mathrm{D1st}},\ \beta_{1\mathrm{h}} = 1\}$ and $\{\alpha_{2\mathrm{h}} = 0,\ \beta_{2\mathrm{h}} = \tau_{\mathrm{D1st}}\}$. All three $\mathrm{LF_{CTDT}}|_i$ form together the DT representation of the nonideal CT modulator in Fig. 5.9. Optimally, this DT loop filter equals the original DT modulator loop filter in (4.7):

$$\sum_i \mathrm{LF_{CTDT}}|_i \stackrel{!}{=} \mathrm{LF_{DT|ideal}}\,. \tag{5.29}$$

Solving the resulting equations yields the scaling coefficients of the modulator with compensated feedback delays τ_{D1st}, τ_{D2nd}. With DT scaling $a_1 = a_2 = 0.5$ for the second-order modulator [6], one gets:

$$\begin{aligned} k_{\mathrm{sig}}^* &= k_1^* = \frac{1}{4}, \quad k_v^* = 1, \quad k_2^* = \frac{6\tau_{\mathrm{D1st}} + 4\tau_{\mathrm{D1st}}\tau_{\mathrm{D2nd}} - \tau_{\mathrm{D2nd}}^2}{8\tau_{\mathrm{D1st}}}, \\ k_{\mathrm{h}}^* &= \frac{3\tau_{\mathrm{D1st}} + 2\tau_{\mathrm{D1st}}\tau_{\mathrm{D2nd}} - \tau_{\mathrm{D2nd}}^2}{4\tau_{\mathrm{D1st}}}\,. \end{aligned} \tag{5.30}$$

Finally note that the scaling coefficients in (5.30) are only directly valid for *one fixed* set of feedback delays. In the case of GBW-induced delays, an iterative calculation has to be done, since the GBW-induced poles and thus delays in (5.25) depend on the scaling coefficients in (5.30) and vice versa [30]. In addition, the coefficients in (5.30) still do not account for the GBW induced gain errors. Thus, the following calculation procedure was proposed [30]:

Step 1 Start the calculation process with the scaling coefficients in (3.21) and $k_{\mathrm{h}}^* = 0$. Correct gain error GE_1 through $k_v^* = 1/\mathrm{GE}_1$.
Step 2 Calculate the resulting GBW-induced delays in (5.22) with (5.25).
Step 3 Use (5.30) to assign a set of compensation coefficients.
Step 4 Repeat steps 2 and 3 up to a certain accuracy.
Step 5 The gain error GE_2 is (ideally) compensated for by the used single-bit quantizer.

Example: Compensation for a second-order modulator
With DT modulator scaling $a_1 = a_2 = 0.5$ and NRZ feedback, the *ideal* CT modulator scaling in Fig. 5.9 becomes:

$$k_{\mathrm{sig}} = k_1 = 1/4, \quad k_2 = 3/8, \quad k_v = 1, \quad k_{\mathrm{h}} = 0\,. \tag{5.31}$$

If an exemplary amplifier GBW is taken with $c = 0.1$, (5.22) yields a first integrator gain error for the original scaling coefficients:

$$\mathrm{GE}_1 = \frac{\mathrm{GBW}_1}{\mathrm{GBW}_1 + f_{\mathrm{S}}(k_1 + k_{\mathrm{sig}})} \overset{(5.24)}{=} \frac{2\pi c}{2\pi c + (k_1 + k_{\mathrm{sig}})} \approx 0.55\,. \tag{5.32}$$

To compensate for the first integrator gain error, set as new value:

$$k_v^* = 1/\mathrm{GE}_1 \approx 1.8 \tag{5.33}$$

Equation (5.31) with (5.33) and (5.22) yields a first set for the GBW-induced feedback time delays in (5.25):

$$\begin{aligned} \omega_{\mathrm{p1}} &= 2\pi f_{\mathrm{S}} c + (k_1^* + k_{\mathrm{sig}}^*) f_{\mathrm{S}} \approx 1.12 f_{\mathrm{S}}\,, \\ \omega_{\mathrm{p2}} &= 2\pi f_{\mathrm{S}} c + (k_2^* + k_v^* + k_{\mathrm{h}}^*) f_{\mathrm{S}} \approx 2.8 f_{\mathrm{S}}\,, \\ \tau_{\mathrm{D1st}} &\approx 0.78\,, \\ \tau_{\mathrm{D2nd}} &\approx 0.33\,, \end{aligned} \tag{5.34}$$

which yields a first set of compensation scalings in (5.30):

$$k_{\mathrm{sig}}^* = k_1^* = 1/4, \quad k_2^* \approx 0.91, \quad k_v^* \approx 1.8, \quad k_{\mathrm{h}}^* \approx 0.69\,. \tag{5.35}$$

With these coefficients, the time delays are recalculated and then the scalings are recalculated, which is repeated in an iterative style. Note that only k_2^* and k_{h}^* alter during the iteration. The *final* values (after five iterations) yield:

$$k_{\mathrm{sig}}^* = k_1^* = 1/4, \quad k_2^* \approx 0.85, \quad k_v^* \approx 1.8, \quad k_{\mathrm{h}}^* \approx 0.57\,. \tag{5.36}$$

This set of coefficients in an architecture as in Fig. 5.9 compensates the influence of finite amplifier GBW with $c = 0.1$. *This is the end of the example.*

Simulation results over various values for GBW and for a second- and third-order modulator with both RZ and NRZ feedback are given in Fig. 5.10. Note that ideally the GBW can have extremely low values without altering

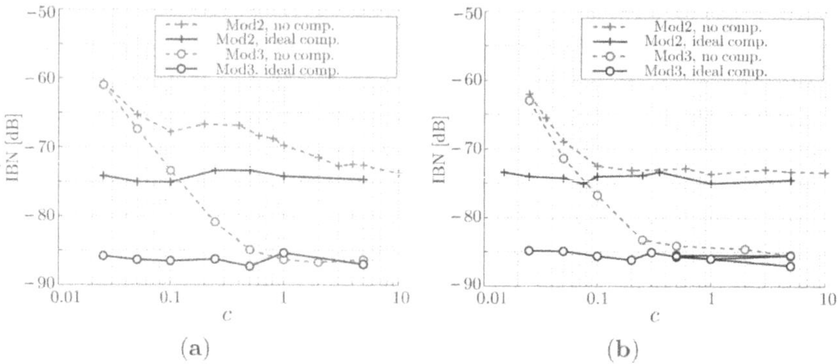

Fig. 5.10. IBN of CT modulators with finite GBW amplifiers and ideal compensation. $c = \text{GBW}/2\pi f_S$, after [30]. (**a**) Second-order and third-order NRZ, 1p model; (**b**) second-order and third-order, RZ, 1p model

the noise shaping behavior of the modulators, and all considered architectures become nearly insensitive to finite GBW. This obviously alleviates by some degree when taking nonidealities, other architectures or different amplifier models into account [30].

Concluding, the presented technique for compensating the influence of finite GBW on single-loop CT ΣΔ modulators promises a way to reduce the required speed requirements significantly. Nonetheless, as already discussed in [30], the ideally seen results cannot be expected in reality: first, reducing the bandwidth goes usually along with also decreasing the slew rate of an amplifier. While the compensation works for low bandwidth, the slew rate requirements of the CT modulator still have to be achieved! Additionally, reducing the bandwidth by reducing the amplifier transconductance may also increase the input referred noise, thus decreasing the SNR. Finally, the reduced signal swings can be ideally canceled by subsequent gains, but in reality this will be limited by the achievable SNR at the different modulator nodes. Thus, actual implementations on circuit level still have to prove this compensation technique!

Cascaded CT ΣΔ Modulators

For cascaded modulators, it turned out in Sect. 5.4.1 that finite GBW induced gain errors dominate the behavior due to the high sensitivity of cascaded structures to it. Nonetheless, in Sect. 5.3.6 a method for digital correction of integrator gain errors in ΣΔ modulators was presented, based on digitally counteracting the disturbance of the analog noise-transfer and signal-transfer functions. Previously, this method was only intended to counteract gain errors caused by variations of the passive components (δ_{RC}) [20, 96, 98]. Nonetheless, it can be directly adapted also for integrator gain errors caused by finite

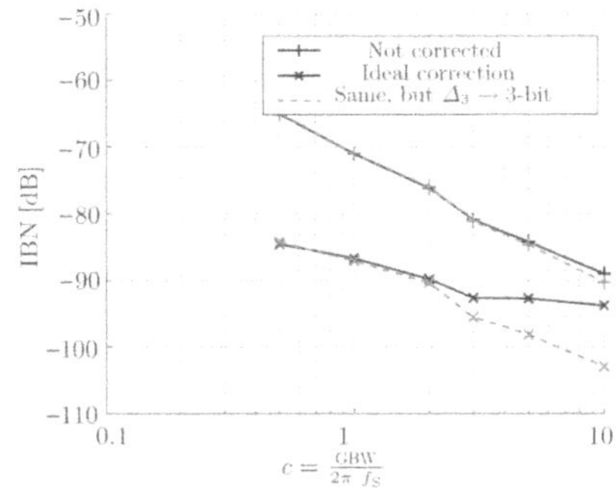

Fig. 5.11. Simulated 2–1–1 CT modulator, NRZ, single-bit and multibit last-stage quantization, under the influence of finite gain bandwidth with and without correction. $P_{\text{sig}} = -9\,\text{dB}$, $f_{\text{sig}} = f_{\text{B}}/3$ and $\text{OSR} = 24$, after [30]

amplifier GBW, when the gain error in (5.22) is canceled by a correction factor (5.19).

This is shown through ideal simulations in Fig. 5.11, where the GBW caused gain errors are canceled with digital correction. In addition, the nonideal results from Fig. 5.5b are given. Obviously, by employing digital correction, the requirements decrease by an order of magnitude for the single-bit implementations, i.e., to a range of $c = 1$–2. In the case of multibit last stage quantization, still a much higher GBW is required to obtain the ideal resolution.

The remaining performance degradation is caused by the influence of the second integrator pole, which is not considered and thus not corrected. Adopting the same correction procedure as for the single-loop modulators with delay compensation is rather not feasible, since the cascaded modulators need too accurate matching as provided with this method. Note that the digital correction for finite GBW induced gain errors could also be useful in DT, switched-capacitor $\Sigma\Delta$ modulators. Again, a possibility to perform the above correction on circuit level will be discussed in Sect. 7.6.1.

5.4.4 Influence on Different Feedback Implementations

Finite amplifier GBW is a typical nonideality concerning the dynamics of a $\Sigma\Delta$ modulator. Thus, by changing the shape and the dynamic behavior of the feedback pulses, also the influence of finite GBW is altered.

Nonetheless, it cannot be stated that the faster the feedback the higher the GBW requirements. This was exemplarily seen in Fig. 5.8, where the RZ feedback pulse requires even a little less amplifier bandwidth than its NRZ counterpart. The reason is the induced feedback delay and the higher sensitivity of the NRZ implementation to it. Similar effects are seen for example for the exponential SCR feedback, whose nonideal influence on the GBW requirements is not as severe as could be expected [31]. Furthermore, note that current feedback differs from resistive feedback, since in first-order approximation the current source supplies current regardless from the voltage (e.g., at the virtual ground node), while the resistive feedback varies proportionally to that voltage.

Another special influence of finite GBW is also seen on the exponentially sloping, passive SCR implementation: this is shown in Fig. 5.12, where the output of an integrator is simulated on circuit level upon an SCR pulse derived from a circuit equivalent to Fig. 4.14 connected to a CT integrator with illustrative values of $T_S = 100\,\text{ns}$, $\tau_{DAC} = 10\% T_S = 10\,\text{ns}$ and $V_{ref} = 1\,\text{V}$. The tangents to the integrator output waveforms show that the SCR pulse response of the ideal integrator has a time constant as chosen, approximately 10 ns, while for smaller bandwidths the integrator pulse response gets slower. Thus, the effective discharge time constant $\tau_{DAC|eff}$ increases. Consequently, the effective jitter sensitivity of the passive SCR DAC implementation increases with decreasing bandwidth. This behavior is clearly understood, since the virtual ground node of the CT integrator in Fig. 5.1 moves upon rising feedback signal and thus shows an influence on the SCR pulse itself. The behavior in Fig. 5.12 has also been considered analytically [124] and through simulations [31], and actually the same behavior can be seen for discrete-time, SC implementations [119]. Alternatively, the passive SC discharge pulse can be generated separated from the main integrator virtual ground node, and then converted to the actual feedback signal by means of a voltage-controlled voltage or current source [200].

None of the above mentioned effects can be precisely predicted by the simple GBW model derived throughout Sect. 5.4.1: there the calculation assumes an *RC*-integrator with voltage mode, constant feedback DAC. Nonetheless, as long as the feedback is generated in a way as to follow the simple linear signal theory, the derived expressions approximate the required GBW rather closely [31].

Concluding, there exist nowadays some methods for the characterization of finite GBW in CT ΣΔ modulators, some of them giving a good and feasible insight into its influence on a special circuit, others simulating themselves out of theory for a special application. Nonetheless, up to now there is *no almighty theory* including all the possibilities for the feedback

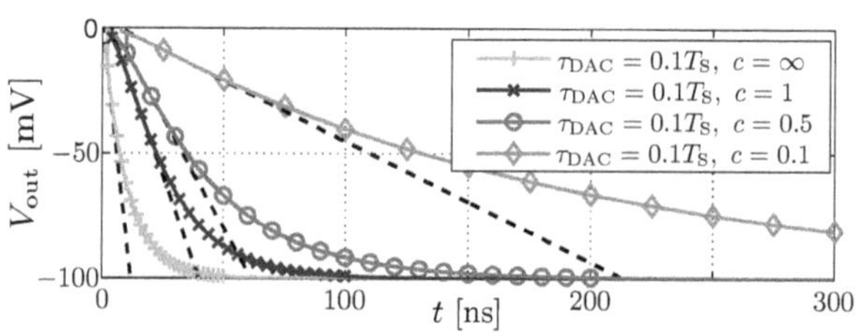

Fig. 5.12. Integrator output with finite GBW upon an SCR-pulse, after [31]

DAC, as rectangular (NRZ, RZ), shaped (SCR [136], sine [130], etc.), current mode [9], resistive [37], and filter implementations as gmC, active RC or bandpass LC.

5.5 Finite Amplifier Slew Rate

The second nonideality due to the integrator dynamics is a finite slew rate (SR) of the output signal arising through limited current capability, which is intended to charge the integrating (or an internal compensation) capacitor. As opposed to errors due to finite GBW, finite SR is a purely nonlinear effect and causes distortion as well as an increase of the noise floor [223, 226]: in DT implementations signal transitions are very fast SC-pulses and finite slew rate yields incomplete settling, which is equivalent to the effect of finite GBW on SC-implementations, which again yields a gain error. More important, the SR-limitation is signal dependent, which makes it nonlinear and results in distortion.

Therefore the requirements on the SR in DT modulators are usually high: especially in the case of cascaded modulators, successful implementations [60, 62, 94] showed $\mathrm{SR} = 5, \ldots, 10 V_{\mathrm{ref}} f_{\mathrm{S}}$, where f_{S} is the sampling frequency and V_{ref} the reference voltage as defined in Sect. 2.2.2. This is, because in addition to the induced distortion, also the noise-floor increases dramatically due to noise leakage [121]. In single-loop implementations, mostly the distortion is the performance limiting factor [110]. The only possibility to reduce the SR requirements in DT modulators is the usage of multibit internal quantization, because then the input signals to the integrators decrease [8].

5.5.1 Slew Rate in CT ΣΔ Modulators

By using CT circuitry, the slew rate specifications can be relaxed even for the single-bit quantizer case. This is because in CT modulators the various signals are changing much more slowly depending on the feedback waveform. This can be shown when considering the maximum output slew rate of a single-bit CT integrator as described in Fig. 5.1 and (5.1) to be:

$$\mathrm{SR}_{\max} = \left| \frac{\mathrm{d}v_{\mathrm{out}}}{\mathrm{d}t} \right|_{\max} = f_{\mathrm{S}} V_{\mathrm{in}}|_{\max} \,, \tag{5.37}$$

$$V_{\mathrm{in}}|_{\max} = \sum_{i=1}^{m} k_i|_{\mathrm{fb}} V_{\mathrm{ref}} + \sum_{i=1}^{n} k_i|_{\mathrm{fw}} V_{\mathrm{fw}}|_{\max} \,,$$

where $k_i|_{\mathrm{fb}}$ are the feedback scaling coefficients of the respective integrator, while $k_i|_{\mathrm{fw}}$ are its forward scaling coefficients. $V_{\mathrm{ref}} = \Delta/2$ is the reference voltage which has been introduced in Sect. 2.2.2. V_{fw} are the forward signals, i.e., the modulator input signal or the outputs of preceding integrators, both of which are usually kept in a $\pm V_{\text{ref}}$-limit through proper scaling coefficients [6]. Independent from the underlying architecture, the most critical integrator is the one at the overall modulator input, i.e., the first integrator of the first stage. Here according to Sect. 3.8.1, input referred errors are not suppressed and thus any SR-induced error will directly appear in the output spectrum of the modulator.

Due to oversampling, the input signal appears almost constant during one sampling period, while the feedback signal is assumed in the following to be rectangular, RZ or NRZ pulses. In the case of an NRZ pulse, the input signal to the first, most critical integrator is constant during a complete sampling period, yielding an integrator output signal with constant slope. Consequently, from (5.37) the required SR for a single-bit modulator can be approximated to:

$$\text{SR}_{\text{max}}|_{\text{Int11}} = f_{\text{S}} \left(k_{\text{sig}} \hat{u} + k_1 \frac{\Delta}{2} \right) , \tag{5.38}$$

where $\hat{u}$ is the amplitude of the input signal. Errors due to finite SR in the integrators after the first one are suppressed by the preceding filters. On the other hand, their scaling coefficients are (much) higher than for the first integrator. This applies especially for the cascaded architectures where more input signals are summed at the integrators of higher stages (see Fig. 3.8b). Therefore it may not be expected that the SR requirements are relaxed compared to one of the first integrator in (5.38), but only compared to their calculated, maximum values in (5.37) are due to error suppression.

5.5.2 Influence of Different Feedback Waveforms and ΣΔ Architectures

As is the case for finite GBW-induced errors in CT ΣΔ modulators, also for finite slew rate requirements the form of the feedback waveform and generation has a significant influence. But in contrast to the influence of finite GBW, in the case of amplifier SR even less published work exists and currently most designs strongly rely on simulations. Therefore, in the following only some considerations are given. Obviously, the use of NRZ-feedback pulses requires the lowest SR requirements, since the scaling coefficients are the lowest and consequently the resulting ideal integrator output slope is small. Additionally, for the NRZ feedback a rather exact limit for integrator SR exists, from where on no nonideal integrator response occurs, see (5.37).

When the scaling coefficients increase, e.g., for RZ and even more for pulse-shaped feedbacks as the exponential SCR feedback, also the required integrator slope and with it the integrator slew rate increase, Fig. 5.13 [31]. Nonetheless, the SR requirements do not increase proportionally to the scaling coefficients, which becomes obvious when considering switched capacitor modulators: here, the SC pulses are without significant parasitic switch resistance, extremely peaking, but the required slew rate is still finite and depends on the required resolution.

Similar expectations can be made for CT modulators with pulse-shaped feedback waveforms, but with one exception: the CT integrator does integrate signals over time, and thus its virtual ground node is of special importance. This again raises concerns about the feedback implementation, e.g., resistive current feedback [37] or current mode feedback [9] and the integrator implementation itself: a current mode feedback onto an active *RC* integrator disturbs the virtual ground node more than a purely resistor feedback. Thus, more signal distortion should be expected.

Additionally, note that the estimation in (5.38) is not useful for multibit $\Sigma\Delta$ implementations, since here the SR requirements are quite relaxed! This is, since the feedback DAC rather closely tracks the input signal, thus keeping the difference of both, which is the input to the first integrator, almost below one least significant bit of the DAC. Consequently, the output slope of the integrator and thus its slew rate can be drastically reduced. Nonetheless, this advantage alleviates if no NRZ feedback is used, since then the feedback signal only tracks the input over time, but not accurately at each instant: thus, the slew rate requirements increase. Subsequently, this behavior is not further considered.

Finally, one can expect that different $\Sigma\Delta$ modulator architectures require different slew rate, at least in the later integrators. This is since the input signals to the integrators in an architecture as exemplarily shown in Figs. 2.14 and 2.16 vary significantly. All of these considerations thus lead to one conclusion: integrator dynamics in CT $\Sigma\Delta$ modulators were not intensively regarded in the past, since they were known and claimed to work with much lower speed requirements than their discrete time, SC counterparts. Nonetheless, especially due to the recently arising modulators with feedback waveform shaping and additional advances in DT modulator speed requirements, also CT modulators are nowadays required to be designed more at the edge of what is possible, while this edge is not well analyzed theoretically up to now. Consequently, as claimed for finite GBW, also the slew rate requirements in CT $\Sigma\Delta$ modulators have to be studied more extensively in the future.

5.5.3 Simulation Results

Due to the strong dependency of the SR on the exact waveforms in a circuit implementation, e.g., the movement of the virtual ground node, etc., simulations were first performed in a circuit simulator (SPECTRE) and then compared to simulation results from MATLAB. This revealed that high-level simulations do not give good results and SR simulations should be performed in a circuit simulator.

Simulation results are given in Fig. 5.13 for two single-loop, low-pass modulators with distributed feedback and OSR = 48 employing RZ and NRZ feedback DAC. The 2–1–1 modulator is considered with an OSR = 24 and last-stage single-bit or multibit quantization. Additionally, the third-order

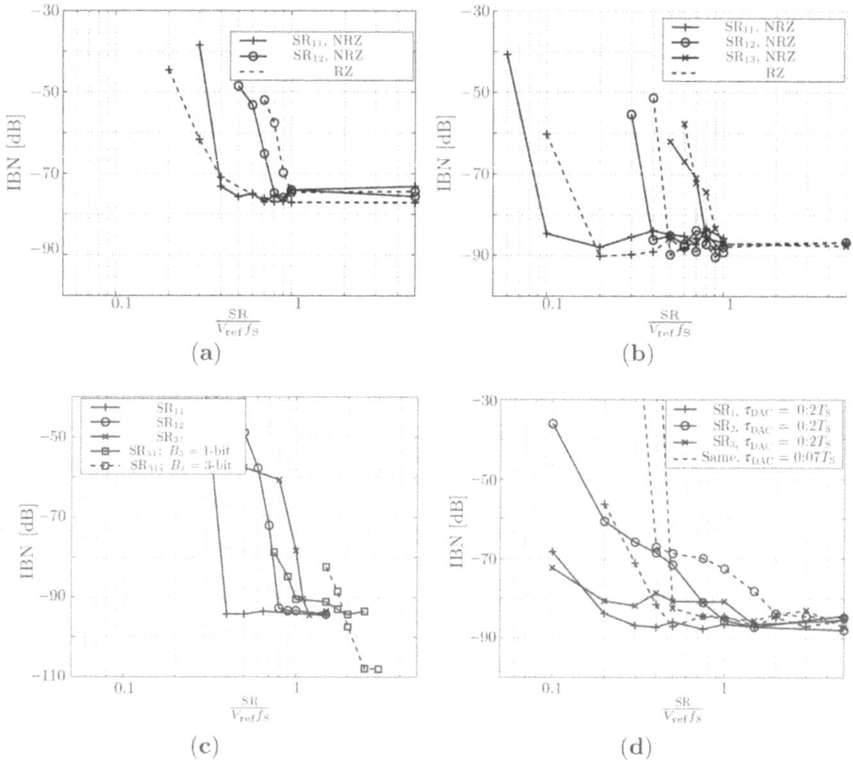

Fig. 5.13. Simulated CT modulators with NRZ, RZ and SCR implementation under the influence of finite slew rate. OSR = 24 or 48, $P_{\text{sig}} = -15\,\text{dB}$, $f_{\text{sig}} = f_{\text{B}}/5$. (**a**) sb, CT second-order, RZ and NRZ; (**b**) sb, CT third-order, RZ and NRZ; (**c**) CT 211, NRZ; (**d**) sb, CT third-order, SCR, [31]

modulator is simulated with exponentially shaped, SCR feedback DAC, and two different feedback time constants $\tau_{\text{DAC}} = (0.2/0.07)T_{\text{S}}$. The IBN is simulated over finite SR for each integrator separately, where SR_{il} is the SR of the lth integrator in the ith stage. Thereby, an exemplary reference voltage of $V_{\text{ref}} = 1\,\text{V}$ is chosen. Thus, a value of $\text{SR}/V_{\text{ref}} f_{\text{S}} = 1$ corresponds to $\text{SR} = 1\,\text{V}/\mu\text{s}$ per MHz of sampling frequency f_{S}.

To see the distortion caused by the finite SR, a reasonably large input signal $P_{\text{sig}} = -15\,\text{dB}$ has been applied, which corresponds to a signal amplitude of $0.25\,V_{\text{ref}}$. To include the majority of the distortion, the signal frequency has been chosen to be $f_{\text{sig}} = f_{\text{B}}/5$. Through these simulations, the SR estimation in (5.38) is roughly confirmed, which is exemplarily discussed for the second-order, the SCR and the 2–1–1 modulator:

– *Second-order.* In the case of the second-order modulator, with an applied input signal amplitude $\hat{u} = 0.25$ and a modulator scaling $k_{\text{sig}} =$

$k_1|_{\mathrm{NRZ}} = 0.25$, $k_1|_{\mathrm{RZ}} = 0.5$ (see Sect. 3.2.1), following (5.38) the required SR of the first integrator is $\mathrm{SR}_1|_{\mathrm{NRZ}} \approx 0.3125\,(f_\mathrm{S} V_\mathrm{ref})$ and $\mathrm{SR}_1|_{\mathrm{RZ}} \approx 0.5625\,(f_\mathrm{S} V_\mathrm{ref})$. This is confirmed in Fig. 5.13a, where the ideal IBN is achieved at $\mathrm{SR}_1|_{\mathrm{NRZ}}/(f_\mathrm{S} V_\mathrm{ref}) \geq 0.3,\ldots,0.4$ and $\mathrm{SR}_1|_{\mathrm{RZ}}/(f_\mathrm{S} V_\mathrm{ref}) \geq 0.5,\ldots,0.6$.
For the second integrator, a much higher slew rate is required because the input swing is also much higher. The forward path of either the RZ and NRZ implementation sees a unity scaling, while the feedback scaling is $k_2|_{\mathrm{NRZ}} = 0.375$, $k_2|_{\mathrm{RZ}} = 0.625$ (see Sect. 3.2.1). In contrast, the simulated slew rate requirement for almost ideal IBN is simulated to be only $\mathrm{SR}_2|_{\mathrm{NRZ}}/(f_\mathrm{S} V_\mathrm{ref}) \geq 0.8$ and $\mathrm{SR}_2|_{\mathrm{RZ}}/(f_\mathrm{S} V_\mathrm{ref}) \geq 1$. Thus, they are lower than expected by (5.38), but still higher than the slew rate of the first integrator, which confirms the above discussion.

- *2–1–1 Modulator.* In the case of the cascaded CT 2–1–1 modulator, the simulated slew rate requirements in the first stage are almost equivalent to those of the second-order modulator. This is obvious, because the architecture of the first stage of the 2–1–1 is equivalent to the second-order CT modulator. Interestingly, the requirements on the slew rate of the integrator in the second stage are the highest. Here, the input signal is large, while the noise-shaping is still only second-order.
  Surprisingly, when using multibit quantization in the third stage, the requirements here increase to a maximum. This is a bit confusing since usually with multibit quantization the feedback signal tracks the input signal well and thus the input to the integrator becomes minimal [8]. In opposite, simulations showed that this is correct for a slowly changing input signal as for the first integrator in the first stage, the input to the higher stages in a cascaded modulator is rapidly moving. Thus, the difference of the feedback and the input signal is anything but low. Thus, since the scaling of the last stage was doubled in the case of multibit quantization, and additionally the ideal resolution increases, the slew rate requirements obviously increase.
- *Third-order modulator with* SCR. Finally, the simulation for the third-order modulator with SCR feedback confirms the discussion in Sect. 5.5.2: in comparison to the third-order modulator with rectangular feedback in Fig. 5.13b, the SR requirements are increased. While the third-order, NRZ modulator shows decreased performance below $\mathrm{SR}_1|_{\mathrm{SCR}}/(f_\mathrm{S} V_\mathrm{ref}) \leq 0.1$, this increases to $\mathrm{SR}_1|_{\mathrm{SCR}}/(f_\mathrm{S} V_\mathrm{ref}) \leq 0.3/0.6$ in the case of the SCR feedback with $\tau_{\mathrm{DAC}} = 0.2/0.07 T_\mathrm{S}$.
  Even if the derivation in Sect. 5.5.1 was intended for constant NRZ feedback, these simulated values for the SCR feedback are approximately predictable by the scaling coefficients in Table 4.2 together with the expression derived in (5.38): for example, with an applied input signal amplitude $\hat{u} = 0.25$ and an SCR modulator scaling $k_{\mathrm{sig|Mod3}} \approx 0.05$ and $k_1|_{\mathrm{SCR|0.2}} \approx 0.27$ or $k_1|_{\mathrm{SCR|0.07}} = 0.71$ (see Table 4.2), following (5.38) the approximately required SR of the first integrator is $\mathrm{SR}_1|_{\mathrm{SCR|0.2}} \approx 0.28\ (f_\mathrm{S} V_\mathrm{ref})$

and $\mathrm{SR}_1|_{\mathrm{SCR}|0.07} \approx 0.72\ (f_S V_{\mathrm{ref}})$. These values correspond to the simulated ones. For the second and third integrator, the SR requirements become higher again, because the input signals increase. But as above, they are lower than expected by (5.38) due to error suppression by the previous loops. Also note that the performance decrease was very abruptly for rectangular feedback DACs, while for the pulse-shaped DAC the loss starts continuously. This is clear since the shaped pulse transiently shows parts, which exceed the integrator slew rate, while others do not. In contrast, the rectangular feedback either completely exceeds the maximum slew rate or not!

Finally, it can be concluded from (5.37) and (5.38) that the simulations in Fig. 5.13 and the discussion above that the SR of the OpAmps in CT modulators with rectangular feedback can be kept in a very low range which is at least half a magnitude lower than in the DT case [60, 62]. Even for pulse-shaped feedback DACs as the exponentially decaying SCR pulse, the requirements increase to values, which are still relaxed compared to what is commonly seen in DT implementations.

As a good rule of thumb, (5.37) and (5.38) can be referred to, while their predictions are too high for the requirements in later integrators and higher stages and only rough approximations, if the DAC implementation is different from the rectangular resistor feedback. Thus, while avoiding slew rate limiting in CT ΣΔ modulators is just as important as in DT modulator implementations, achieving it is usually by far not as difficult.

5.6 Other Integrator Nonidealities

There exist other nonidealities, some having similar influence in discrete-time and continuous-time ΣΔ modulators, or even do not affect CT implementations as severe, e.g., nonzero switch resistance or signal-dependent charge injection. Most of them are covered throughout the published literature, e.g., [4, 5, 8, 60], while some are discussed shortly in the following.

5.6.1 Limited Output Swing

The limitation of the integrator outputs, also known as clipping, is principally a signal-dependent variation of the system states from their ideal values. Thus, the modulator behavior will be altered constantly or even signal dependent. Among others, Northworthy, Schreier, Hauser, Brodersen, Boser and Wooley [4, 66, 110] illustrate that the limitation of the integrator output signals results in severely increasing baseband noise as well as distortion. Nevertheless, in contrast to many other nonidealties, which are intrinsic or process dependent, power consuming or otherwise negatively affected, the clipping can be mostly avoided in the design process by proper parameter scaling, e.g.,

[6, 214], by which all integrator outputs are usually limited to the modulator reference voltage, or by choosing an optimized loop filter architecture, as e.g., the feed-forward topology.

5.6.2 Circuit Noise

In an ideal ΣΔ modulator, the only noise contribution arises from the quantization noise, which is suppressed by oversampling and high-pass noise shaping. Taking nonidealities into account, other noise sources will also appear, which can be white or frequency dependent. In general a ΣΔ modulator is designed in such a way that the overall noise power is dominated by circuit noise [72]. Consequently, the ADC resolution can be specified by means of calculating the total input referred noise. If required, chopper techniques [227] can be applied to reduce these effects in the passband, but at the expense of a sampled integrator.

In CT ΣΔ modulators, in fact only the in-band noise components decrease the performance, while high-frequency parts are filtered by the subsequent decimation filter. As discussed in Sect. 3.8.1 and shown for example in [4, 60], it strongly depends where in the modulator the error sources are located: while the highest degree of noise shaping is available at the quantizer input, the most critical error source is at the input of the modulator, where no noise shaping takes place. Thus, the first integrator has to fulfil the noise requirements of the entire ΣΔ modulator [4, 113].

The actual noise contribution and the main performance limitation depends on the circuit architecture: in DT modulators, implemented in the SC-technique, sampling of the switch resistances results in the well known KT/C noise [222], where T is the absolute temperature and K is Boltzman's constant. Depending on the required resolution of the modulator, this can result in large sampling capacitors at the input of the modulator, increasing chip area [189] and power consumption [60].

In contrast, in purely CT ΣΔ modulators there exists no sampling capacitor, which is an often cited advantage. Thus, high-frequency noise components are filtered (see Sect. 3.5) before they see the sampler, and remaining aliased noise components are attenuated by the complete noise shaping.

Fig. 5.14 shows the dominant noise sources in an exemplary active RC-integrator. The input-referred amplifier noise $\overline{v}^2_{n,\mathrm{OTA}}$ has two main contributions: flicker and thermal noise, whereas the noise caused by the resistors is inherently white. Determining the total input-referred noise power yields:

$$\begin{aligned}\overline{v}^2_{\mathrm{n,in}} \approx\ & 2\left(\overline{v}^2_{\mathrm{n},R} + \overline{v}^2_{\mathrm{n},R_{\mathrm{DAC}}}\frac{R^2}{R^2_{\mathrm{DAC}}} + \overline{v}^2_{\mathrm{n},R_Z}\frac{R^2}{Z^2_{\mathrm{F}}}\right) \\ & + \overline{i}^2_{\mathrm{n,OTA}}R^2 + \overline{v}^2_{\mathrm{n,OTA}}\left(1 + \frac{R}{R_{\mathrm{DAC}}} + \frac{R}{Z_{\mathrm{F}}}\right)^2,\end{aligned} \tag{5.39}$$

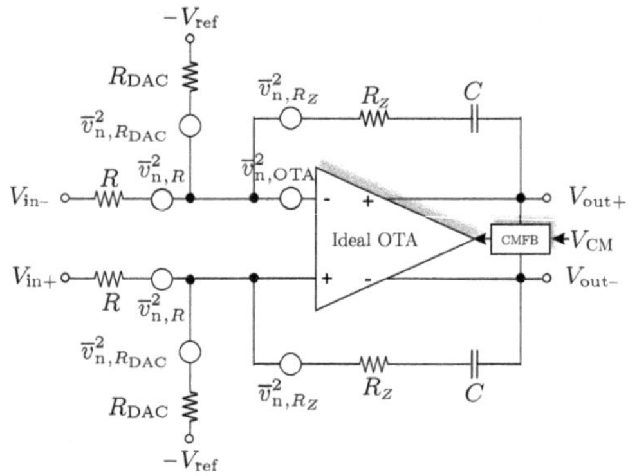

Fig. 5.14. Schematic of a fully differential active RC-integrator with additional noise sources caused by the resistors and the amplifier, after [37]

where Z_F denotes the feedback impedance $R_Z + 1/sC$. Each resistance R_i generates thermal noise, given by:

$$\frac{\overline{v}^2_{n,R}}{\Delta f} = 4KTR, \quad \frac{\overline{v}^2_{n,R_{DAC}}}{\Delta f} = 4KTR_{DAC}, \quad \frac{\overline{v}^2_{n,R_Z}}{\Delta f} = 4KTR_Z\,. \tag{5.40}$$

Additionally, the amplifier introduces beside thermal noise also $1/f$ noise, which yields:

$$\frac{\overline{v}^2_{n,OTA}}{\Delta f} = \frac{8}{3}\frac{KTn_{e,th}}{gm_{OTA}} + \frac{K_f n_{e,f}}{C^2_{ox}WL}\frac{1}{f_f^a}\,. \tag{5.41}$$

Here, $n_{e,th}$ and $n_{e,f}$ describe the thermal and flicker noise excess factors, gm_{OTA} represents the amplifier input transconductance, K_f and a_f are the flicker noise parameters, which strongly depend on both the transistor type and the employed technology. The equivalent thermal input noise current $\overline{i}^2_{n,OTA}$ is heavily frequency dependent and becomes important only at very high frequencies. Thus, it is neglected in the following. Accordingly one obtains:

$$\begin{aligned}
\overline{v}^2_{n,in} &\approx \overline{v}^2_{in,thermal} + \overline{v}^2_{in,1/f}\,,\\
\frac{\overline{v}^2_{in,thermal}}{\Delta f} &= 8KT\left[R + R_{DAC}\frac{R^2}{R^2_{DAC}} + R_Z\frac{R^2}{(R_Z+1/sC)^2}\right.\\
&\quad\left. + \frac{n_{e,th}}{3\,gm_{OTA}}\left(1+\frac{R}{R_{DAC}}+\frac{R}{R_Z+1/sC}\right)^2\right],\\
\frac{\overline{v}^2_{in,1/f}}{\Delta f} &= \frac{K_f n_{e,f}}{C^2_{ox}WL}\left(1+\frac{R}{R_{DAC}}+\frac{R}{R_Z+1/sC}\right)^2.
\end{aligned} \tag{5.42}$$

To calculate now the total in-band noise P_N, (5.42) has to be integrated over the passband:

$$P_{\mathrm{N}} = 2\int_0^{f_{\mathrm{B}}} \left| \frac{\overline{v}^2_{\mathrm{in,thermal}}}{\Delta f} \right| \mathrm{d}f + 2\int_{f_u}^{f_{\mathrm{B}}} \left| \frac{\overline{v}^2_{\mathrm{in},1/f}}{\Delta f} \right| \mathrm{d}f\,. \tag{5.43}$$

In (5.43) the lower integration boundary is set to $f_u > 0$ due to the logarithmic characteristic of the $1/f$ noise term. By applying additionally:

$$C = \frac{1}{2Rf_{\mathrm{B}}\mathrm{OSR}} \tag{5.44}$$

for the total input referred in-band noise

$$\begin{aligned} P_{\mathrm{N}} &= 16KTf_{\mathrm{B}}\Bigg[R + R_{\mathrm{DAC}}\frac{R^2}{R^2_{\mathrm{DAC}}} + R_Z\frac{\pi^2}{3\mathrm{OSR}^2} \\ &\quad + \frac{2n_{\mathrm{e,th}}}{3gm_{\mathrm{OTA}}}\left(\left(1+\frac{R}{R_{\mathrm{DAC}}}\right)^2 + \frac{\pi^2}{3\mathrm{OSR}^2}\right)\Bigg] \\ &\quad + \frac{2K_{\mathrm{f}}n_{\mathrm{e,f}}}{C^2_{\mathrm{ox}}WL}\left[\ln\left(\frac{f_{\mathrm{B}}}{f_u}\right)\left(1+\frac{R}{R_{\mathrm{DAC}}}\right)^2 + \frac{\pi^2}{\mathrm{OSR}^2}\left(1-\left(\frac{f_u}{f_{\mathrm{B}}}\right)^2\right)\right] \end{aligned} \tag{5.45}$$

can be obtained. Expression (5.45) illustrates the complete input referred in-band noise. However, it is difficult to obtain the desired design guidelines regarding the noise contribution for the active-RC integrator. Thus, by simplifying (5.45):

$$\begin{aligned} P^*_{\mathrm{N}} &= 16KTf_{\mathrm{B}}\left[R + R_{\mathrm{DAC}}\frac{R^2}{R^2_{\mathrm{DAC}}} + \frac{2n_{\mathrm{e,th}}}{3gm_{\mathrm{OTA}}}\left(1+\frac{R}{R_{\mathrm{DAC}}}\right)^2\right] \\ &\quad + \frac{2K_{\mathrm{f}}n_{\mathrm{e,f}}}{C^2_{\mathrm{ox}}WL}\left[\ln\left(\frac{f_{\mathrm{B}}}{f_u}\right)\left(1+\frac{R}{R_{\mathrm{DAC}}}\right)^2\right]. \end{aligned} \tag{5.46}$$

Obviously, all parts depending on the OSR are neglected. Still, the resulting inaccuracy is absolutely tolerable, since the deviations are $1.3\%, 0.19\%$ and 0.05% for oversampling ratios of OSR = 4, 8, and 16, respectively.

In addition, (5.46) can be further simplified, if it can be realized that in the CT $\Sigma\Delta$ modulators the feedback resistor resistor equals $R_{\mathrm{DAC}} = (1/\kappa)R$. Thus, approximatley ($\kappa \approx 1$):

$$P^{**}_{\mathrm{N}} = 32KTf_{\mathrm{B}}\left(R + \frac{4n_{\mathrm{e,th}}}{3gm_{\mathrm{OTA}}}\right) + \frac{8K_{\mathrm{f}}n_{\mathrm{e,f}}}{C^2_{\mathrm{ox}}WL}\ln\left(\frac{f_{\mathrm{B}}}{f_u}\right). \tag{5.47}$$

In order to achieve a certain noise performance, the sum of all three noise contributions have to meet the demanded noise floor. Yet, for low or medium signal bandwidths, the $1/f$-noise term is the dominant part, while for higher passbands the white noise term dominates.

Additionally, (5.46) gives a design guideline concerning the maximum integrator resistance and the minimum amplifier transconductance for a given modulator resolution. The major benefit of using the maximum resistor size is the reduction of the overall integrator power consumption, due to the fact that these are these are specified by $f_S = 1/RC$. Thus, both and the amplifier transconductance can be reduced down to a limit given in (5.46). Another advantage resulting from a maximized input resistor is the improved linearity behavior of an active RC-integrator as subsequently presented in Sect. 5.6.3. Note that the above calculation procedure is done for the special architecture of active RC, first integrator. Changing the feedback to current source or the filter to gmC obviously would alter the expressions!

5.6.3 Integrator Nonlinearity

The most commonly used devices presented in analog integrated circuits such as transistors, resistors, capacitors, and diodes are generally nonlinear [228]. By taking this (usually weakly) nonlinear behavior into account, a significant deviation of the (wanted) output signal caused by the amount of harmonic distortion can be obtained. Fig. 5.15 indicates those sources of harmonics that limit the linearity of an active RC-integrator employed in a CT ΣΔ modulator. Concerning distortion, primarily the first integrator has to be considered, because of the fact that nonlinearities of the subsequent stages are suppressed by the respective loop gain: any nonlinearity of the second or third integrator is attenuated proportionally to $(\pi/\mathrm{OSR})^2$ or $(\pi/\mathrm{OSR})^4$, respectively.

Thereby, the voltage dependency of the amplifier dc gain, integrator resistances and capacitors may introduce substantial distortion especially at high-input amplitudes. Because monolithic capacitors feature high linearity $C(V) = C(1 + DV_1V + DV_2V^2)$, with DV_1, DV_2 a few ppm/V and ppm/V^2, respectively), mainly the nonlinearity of the monolithic resistors such as of the used amplifiers are of concern.

In general, linear (negative) feedback is widely used in analog circuits, because of its important benefits, e.g., stabilization of the gain of an amplifier.

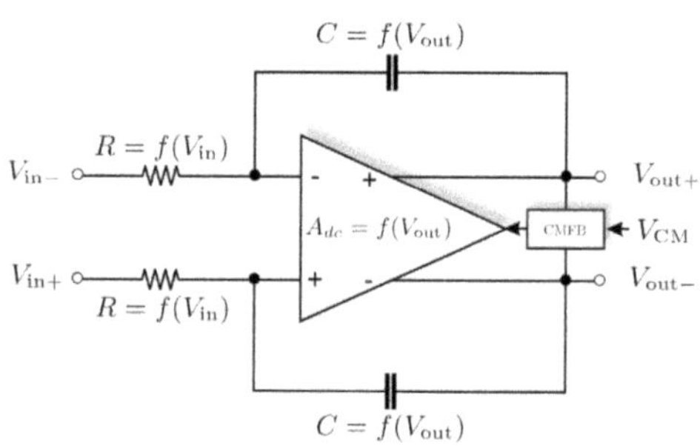

Fig. 5.15. Schematic of a fully differential integrator with nonlinear error sources

But there is still another considerable performance gain; the reduction of nonlinear distortion [228, 229]. Thus, the linearity of the implemented amplifier as well as the integration capacitor reside in a feedback loop, thus the impact of both nonlinearities is reduced.

Additionally note that it is not feasible trying to give general expressions for how much distortion can be allowed in CT $\Sigma\Delta$ modulators for two reasons: first, this heavily depends on the circuit level implementation of the integrators, and additionally the architectural alternatives combining modulator scaling and loop filter realization are too manifold and have a strong influence on the tolerable distortion [230]. Thus, only exemplary considerations are given.

Nonlinear Amplifier Gain

Figure 5.16 exemplarily shows the simulated open loop gain of a single-stage folded-cascode amplifier [34]. The maximum gain of almost 78 dB is obtained at the common mode output voltage and decreases rapidly as the output voltage approaches the saturation region. By estimating the nonlinearity, a THD of less than −99 dB according to [229] can be obtained. Comparable results are obtained in [213]. Furthermore, the bias current and the transconductance of the first integrator are determined by the linearity requirements of the overall ADC [128]:

$$\mathrm{HD3} \approx \frac{V_{\mathrm{in}}^2}{64 gm R^3 I_{\mathrm{DS}}^2}\left(1+\frac{R}{R_{\mathrm{DAC}}}\right), \tag{5.48}$$

where R is the input resistor, R_{DAC} the feedback resistor, while gm and I_{DS} are the transconductance and bias current of the input transistors of the respective differential amplifier. From this equation yields that by increasing R up to the allowable thermal noise limit, the linearity can be improved. The same holds true for increasing the input transconductance, but with lower efficiency due

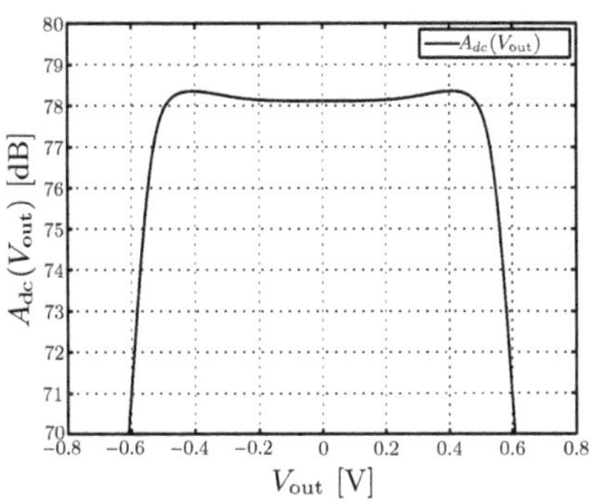

Fig. 5.16. Simulated amplifier gain nonlinearity

to the smaller exponential effect. Calculating the power consumption from (5.48), yields:

$$P \approx \frac{V_{\mathrm{dd}}\,\mathrm{FS}}{8R}\sqrt{\frac{1}{gm_{\mathrm{OTA}}\mathrm{HD}_3}\left(\frac{1}{R}+\frac{1}{R_{\mathrm{DAC}}}\right)}, \tag{5.49}$$

where V_{dd} is the supply voltage, FS the full-scale input voltage range, HD_3 third harmonic distortion, gm_{OTA} the transconductance of the input transistors and R and R_{DAC} are the input and feedback (DAC) resistances, respectively. Power minimization can be achieved by increasing R up to the thermal noise limit in the same way as R_{DAC}, see (5.46). Moreover, the power consumption is reduced by a large gm_{OTA}.

Nonlinear Resistance/Transconductance

The first voltage-to-current converter implemented as a resistor (in an *RC*-integrator) or a transconductance amplifier (e.g., in a *gmC*-integrator) is the main source of nonlinearity, because its harmonic distortion adds directly to the input signal. Accordingly, the demands on the linearity of the first resistor, at the maximum input level, has to be as high as the overall accuracy of the ΣΔ modulator. This statement is confirmed by simulation results of a second-order and third-order modulator, presented in Fig. 5.17a and b. In both cases the required linearity of the first integrator (denoted with Int1) meets nearly that of the overall converter. Besides, it is evident that the requirements on the linearity of the succeeding stages, here the second integrator (Int2), are highly reduced. If the attenuation of the second-stage nonlinearity transferred

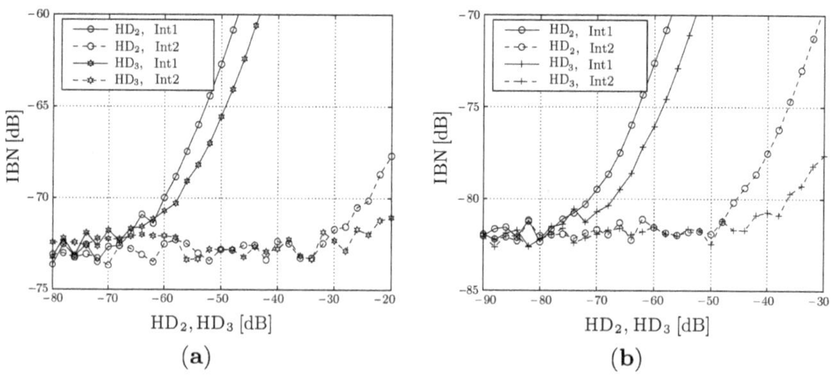

Fig. 5.17. Effect of second-harmonic and third-harmonic distortion caused by the first and second transconductor on the (**a**) second-order ΣΔ modulator with $\{k_1, k_2\} = \{1/4, 3/8\}$ and (**b**) third-order ΣΔ modulator with $\{k_1, k_2, k_3\} = \{0.05, 0.3, 0.64\}$. $f_{\mathrm{B}} = 25\,\mathrm{kHz}$, $\mathrm{OSR} = 48$, $P_{\mathrm{sig}} = -7\,\mathrm{dB}$

to the modulator output is determined in detail, $4\pi^2 f^2/f_S^2$ will be obtained. Therefore, the worst case attenuation takes place at the passband edge ($f = f_B$) which results in a $4\pi^2 f_B^2/f_S^2 = \pi^2/\mathrm{OSR}^2 \approx -25\,\mathrm{dB}$ attenuation for an oversampling ratio of 48.

Even-Order Distortion due to Mismatch in Differential Integrators

Ideally, fully differential circuits are free from even-order distortion by virtue due to symmetry. In reality however, mismatches degrade the symmetry, introduce thereby a finite even-order nonlinearity [210, 229]. Through relative component mismatch, two resistors can be merely integrated up to a certain accuracy. Therefore, the second harmonic has an amplitude proportional to the difference between the second-order nonlinear term [228, 229].

6

Quantizer Nonidealities in Continuous-Time $\Sigma\Delta$ Modulators

This section shortly covers the influence of the nonideal behavior of the internal quantizer in a continuous-time $\Sigma\Delta$ modulator. Usually, the quantizer is the most uncritical building block, and therefore this chapter only covers its behavior in a glance. Nonetheless, the specialities of the quantizer in CT $\Sigma\Delta$ modulators is shortly reviewed and discussed.

The quantizer is located at the most insensitive place of the loop, i.e., preceded by the complete loop filter. This is why $\Sigma\Delta$ modulators are mostly insensitive to errors induced by the internal quantizer, which usually are offset, hysteresis, and nonlinearities. Nonetheless, in CT $\Sigma\Delta$ modulators special attention has to be paid for all timing issues related with the quantizer!

In the case of a single-bit quantizer, the comparator offset is largely attenuated by the overall dc gain of all preceding integrators. Thus, single-bit $\Sigma\Delta$ modulators are almost insensitive to these errors [5]. The effect of comparator hysteresis on the overall $\Sigma\Delta$ modulator was considered in [110]; here, also an analytical expression was derived, and comparator hysteresis was found to cause additional noise, but which is fortunately shaped in the same way as quantization noise [8, 110, 121]. Thus, its nonideal influence is mostly negligible, provided an adequate reset of the comparator is accomplished [5, 110]. The proper reset is needed to cancel the *memory* of the comparator circuit [23].

Nevertheless, note that both offset and hysteresis start to be a more severe design issue in high-resolution implementations and especially if multibit internal quantization is applied [4, 63, 231]. In the latter case, commonly internal flash ADCs are employed due to speed reasons. Consequently, the exponentially increased input capacitance imposes a very dominant load for the driving stage, which therefore can even become the dominant power consumer in the overall $\Sigma\Delta$ modulator [21, 126]. Furthermore, the flash architecture needs a reference resistor ladder. In order to minimize capacitive feedthrough from the comparators into the reference ladder, the resistors value need to be small, which significantly adds to the overall power consumption [36, 232]. As already indicated in Sect. 3.1.3, a possibility to circumvent this is the implementation of the multibit internal quantizer by an alternative low resolution architecture, e.g., by a tracking [22] or a successive approximation ADC [129].

The just discussed commonly seen low influence of the quantizer nonidealities is somewhat alleviated in the case of CT $\Sigma\Delta$ modulators: this is caused by timing-induced errors like propagation and signal dependent delay: constant excess loop delay can be caused by the internal quantizer just as it was seen in the feedback DAC. More severely, the delay of the decision is dependent on the signal amplitude and slope: obviously, a differential input stage will settle faster with large input overdrive. It has been shown that quantizer metastabiliy yields equivalent performance degradation as clock jitter, because due to the statistically varying quantizer delay, also the length of the feedback pulses will vary [12]. This is known as signal dependent timing jitter!

Nonetheless, the signal dependency of the quantizer delay is easy to circumvent by inserting, for example, a latch between the quantizer and the feedback DAC, which is clocked differently than the quantizer itself. Thus, the quantizer has a constant time to decide and settle, and subsequently the result is transferred to the latch and the DAC [37, 113, 126].

Consequently, it can be concluded that the quantizer performance is of minor importance in CT ΣΔ modulators, when compared to the influence of the nonidealities of the feedback DAC or the filters, just as it was in DT modulators. Nonetheless, since transient behavior and delays are much more of an issue in continuous-time modulators, care has to be taken on an architectural level to circumvent signal dependent delay and also constant quantizer delay has to be taken into account in addition to the DAC excess loop delay. For both tasks, published solutions exist.

7

CT $\Sigma\Delta$ Modulator Design Examples

This chapter presents several design examples, which correspond to the topics discussed in the previous chapters. Therefore, first a design strategy for CT $\Sigma\Delta$ ADCs is proposed. Using a Figure of Merit, the optimal $\Sigma\Delta$ modulator topology with respect to a minimal power consumption for the desired design constraints such as accuracy and bandwidth is found. Based on the design strategy, two power-efficient CT $\Sigma\Delta$ modulators are introduced. Several power optimizations, design requirements, and performance limitations are specified. Additionally, based on the first implementation example, a modification with shaped feedback waveform is shown, which improves the jitter behavior. Finally, a breadboard prototype cascaded CT modulator is shown, which was implemented to prove the possibility for gain-error correction.

7.1 FOM Based Design Strategy for CT $\Sigma\Delta$ Modulators

In general, the design flow of an integrated circuit consists of a number of hierarchical levels. Here, several techniques and approaches have been proposed, to obtain low-power $\Sigma\Delta$ modulators, e.g., [19, 35–37, 48, 71, 117, 176, 233–237]. However, most of them are optimizations on circuit level. Principally, these techniques originate from a selected or given $\Sigma\Delta$ modulator topology and optimize the overall power drain by means of calculating and optimizing the required DR, KT/C-noise, just as the resulting capacitive load and adapt subsequently the analog building blocks on the determined requirements.

A power optimization regarding modulator topologies was presented in [71]. The authors perform a coarse estimation for single-loop modulators with regard to the SNR and noise benefits as well as power savings by increasing the modulator order from 2 to 4. As a result, power savings of approximately 10% can be obtained for each modulator order increment, by assuming a constant modulator performance. A more effective way is to use a higher abstraction level as an optimal starting point. Thus, a global optimization has also to consider a modulator topology selection. An improved software tool SDOPT that explores the different modulator architectures by using simple design equations to find a best modulator topology with regard to high-level specifications (e.g., signal bandwidth, resolution, etc.) was presented in [5]. In [38, 48], an approach has been presented that determines the power-efficiency of different modulator structures subject to the required converter resolution and signal bandwidth. This method will be illustrated in detail in the following. Based on these results, additional power reduction techniques on circuit level can be applied [34, 71, 120].

7.1.1 Generic Figure of Merit Calculation

To find the optimal ΣΔ modulator implementation with respect to a minimal power consumption for a given modulator resolution (B-bit) and oversampling ratio ($\mathrm{OSR} = f_\mathrm{S}/2f_\mathrm{B}$) on the one hand, and to fulfill also a rapid prototyping approach on the other hand, a Figure of Merit was used [5]. The FOM estimation comprises a mixed bottom-up top-down approach. Further parameters that are essential for the FOM estimation (V_dd, etc.) could be extracted from different design abstraction levels. The fundamental equation employed in this approach is

$$\mathrm{FOM} = \frac{P}{2^B 2 f_\mathrm{B}}\,. \tag{7.1}$$

Based on (7.1), the parameters such as the modulator power consumption P and the dynamic range DR are replaced by a set of equations that possess a minimal number of degrees of freedom. Furthermore it will be shown, that – besides the prescribed signal bandwidth and the resolution – these are mostly technology dependent parameters, such as the supply voltage V_dd, the PMOS transistor transconductance parameter β_p, the CMOS inverter power consumption P_inv as well as the possible signal swing (which depends again on the supply voltage and the employed technology) and the size of the input transistors.

To simplify the entire estimation on the one hand, but to consider the most important error sources on the other hand, the dynamic range calculations comprise both quantization (IBN_Q) and thermal noise (P_th). According to [110], the DR results to

$$\mathrm{DR} \approx \frac{\frac{\mathrm{FS}^2}{2}}{\mathrm{IBN}_Q + P_\mathrm{th}} = 3 \times 2^{2B-1}\,, \tag{7.2}$$

where FS denotes the full-scale input signal range. Further noise and error contributions such as thermal and $1/f$ noise of the amplifiers as well as harmonic distortion will be neglected. Thus, the in-band quantization noise of an ideal ΣΔ modulator of the order N is calculated as follows [4]:

$$\mathrm{IBN}_Q = \frac{\mathrm{FS}^2}{12} \frac{\pi^{2N}}{(2N+1)\,\mathrm{OSR}^{2N+1}}\,. \tag{7.3}$$

The corresponding thermal noise of a CT integrator results to

$$P_\mathrm{th} \approx 16KTRf_\mathrm{N} = \frac{16KT}{\mathrm{OSR} \times C}\,, \tag{7.4}$$

where R is the resistance of the fully differential voltage-to-current converter of the CT integrator, as shown in Fig. 7.1a. The overall power consumption P of a ΣΔ modulator is made up of a static and dynamic portion of the analog part (P_static, P_dynamic) and a portion of the digital part (P_digital):

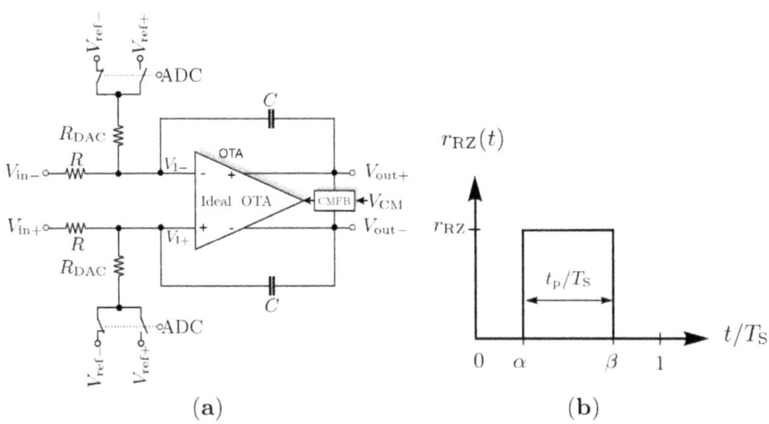

Fig. 7.1. Active *RC*-integrator schematic with its DAC feedback pulse waveform. (**a**) Integrator schematic; (**b**) DAC feedback pulse

$$P \approx P_{\text{static}} + P_{\text{dynamic}} + P_{\text{digital}} \,. \tag{7.5}$$

The static power consumption caused by the amplifier, depends substantially on the amplifier topology. Thus, P_{static} is proportional to the number of current branches n_I times the current I_{B}. We assume, for reasons of simplicity, that the current is the same in every branch. Furthermore, every integrator features the same power, so that the overall static modulator power consumption is N-times to that of the first integrator. This simplification could be eliminated by considering the noise-shaping effect at every filter stage, which allows an increased resistance and accordingly a scaled capacitive load and power consumption:

$$P_{\text{static,OpAmp}} = n_{\text{I}} I_{\text{B}} V_{\text{dd}} N \,. \tag{7.6}$$

The gain-bandwidth product of the amplifier can be approximated by GBW $= gm/C \approx 1.5 f_{\text{S}}$ to ensure an adequate integrator behavior [35]. The current I_{B} again depends on the transconductance as $I_{\text{B}} = gm^2/(2\beta_p)$ where β_p characterizes the transconductance parameter of the input transistors.

On the other hand, the overall integrator power consumption consists of two parts; the first because of the input signal ($P_{\text{dynamic,input}}$) the second because of the DAC feedback ($P_{\text{static,DAC}}$). The first part can be estimated as:

$$P_{\text{dynamic,input}} = 2n_{\text{R}}\ \text{FS}^2\ C f_{\text{S}} N = 4\text{FS}^2\ C f_{\text{N}}\ \text{OSR}\ N \,, \tag{7.7}$$

where n_{R} is the number of resistors applied to the input signal. Using full-differential circuits, there exist two input resistors as illustrated in Fig. 7.1a.

Additionally, the feedback current DAC has to be considered. The DAC power consumption depends on the one hand on the magnitude (r_{RZ}) and time response (α, β with $0 \leq \alpha < \beta \leq 1$) of the DAC pulse (Fig. 7.1b) and on the other hand on the voltage V_{R} across the feedback resistor R_{DAC}. In general, the amount of feedback charge during a sample period is constant, i.e., if the rectangular DAC feedback pulse is only half the sampling period T_S high (return-to-zero DAC pulse), the magnitude r_{RZ} is doubled. Furthermore, V_R can be simplified as $V_{CM} - (\pm V_{ref}) \approx$ FS (see Fig. 7.1a). Likewise, the feedback path consists of two resistors, leading to

$$P_{\text{static,DAC}} = 2n_R r_{RZ}(\beta - \alpha)\frac{\text{FS}^2 M}{R} \approx 4\text{FS}^2 C f_N \text{ OSR } N\,. \tag{7.8}$$

Thus, the overall integrator power results to

$$P_{\text{dynamic}} = 8\text{FS}^2 C f_N \text{ OSR } N\,. \tag{7.9}$$

The digital power dissipation resulting from the comparators, flip-flops and additional logic yields:

$$P_{\text{digital}} = n_Q P_{\text{inv}} f_N \text{ OSR}\,, \tag{7.10}$$

where n_Q represents the size of a quantizer and additional digital logic converted in CMOS gate-equivalent inverters each with a standardized power consumption of P_{inv}/MHz. This way, the entire power consumption results to

$$P \approx n_I I_B V_{dd} N + 8\text{FS}^2 C f_N \text{ OSR } N + n_Q P_{\text{inv}} f_N \text{ OSR}\,. \tag{7.11}$$

With (7.2) to (7.4) the required integrator capacitance C can be calculated as a function of the modulator order N and the oversampling ratio OSR:

$$C(N, \text{OSR}) = \frac{48KT}{\text{OSR}}\left(\frac{2^{2B-1}}{\text{FS}^2 - 3\text{IBN}_Q \times 2^{2B}}\right) \tag{7.12}$$

and accordingly with (7.11) the FOM(OSR). The optimal ΣΔ modulator implementation with respect to a minimal power consumption yields:

$$\frac{\partial \text{FOM}}{\partial \text{OSR}} = 0 \quad \text{and} \quad \frac{\partial^2 \text{FOM}}{\partial \text{OSR}^2} > 0\,. \tag{7.13}$$

During the last 25 years, a large amount of ΣΔ modulator architectures has been proposed, e.g., [2, 19, 36, 54, 55, 107, 124, 175, 185]. The main objective of all these schemes is the reduction of the quantization noise within the baseband. In the following, the most common architectures will be analyzed and the results are used to adapt the FOM computation.

7.1.2 Single-Loop Architectures

As shown in Sect. 2.7, single-loop structures can be implemented in different ways. However, the overall performance of the modulators with distributed feedback shown in Fig. 2.14 and with feed-forward summation Fig. 2.16 is quite comparable. The disadvantage of single-loop single-bit ΣΔ modulators with orders $N > 2$ is the tendency to instability. Therefore, a proper selection of the scaling coefficients has to be made [81, 238–240]. The optimal scaling coefficients are a trade-off between stability, maximum stable input amplitude and peak signal-to-noise ratio, see Sect. 2.7.

A more realistic FOM estimation takes also this scaling of the noise transfer function into account. Therefore as a rule-of-thumb, an NTF out-of-band gain $\mathrm{NTF}_\infty \approx 1.5$ was used [226] (note, ideally $\mathrm{NTF}_\infty = 2^N$). This scaling of the ΣΔ modulator involves that the resulting in-band noise is rather higher than the one, calculated with (7.3). Consequently, the oversampling ratio has to be increased which results in a higher power drain and accordingly in a higher FOM:

$$\mathrm{IBN}_\mathrm{Q}^* \approx \left(\frac{2^{N+1}}{3}\right)^2 \mathrm{IBN}_\mathrm{Q}\,. \tag{7.14}$$

A more efficient way to stabilize high-order ΣΔ modulators is to increase the number of levels of the internal quantizer. Furthermore, by using a small negative feedback around pairs of integrators (as specified in Sect. 2.7.2) it is possible to move the NTF zeros away from dc. As a result, the noise attenuation is spread over the baseband and thus the in-band quantization noise is further reduced. This technique was not considered during this estimation, but it can be directly incorporated.

7.1.3 Multibit Single-Loop Architectures

The use of a multibit ADC as the internal quantizer offers beside a better loop stability also the advantage that the in-band quantization noise is a factor of $(2^{B_\mathrm{int}} - 1)^2$ lower in comparison with a simple two level comparator. Thus, (7.3) yields:

$$\mathrm{IBN}_\mathrm{Q}^{**} \approx \frac{\mathrm{IBN}_\mathrm{Q}}{(2^{B_\mathrm{int}} - 1)^2}\,, \tag{7.15}$$

where B_int indicates the number of bits of the internal ADC. Consequently, the power consumption of the digital part is approximately increased with the resolution of the internal ADC. Therefore, (7.10) extends to

$$P_\mathrm{digital}^* \approx 2^{B_\mathrm{int}} n_\mathrm{Q} P_\mathrm{inv} f_\mathrm{S}\,. \tag{7.16}$$

Circuit nonidealities due to the ADC are mostly suppressed by the modulator loop gain. In contrast, the digital-analog conversion errors, are not attenuated, but even add directly to the modulator input. These DAC errors appear at the modulator output as additional noise and distortion. Thus, the DAC linearity has to be as high as the overall modulator resolution. To ensure the desired modulator resolution, many DAC linearisation techniques have been employed, Sect. 4.5. All of these methods consume additional power and silicon area. For reasons of simplicity, we neglect these factors during the FOM calculation.

7.1.4 Cascaded Architectures

Continuous-time cascaded or multistage ΣΔ modulators consist of a cascade connection of low-order modulators $N \leq 2$, which are inherently stable. Consequently, these topologies demand almost no scaling, thus they achieve almost ideal noise shaping performance [239].

In Figs. 3.7b and 3.8b the architectures of a third and fourth-order cascaded continuous-time system are given. A considerable disadvantage of multistage ΣΔ modulators is the high sensitivity to the imperfections of the analog components as well as coefficient mismatch against the digital filter [241]. Nevertheless the design of cascaded CT modulators is possible, if an error correction system is used and thus the most important errors can be automatically detected and canceled out [96, 98, 242]. Thus, these errors were not taken into account during the FOM computation.

In general, the scaled quantization noise generated in the first stage will be remodulated in the next one, and later canceled by digital filters. Hence, the characterization of the digital power consumption in (7.10) has to be adjusted. The number of quantizers is proportional to the number of modulator loops M. Thus, (7.10) results to

$$P^{**}_{\text{digital}} = (n_{\text{Q}} M) P_{\text{inv}} f_{\text{S}} \,. \tag{7.17}$$

Furthermore, the additional demanded digital filters, which remove the quantization error of the first stage, have to be considered [6]. A coarse approximation of the additional hardware effort for different modulator orders and structures is shown in Table 7.1. n_{Adder} and n_{Delay} represent the size of the adders and flip-flops converted in CMOS gate-equivalent inverters, whereas k_{Adder} and k_{Delay} specify the number of adders and flip-flops (7.10). Hence, the adjusted entire digital power consumption yields:

$$P^{**}_{\text{digital}} = (n_{\text{Q}} M + n_{\text{Adder}} k_{\text{Adder}} + n_{\text{Delay}} k_{\text{Delay}}) P_{\text{inv}} f_{\text{S}} \,. \tag{7.18}$$

To estimate the digital hardware effort of the cancellation logic several assumptions are made:

Table 7.1. Additional minimal hardware effort in the digital part of the multistage ΣΔ modulators

Structure	Order N	M	k_{Adder}	k_{Delay}
2–1 modulator	3	2	4	3
2–2 modulator	4	2	4	4
2–1–1 modulator	4	3	9	7

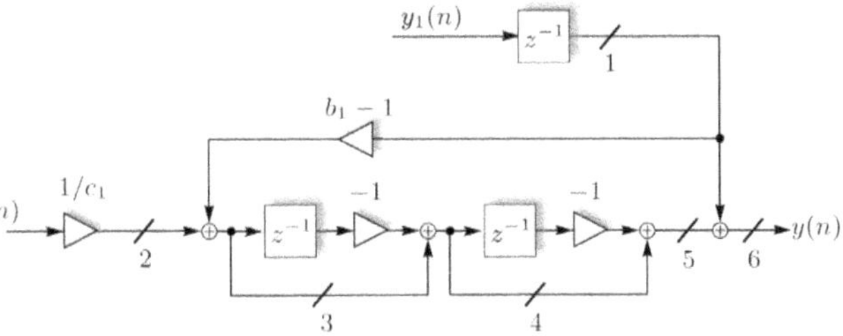

Fig. 7.2. Generic architecture of the cancellation logic for a third-order 2–1 ΣΔ modulator with a coarse estimation of the bit width

- In all stages the employed internal quantizer is a comparator. This limitation can easily be removed, merely (7.3) and (7.18) have to be adapted.
- No digital error correction was taken into account e.g., [125, 242, 243].
- The estimation of the additional digital hardware effort in multistage ΣΔ modulators is pointed out in case study for the cancellation logic of a 2–1 modulator as shown in Fig. 7.2. As proposed in [6], a good choice for the coupling coefficient c_1 and the gain factor b_1 is 0.5 and 2, respectively. Thus, the first adder is 2×1-bit wide, leading to an output width of three bit. To simplify matters, the output of every sequencing adder is increased by one bit. In this way one obtains the values shown in Table 7.1.
- Other values for b_1 and c_1 increase the hardware effort, in particular c_1 due to the fact that c_1 resides at the beginning of the cancellation logic.

7.1.5 FOM Based Design Example: A 12-Bit 25 kHz ΣΔ Modulator

The aim of this section is to exemplarily find a power-efficient modulator implementation for a given set of specifications using the above presented FOM and by assuming that the amplifier is a folded-cascode OTA, with a total current of $4I_\mathrm{B}$. The supply voltage is 1.5 V with a resulting full scale input amplitude of $\mathrm{FS}/2 = V_\mathrm{dd}/3$. The 3.3 V, 0.5 µm CMOS technology used provides a transconductance parameter β_p of 40 µA/V^2 with a W/L ratio of 400. The following FOM estimations will be made to guarantee a modulator resolution of B = 12-bit in a signal bandwidth of 25 kHz. Figure 7.3a illustrates the FOM for a single-loop (SL) single-bit quantizer according to (7.11) and (7.12). The best FOM can be obtained by a third-order modulator with an OSR close to 30.

A more realistic FOM estimation takes also the scaling of the noise transfer function into account, see (7.14). This results in Figure 7.3b. The minimal FOM, as well as the minimal OSR at which the desired performance is achieved, increases. In other words, the minimal power consumption increases for a fixed resolution and signal bandwidth. Again, the best FOM can be

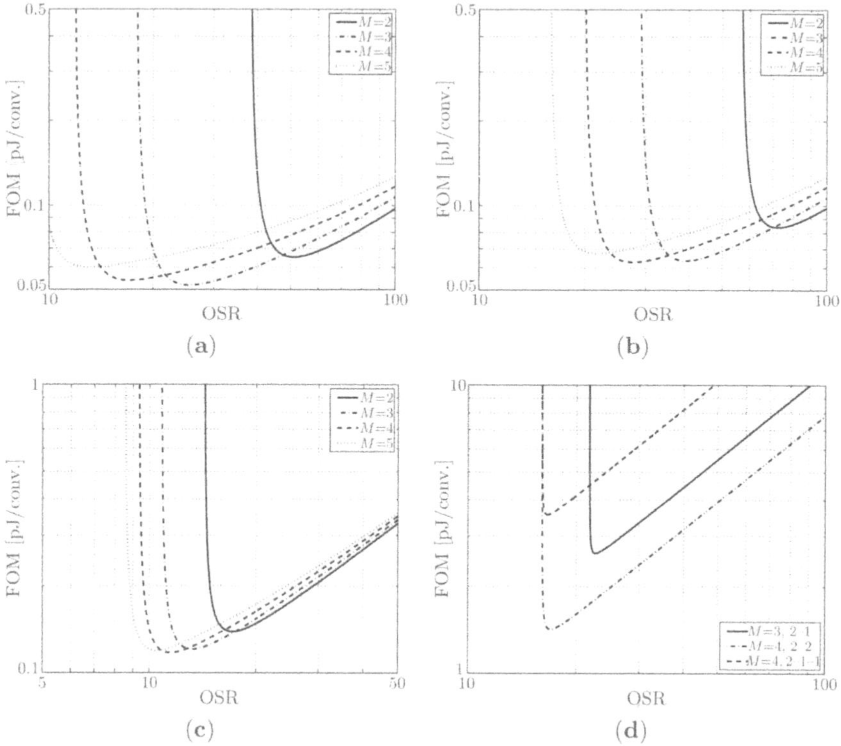

Fig. 7.3. Estimated FOM as a function of the oversampling ratio OSR corresponding to the design example 1 (12-bit, 25 kHz) for different modulator topologies and orders [48]. (**a**) Ideal SL; (**b**) scaled SL; (**c**) SL $B_{\mathrm{int}} = 3$; (**d**) 2–1, 2–2 and 2–1–1

obtained with a third-order modulator with an OSR of approximately 40. An almost equal power efficiency is exhibited by the fourth-order modulator with an OSR around 30. Figures 7.3c, d show the estimated FOM for single-loop 3-bit modulators with respect to (7.15) and (7.16) and for multiloop (2–1, 2–2, 2–1–1) ΣΔ modulators considering (7.18). As expected, the required OSR is much lower in comparison with a single-loop single-bit modulator, *but* the power efficiency is much lower.

These FOM results were considered during the design of the 1.5 V low-power third-order ΣΔ modulator proposed in [34, 37]. Detailed implementation issues and performance details are presented in Sect. 7.3.

7.1.6 Expansion Features of the FOM Based Design Strategy

As presented in Sect. 2.5, the entire ΣΔ ADC includes also an antialiasing filter and a back-end decimation filter. Both filters require a significant amount of power and silicon area; thus, the integration of both filter requirements and parameters in the FOM estimation is particularly desirable to complete the entire design strategy:

- The number of poles of an antialiasing filter is directly related to its entire power dissipation. As a result, if the number of poles in the AAF filter can be decreased by increasing the oversampling ratio, the overall ADC power dissipation decreases, although the $\Sigma\Delta$ modulator power will increase.
- A decimation filter is a power and area consuming building block of the overall $\Sigma\Delta$ ADC whose architecture, parameters and hardware effort strongly depend on the OSR.
- From the sytem point of view, without including the AAF filter nor the decimation filter parameters within the design strategy, topologies with the same FOM but different OSR show the same power efficiency.

7.2 Low-Power Limits in Analog Circuits

In literature, a large number of approaches can be found to define the absolute minimum power consumption of an analog circuit or system. In the following a short review of the different approaches is presented. Finally, the low-power limit for CT $\Sigma\Delta$ modulator is calculated and compared with the DT counterpart proposed in [120].

7.2.1 Low-Power Limits in Noise-Dominated Circuits

The well-known expression for the minimum power consumption in a class-B operated analog system that satisfies both the signal bandwidth f_{B} and the SNR was proposed [244]:

$$P_{\mathrm{min,SNR}} = 8KT\mathrm{SNR}\, f_{\mathrm{B}} \,, \tag{7.19}$$

where the Boltzmann's constant is defined as K and the absolute temperature as T. First of all, (7.19) starts out from the fact, that the peak-to-peak output signal swing corresponds to the supply voltage V_{dd} and secondly it only takes thermal noise into account. Furthermore, it was expected that the level of harmonic distortion due to either voltage or current limitations is (close to) zero. In fact, the minimum achievable power consumption is in general more than two to three orders of magnitude higher, as real implementations show, due to additional noise sources like $1/f$-noise as well as parasitic capacitances [59].

Additionally, the output swing assumed in (7.19) is generally smaller by the magnitude ΔV, which is the part of the supply voltage not used for the signal swing. Furthermore, this introduces some supply voltage dependency, as (7.20) indicates:

$$P_{\mathrm{min,SNR}} = 8KT\mathrm{SNR} f_{\mathrm{B}} \frac{V_{\mathrm{dd}}}{V_{\mathrm{dd}} - \Delta V} \,. \tag{7.20}$$

The following context will indicate the trade-off between low-voltage and low-power analog design in noise dominated systems. Obviously, a reduction of the supply voltage V_{dd} also shrinks the signal amplitude. This implicates for a constant SNR a lower noise contribution $\overline{v_{\mathrm{n}}}^2$ which is inversely proportional to the transconductance $\overline{v_{\mathrm{n}}}^2 \propto gm^{-1}$. The low-voltage approach causes again:

$$\left.\begin{aligned} gm &= \frac{2I_{\mathrm{DS}}}{V_{\mathrm{DS,sat}}} \\ V_{\mathrm{DS,sat}} &= \mathrm{const.} \end{aligned}\right\} gm \propto I_{\mathrm{DS}} \, . \tag{7.21}$$

Therefore, the bias current has to be quadratically increased to maintain the overall performance. Accordingly, the overall power consumption $P = V_{\mathrm{dd}} I_{\mathrm{ges}}$ increases as the supply voltage decreases.

7.2.2 Low-Power Limits in Noise-Dominated and Distortion-Dominated Circuits

An extension of (7.19) and (7.20) was presented in [245]. This approach considers beside thermal noise also the effect of the nth-order harmonic. Thus, it employs the SNDR in contrast to the SNR. Therefore, the minimal power consumption of a class-A biased system yields:

$$\begin{aligned} P_{\mathrm{min,SNDR}} &= 16\pi K^2 T^2 f_{\mathrm{B}} \left(\frac{2\alpha_n}{V_{\mathrm{dd}} - \Delta V} \right)^{(1/n)} \\ &\quad \times \left[(2n+1)\, \mathrm{SNDR} \right]^{1+1/(2n)} \frac{V_{\mathrm{dd}}}{nq \left(V_{\mathrm{dd}} - \Delta V \right)} \, , \end{aligned} \tag{7.22}$$

assuming the transistors operate in (ideal) weak-inversion $gm = qI_{\mathrm{DS}}/4KT$. The factor α_n denotes the magnitude of the nth-harmonic. Equation (7.22) looks similar to (7.20). The major differences are that the power consumption is more than linear with the SNDR and a multiplication factor which depends among other things on the circuit nonlinearity [245].

7.2.3 Low-Power Limits in Matching-Dominated Circuits

Another important expression concerning the minimal power consumption in analog circuits dominated by transistor mismatch was proposed in [246, 247]:

$$P_{\mathrm{min},A_{V_{\mathrm{T}}}} = 24 C_{\mathrm{ox}} A_{V_{\mathrm{T}}}^2 \mathrm{SNR} f_{\mathrm{B}} \, , \tag{7.23}$$

where C_{ox} represents the gate-oxide capacitance per unit area and $A_{V_{\mathrm{T}}}$ the technology-dependent mismatch constant concerning the threshold voltage. As shown in [246], this improved power limitation (7.23) matches considerably better to power drains of actual filter and ADC implementations. Moreover, expression (7.23) clarifies that the *Speed–Accuracy–Power* figure of merit is just a function of technological constants.

Estimating again the trade-off between low-voltage and low-power analog design, but now in matching dominated systems, leads again to an overall power drain increase. The reduction of the supply voltage V_{dd} diminishes the reference voltage V_{ref}, e.g., in an ADC. This leads to a smaller LSB and accordingly it requires a higher accuracy or rather a better matching for a constant resolution:

$$\sigma(\Delta V_T) = \frac{A_{V_{\mathrm{T}}}}{\sqrt{\mathrm{WL}}} \propto \frac{1}{\sqrt{C_{\mathrm{GS}}}} . \tag{7.24}$$

According to (7.24) both the overall transistor area as well as the gate–source capacitance have to be quadratically magnified. Accordingly, the currents have to be likewise increased to maintain the overall speed. As a result, the power consumption increases as the supply voltage is decreased.

7.2.4 Low-Power Limits in ΣΔ Modulators

In [120] the minimal power consumption of discrete-time modulators was evaluated:

$$P_{\mathrm{min},\Sigma\Delta_{\mathrm{DT}}} = 8k_{\Sigma\Delta}k_{\mathrm{OTA}}V_{\mathrm{dd}}f_{\mathrm{N}}\frac{\Delta V}{4}\frac{k_{\mathrm{noise}}KTk_{\mathrm{C_{DT}}}}{2d_{\mathrm{I}}}\frac{\ln(2/3\mathrm{DR})\mathrm{DR}}{(V_{\mathrm{dd}}-\Delta V)^2} . \tag{7.25}$$

Here, the coefficients denote:

$k_{\Sigma\Delta}$: sum of the relative integrator scaling coefficients
k_{OTA} : number of current branches in an OTA
d_{I} : duty cycle of the integration phase
$k_{\mathrm{C_{DT}}}$: capacitive excess factor
k_{noise} : noise excess factor

Furthermore, the validation and conformance were shown by a comparison with the power limit concerning the mismatch (7.23). In the following, the used speed–accuracy–power trade-off for a continuous-time ΣΔ modulator $P_{\mathrm{min},\Sigma\Delta_{\mathrm{CT}}}$ is determined [38]. The key circuit building block in a CT ΣΔ modulator is the integrator, typically implemented as active *RC*-filters especially in the first stage [17, 19, 34, 36, 37, 176]. Moreover, the integrators will dominate the power dissipation of the whole modulator. The minimum achievable power dissipation is set by the necessity to follow signal speeds equal to the sampling frequency. In practice, active *RC*-circuits are implemented using class-A amplifiers which have static bias currents and dissipate significantly more power than the dynamic limit. In this section, the power limit for active *RC*-integrators is derived in the context of an oversampled system.

The power consumption of a differential modulator corresponds to the minimum supply voltage multiplied by the total bias current, which is calculated by $k_{\Sigma\Delta}k_{\mathrm{OTA}}$ times the current in a branch I_{DS}. The transconductance to current relationship is given by:

$$gm = \frac{2I_{\mathrm{DS}}}{V_{\mathrm{GS,eff}}} . \tag{7.26}$$

As stated in [35, 37] the OTA's gain-bandwidth product can be modeled by a factor $2\pi c$ times the sampling frequency f_S, see (5.24):

$$\mathrm{GBW} = \frac{gm}{C_\mathrm{L}} = 2\pi c f_\mathrm{S} \quad \Leftrightarrow \quad gm = 2\pi c f_\mathrm{S} C_\mathrm{L} \,. \tag{7.27}$$

The modulator's dynamic range is given by:

$$\mathrm{DR} = \frac{P_\mathrm{sig}}{P_\mathrm{N}} \tag{7.28}$$

if only circuit noise is considered. Please note that generally the accuracy of a ΣΔ modulator is dominated by circuit noise, which is why the quantization noise is neglected in (7.28). Additionally, in the following only thermal noise of both the amplifier and the resistor are taken into account. The signal power P_sig can be approximated according to [120] with $(2V_\mathrm{ref}\mathrm{OL})^2/2$. The overall circuit noise contribution P_N yields:

$$P_\mathrm{N} \approx 16KTf_\mathrm{B}\left(R + \left(1 + \frac{R}{R_\mathrm{DAC}}\right)^2 \frac{2}{3}\frac{k_\mathrm{noise}}{gm}\right) \tag{7.29}$$

$$\approx 8KTf_\mathrm{N}\left(R + \frac{8}{3}\frac{k_\mathrm{noise}}{\mathrm{gm}}\right) \tag{7.30}$$

if beside the resistive noise of the RC-integrator also the thermal noise of the OTA is considered. R and R_DAC denote the integrator resistance and the feedback (DAC) resistance, which are almost equal. Furthermore, k_noise represent the noise excess factor. Finally, the sampling frequency is determined by:

$$f_\mathrm{S} = \frac{1}{RC} = \frac{k_{C_\mathrm{CT}}}{RC_\mathrm{L}}, \tag{7.31}$$

where k_{C_CT} represents the capacitive load excess factor. Combining equations (7.26–7.31), give the total power consumption of a CT ΣΔ modulator:

$$\begin{aligned} P_{\mathrm{min},\Sigma\Delta_\mathrm{CT}} &= 32k_{\Sigma\Delta}k_\mathrm{OTA}2\pi c\,V_\mathrm{dd}f_\mathrm{N}\frac{\Delta VKT\mathrm{DR}}{(V_\mathrm{dd}-\Delta V)^2} \\ &\times\left(k_{C_\mathrm{CT}} + \left(1 + \frac{8}{3}\frac{k_\mathrm{noise}}{2\pi c}\right)\right). \end{aligned} \tag{7.32}$$

A comparison concerning the minimal power consumptions of both the DT and CT ΣΔ modulator is shown in [38]. The illustrated ratio $P_{\mathrm{min},\Sigma\Delta_\mathrm{DT}}/P_{\mathrm{min},\Sigma\Delta_\mathrm{CT}}$ displays a weak dependency on the overall resolution.

$$\frac{P_{\mathrm{min},\Sigma\Delta_\mathrm{DT}}}{P_{\mathrm{min},\Sigma\Delta_\mathrm{CT}}} = \left(\frac{k_\mathrm{noise}k_{C_\mathrm{DT}}\ln(2/3\,\mathrm{DR})}{32d_\mathrm{I}\,2\pi c(k_{C_\mathrm{CT}} + \frac{8}{3}\frac{k_\mathrm{noise}}{2\pi c})}\right). \tag{7.33}$$

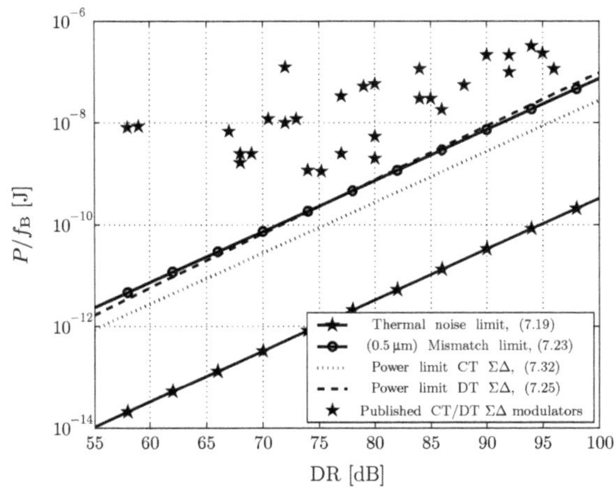

Fig. 7.4. Power consumption over the signal bandwidth vs. dynamic range of recently published CT and DT ΣΔ ADCs in comparison with the presented speed–accuracy–power trade-offs

The made assumptions taken for the best case power limits in a DT modulator, are given in [120]. This leads to capacitive and noise excess factor of eight and four, respectively, in fully differential implementations. The integration phase duty cycle δ_I is at best 2. As proposed in [35], a good rule of thumb of the CT gain bandwidth surplus factor c is 1.5. Finally, the amplifier noise and load capacitance excess factor are 2 and 1. It can be calculated from (7.33) that the DT modulator requires approximately a factor of three higher power drain (Fig. 7.4).

7.3 Implementation Example I: A 12-Bit 25 kHz 1.5 V CT ΣΔ Modulator

For low-voltage low-power applications single-loop architectures are the preferred structures over cascaded stages, because they do not put stringent requirements on the amplifiers [34, 235] and coefficient matching [37, 91], they require only a few critical analog components [48, 72] and are very robust if the modulator order is $N \leq 2$. But however, low-order single-loop modulators demand high oversampling ratios to achieve the desired resolutions, which accordingly leads to a higher power consumption and increases constraints due to clock jitter and excess loop delay [3, 153].

Obviously, by using high-order single-loop modulators with modest OSR, a larger amount of error power is shaped out of the band of interest, which reduces the previous requirements and the presence of pattern noise at the expense of occurring loop stability problems. An appropriate way to stabilize these modulators is the scaling of the NTFs, so that the usable input range increases on the one hand, but the NTFs are not so aggressive in the context of noise shaping because the ideal modulator performance degrades, depending on the implemented scaling.

According to Sect. 7.1.5, the third-order single-loop modulator topology combines both high power efficiency and a low mismatch sensitivity. Therefore, a third-order structure with distributed feedback as in Fig. 3.11a was used.

7.3.1 Loop Filter Design

The design of the loop filter for a CT ΣΔ modulator can be done in discrete time domain, as mentioned in Sect. 3.1, to speed-up the overall design procedure [3, 105]. The used third-order DT modulator is shown in Fig. 7.5, whose loop filter LF(z) yields:

$$\mathrm{LF}(z) = -a_3 I_3(z) - a_2 a_3 I_2(z) I_3(z) - a_1 a_2 a_3 I_1(z) I_2(z) I_3(z), \tag{7.34}$$

where a_i denotes the DT scaling factors. In [6] an optimal implementation of this third-order DT modulator has been found to incorporate three DT integrators $\{I(z) = z^{-1}/1 - z^{-1}\}$, while the integrators are scaled with $\{a_1 = 0.2,\ a_2 = a_3 = 0.5\}$. But the architectures in [6] were derived for DT switched capacitor implementations, whose scalings are precisely matched in the integrated circuit. Therefore, the scaling and the noise shaping can be performed more aggressively, i.e., close to the stability limit.

On the other hand, most CT modulators are realized with active *RC* or *gmC* integrators, whose scaling and gain clearly are subject to a large process variation, as shown throughout Sect. 5.3. Therefore a possible shift into instability has to be considered. As a prevention, the scaling can be done less aggressively. Alternatively, the original DT modulator is chosen to be different than the ideal one described in [6], i.e., with the same scaling coefficients but with two DT resonators $\{I_{1,2}(z) = 1/1 - z^{-1}\}$ as first two filters and thereafter only one DT integrator as last filter $I_3(z)$. A corresponding DT simulation is shown in Fig. 7.6, where the in-band noise of the original ideal DT modulator as well as the one with two DT resonators is simulated over an integrator gain error: $a_i \rightarrow a_i/(1 + \delta_{\mathrm{GE}})$. (Note that the variable δ_{GE} chosen for the DT simulation corresponds to CT modulator's *RC*-variation δ_{RC}!)

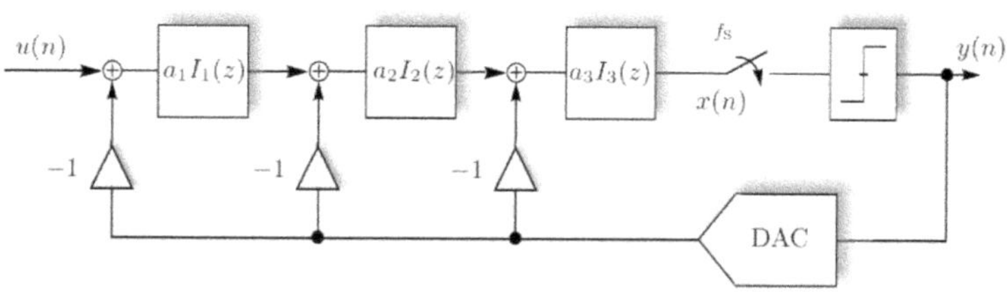

Fig. 7.5. Original third-order single-bit DT ΣΔ modulator consisting of a chain of integrators with weighted feedback a_i, after [6, 37]

It is seen from Fig. 7.6 that the third-order modulator employing three integrators gets unstable for negative variation of approximately $\delta_{\mathrm{GE}} \le -5\%$, while the other is stable down to >20%. The price for the gain-mismatch insensitivity is a 3 dB IBN penalty compared to the ideal architecture [6]. An alternative possibility would be to use a compensation or tuning approach as outlined in Sect. 5.3.5.

Due to this performance loss, in order to achieve a dynamic range of more than 80 dB, the OSR should be at least 48 when using $\{a_1, a_2, a_3\} = \{0.2, 0.5, 0.5\}$ for the scaling coefficients. Accordingly, for the desired passband of 25 kHz, this results in a sampling frequency of $f_S = 2.4\,\text{MHz}$. The corresponding noise and signal transfer functions are shown in Fig. 7.7. Furthermore, Fig. 7.7 indicates, that the total error attenuation and the maximal

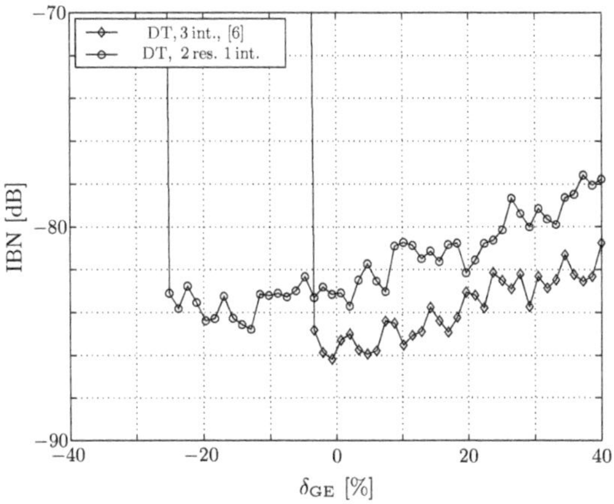

Fig. 7.6. Simulated IBN of a third-order DT modulator under gain-error influence

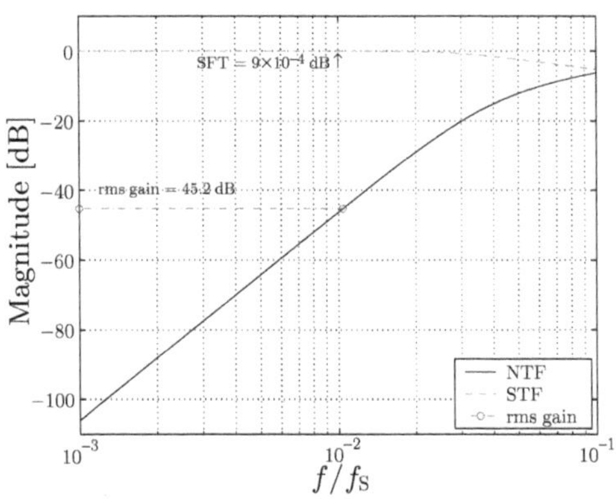

Fig. 7.7. The passband magnitude response of both the NTF and STF of the modified third-order DT modulator with $\{a_1, a_2, a_3\} = \{0.2, 0.5, 0.5\}$. The passband is flat within 9×10^{-4}dB

ripple of the STF within the passband yields −45.2 dB and $\varepsilon_{\text{SFT}} = 9\times10^{-4}$dB, respectively.

Now the modified original DT loop filter function LF(z) can be replaced by a CT equivalent LF(s) with respect to the feedback DAC impulse response $R_{\text{DAC}}(s)$. The coefficients k_i of the CT modulator are calculated as a function of the DT integrator coefficients a_i by using either the modified $\mathcal{Z}$-transform [134, 139, 248] or the impulse-invariant transform [3, 105]. The CT third-order loop filter can be written as:

$$\text{LF}(s) = -\frac{k_1}{T_{\text{S}}^3 s^3} - \frac{k_2}{T_{\text{S}}^2 s^2} - \frac{k_3}{T_{\text{S}} s} \tag{7.35}$$

if a modulator as in Fig. 3.11a is considered. For the feedback DAC an NRZ pulse form has been chosen according to Fig. 3.2a. The transformation according to Sect. 3.2 in combination with a subsequent coefficient comparison yields the wanted CT scaling coefficients k_i. The resulting numerical values using the DT scaling coefficients $\{a_1, a_2, a_3\} = \{0.2,\ 0.5,\ 0.5\}$ are presented in (7.36):

$$\begin{aligned} k_1 &= a_1 a_2 a_3 = 0.05\,, \\ k_2 &= a_1 a_2 a_3 + a_2 a_3 = 0.3\,, \\ k_3 &= \frac{a_1 a_2 a_3}{3} + \frac{a_2 a_3}{2} + a_3 = 0.641\,. \end{aligned} \tag{7.36}$$

The result of the transformation is shown in Fig. 7.8. Obviously, both SNDR curves match very well and exhibit the same DR and peak SNDR of −83 dB and 76 dB, respectively. The maximal stable input amplitude equals −6 dB.

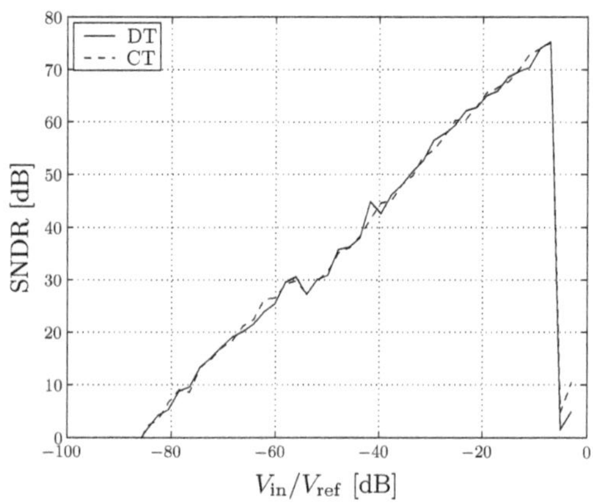

Fig. 7.8. Simulated SNDR vs. input power of both the DT and CT ΣΔ modulator

Stability Considerations

ΣΔ modulators with $N \geq 3$ are not inherently stable. Stability in this context is if the internal modulator states are bounded and the modulator output is free of large limit cycles for a certain input range.

Several stability criteria exist to predict a stable behavior but either they give no guarantee or are over conservative [81, 86, 226]. Some of these techniques were presented in Sect. 2.7.4. Consequently, here a more practical approach is used to get a kind of finger print concerning the stability and reliability of the third-order modulator. In general the onset of instability occurs abruptly when the input amplitude is increased. Therefore, the stability of the modulator was investigated by simulations using a slowly increasing input ramp and a search for long transients [81].

Figure 7.9 denotes the maximum transient length, which uses 2,000 input values equally spaced between 0 and 1. For each input value, 1,000 simulations were performed with randomly selected initial conditions according to [81]. A maximum simulation length of 2,000 points was used. The simulations were stopped either when the quantizer input exceeds an upper bound of 10 or after the maximum 2,000 time steps. The current number of time steps simulated is regarded as a transient length. The maximum transient length is found within each of the 2,000 input amplitude values. The third-order modulator operates reliably and stable for input magnitudes up to approximately 0.7 ($\triangleq -6$ dB below full scale) as Fig. 7.9 indicates. For magnitudes above 0.75 ($\triangleq -5.5$ dB below full scale) the modulator is definitely unstable. As a result, the selected coefficients are a good choice concerning stability and maximum DR of the third-order ΣΔ modulator. After the overall system design has been made, the focus is placed in the following on circuit level design and implementation issues.

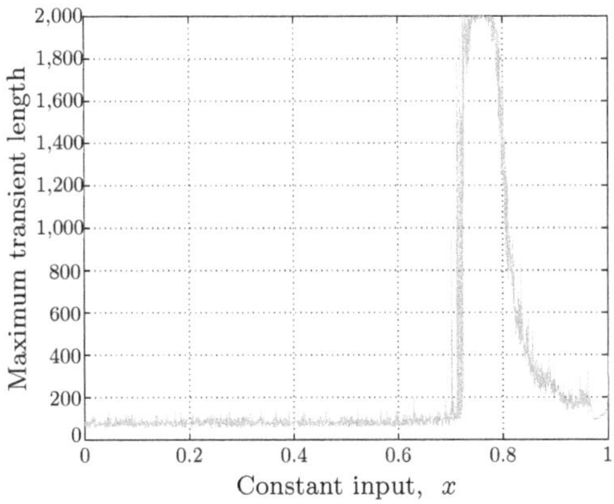

Fig. 7.9. Stability analysis of the used third-order ΣΔ modulator reliably operating for input magnitudes up to 0.7

7.3.2 Circuit Blocks

This section presents the circuit design and implementation of the three building blocks: the CT integrator, the quantizer, and finally the DAC, which were adopted for the integrated 1.5 V–25 kHz third-order modulator prototype.

Integrator

In Chap. 5, the demands on the CT integrator were examined. The strictest requirements on the design of low-voltage low-power single-loop ΣΔ modulators are the gain bandwidth product and a sufficient voltage swing in particular at low supply voltages. The required dc gain and slew rate are only of minor constraint. In order to get a high gain-bandwidth product for a given current drain, a single-stage folded-cascode amplifier has been chosen, Fig. 7.10. For differential amplifiers, dropping the supply voltage means a loss of headroom and dynamic range which is critical to any active analog circuit [249–251]. Besides, having multiple devices stacked on top of each other (as Fig. 7.10 depicts), the output voltage swing is additionally reduced. To enhance the output swing of the used amplifier, all output transistors ($M3$–$M10$) have the same low $V_{\mathrm{DS,sat}} = 100\,\mathrm{mV}$, whereas the input transistors are biased slightly in weak inversion with $V_{\mathrm{DS,sat}} \approx 0$. Consequently, the maximum fully differential input and output swings are:

$$\mathrm{IS}_{\mathrm{diff,max}} \leq 2V_{\mathrm{dd}} - 2V_{\mathrm{GS},M1} - 2V_{\mathrm{DS,lin}}\,, \tag{7.37}$$

$$\mathrm{OS}_{\mathrm{diff,max}} \leq 2V_{\mathrm{dd}} - 8V_{\mathrm{DS,sat}}\,, \tag{7.38}$$

where $V_{\mathrm{DS,lin}}$ is the voltage drop over the linear common mode feedback (CMFB) transistors. Thus, input signals of the amplifier are referenced to a common mode voltage $V_{\mathrm{CM}} = 650\,\mathrm{mV}$ to achieve maximum signal swing at the supply voltage of $1.5\,V$.

According to Sect. 5.4, for the single-loop CT modulator as a rule of thumb the gain-bandwidth product (in Hz) of the employed amplifiers should be in

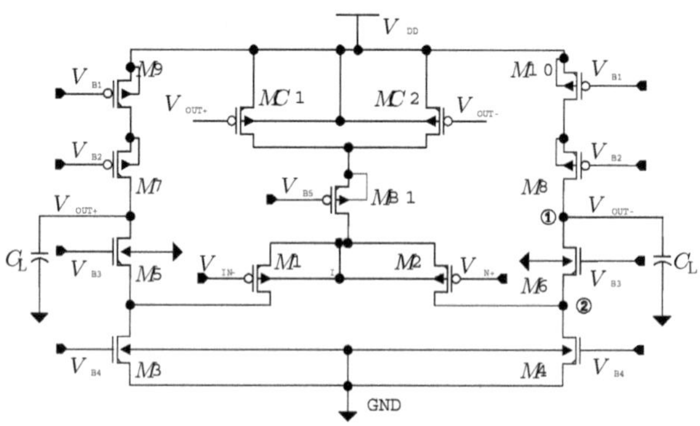

Fig. 7.10. Schematic of the implemented amplifier

the range of the sampling frequency. Taking further parasitic capacitances into account as well as avoiding any problem originated from the limited gain-bandwidth, the implemented GBW of the first amplifier is $\approx 2f_S$. Furthermore, the bias current through $M1, 2$ and each output branch has been designed identically to maximize the slew rate for a certain current drain. The design constraints of the amplifier are given in [36]. Measurement results showed that the bias current can be reduced (i.e., the GBW can be reduced) without significant performance loss. The overall amplifier noise power suffering from the large $1/f$ noise contribution was minimized by employing large input devices ($M1$, $M2$) of $WL = 400\,\mu\text{m}^2$. Simulations of the first amplifier exhibit an integrated input referred in-band noise of approximately $V_{\text{rms}} \approx 46\,\mu\text{V}$.

The common mode feedback circuit (CMFB) of an amplifier is a critical element in CT low-voltage low-power applications, because it should not reduce the output swing and also require as little power as possible. Further important objectives are the gain, sensitivity to mismatch as well as nonlinearity of the common-mode signal detector. Accordingly, a simple CMFB with two transistors (MC1, MC2) biased in the triode region has been chosen [34, 252, 253].

The main drawback of this concept is the low CMFB loop gain A_{CMFB} which is proportional to $V_{\text{DS,lin}}$ [253]. Increasing $V_{\text{DS,lin}}$ to improve the common mode output voltage stability, reduces the input common mode range. As a result a compromise has to be found. Thus, $V_{\text{DS,lin}} = V_{\text{DS,MC1}}$ is designed for 50 mV to provide enough CMFB loop gain and concurrently to decrease the input range merely by 10%. The relative performance between the differential gain of the folded cascode A_{dc} and the common mode gain A_{CMFB} is given by:

$$\frac{A_{\text{CMFB}}}{A_{\text{dc}}} = \frac{\text{GBW}_{\text{CM}}}{\text{GBW}_{\text{DM}}} = \frac{\beta_{\text{MC1}} V_{\text{DS,MC1}}}{\beta_{M1}(V_{\text{GS},M1} - V_{\text{T}})} \approx 0.5\,. \tag{7.39}$$

In order to implement this, both devices MC1, 2 are designed to be quite wide (here $W/L = 70 : 1$), which introduces a substantial parasitic capacitance at the amplifier output. But as long as $M1, 2$ are not strongly unbalanced by a large differential signal, the noise generated by MC1, 2 is of common mode type, which is fortunately rejected by the differential topology. The amplifier was optimized with respect to stability, noise, linearity as well as to power [34]. The subsequent integrators employ the same amplifier structure except for the cascode transistors $M7, 8$ which enhances the output signal swing. Furthermore, both are optimized regarding the power drain.

Internal ADC

ΣΔ modulators are mostly insensitive to errors induced by the internal quantizer, which are mainly offset, hysteresis and nonlinearities. Because the ADC resides within the feedback loop, these nonidealities are attenuated by the loop

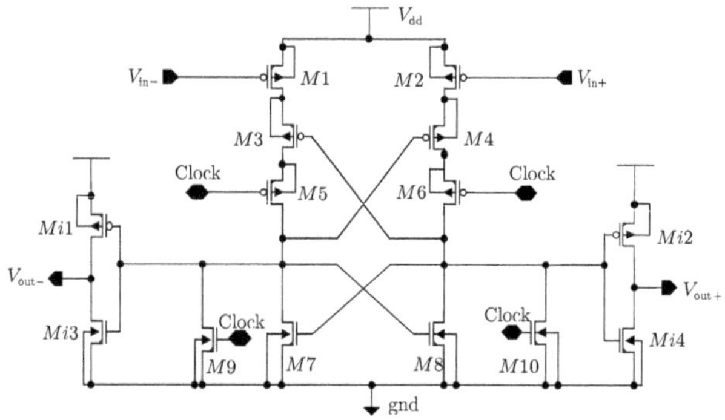

Fig. 7.11. Implemented regenerative comparator with output drivers, after [37]

gain. But still, the internal ADC can cause significant performance degradations in a continuous-time ΣΔ modulator due to timing induced errors like propagation and signal dependent delay, which both depend on the quantizer input amplitude and slope.

For the low-voltage comparator a fully symmetrical regenerative comparator (see Fig. 7.11) proposed in [254] has been used. The PMOS input devices ($M1$, $M2$) operate in the triode region. The cross-coupled inverter-latch regenerates when the signal Clock goes low. Then the output nodes rise simultaneously from gnd to the meta-stability point, here 650 mV, and diverge toward the positive and negative supply rail because of the difference in the input voltages ($V_{\text{in}+}$, $V_{\text{in}-}$) and the resulting asymmetry in the currents of the input devices. The final state is obtained due to the fact that the drain currents of the active switching PMOS devices ($M3$, $M4$) are steered. Thus, in the first part of the decision process a linear rising of the outputs appears while an exponential component is seen after reaching the meta-stability point [7].

The effect of comparator hysteresis on the overall ΣΔ modulator is negligible, provided an adequate reset of the comparator is accomplished [110]. When the Clock signal goes high, the output nodes are shorted to gnd and the comparator is reset, which eliminates memory effects. The quantizer offset voltage, which is generally in the range of a few tens of millivolts for such dynamic comparators, is fortunately also attenuated by the modulator loop gain, in contrast to the previously mentioned delay and metastability [3].

To get an impression of the occurring quantizer input voltages, electrical simulations of the third-order modulator were performed. Figure 7.12a shows the histogram of the quantizer input signal with 5,000 points for a modulator input voltage of $V_{\text{in}} = V_{\text{ref}} = 650\,\text{mV}$. Figure 7.12a depicts that the appearing quantizer input voltages vary between a few microvolt and several hundred millivolt. A corresponding simulation with regard to the signal dependent delay concerning the implemented comparator, is presented in comparator input amplitude V_{in} was swept from 0.75 V amplitude V_{in} was swept from 0.75 V down to 10 μV while using an input slope of 2.4 V/μs, which is given

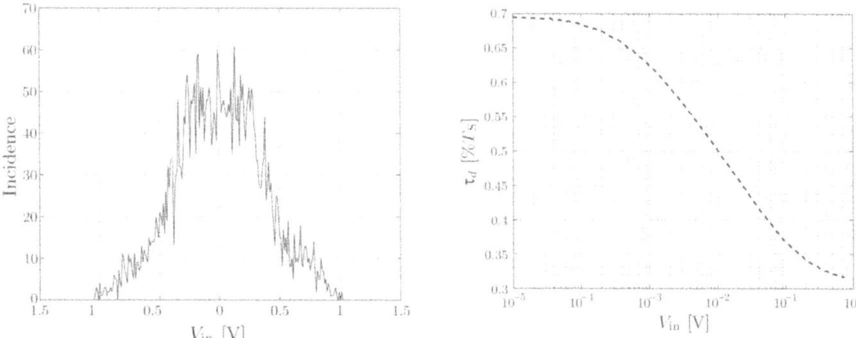

(a) Input signal histogram (b) Signal dependent delay

Fig. 7.12. (**a**) Histogram of the quantizer input signal of the third-order modulator with an input amplitude of $V_{\mathrm{in}} = 650\mathrm{mV}$. (**b**) Signal dependent delay τ_{d} as a function of the comparator input amplitude V_{in} employing a w.c. input signal slope of 2.4 V/µs

by the gain of the last integrator times the reference voltage ($\leq V_{\mathrm{ref}} f_{\mathrm{S}}$). Additionally, this value was verified by behavioral simulations. As a result, the worst case signal dependent delay τ_{d} remains below $0.7\% T_{\mathrm{S}}$ for small input amplitudes, which has no significant influence on the overall modulator performance, see Sect. 4.2. In order to hold the value of the comparator decision for the complete next sampling period, following the quantizer a latch has been introduced, which takes the quantizer decision slightly after it has been settled and holds it until the next sampling instant!

DAC

The main performance objectives on the feedback DACs in CT ΣΔ modulators are besides the linearity [4, 36] again any emerging propagation delay as well as intersymbol interference [37]. Intersymbol interference is mainly caused by mismatches among the positive and negative slopes of the NRZ-DAC feedback signals. Therefore, this mismatch effect depends on the modulator output bit sequence and therefore creates signal dependent distortion.

Fortunately, the used one bit DAC is inherently linear. Thus, the design objectives are to avoid any performance losses due to the remaining circuit nonidealities. The implemented one bit NRZ feedback digital-to-analog converter consist of a feedback resistance, a positive and negative reference voltage ($V_{\mathrm{ref}\pm}$) and a switch. Depending on the output state of the flip-flop OUT, the positive or negative reference voltage will be switched via the non-overlapping clock generator, the switching block and the feedback resistances ($R_{\mathrm{DAC1}}, R_{\mathrm{DAC2}}, R_{\mathrm{DAC3}}$) to the virtual ground of each amplifier [37]. The value of the feedback resistors yield from the integrator scaling coefficients:

$$R_{\mathrm{DAC}i} = \frac{1}{k_i f_{\mathrm{S}} C_i}, \tag{7.40}$$

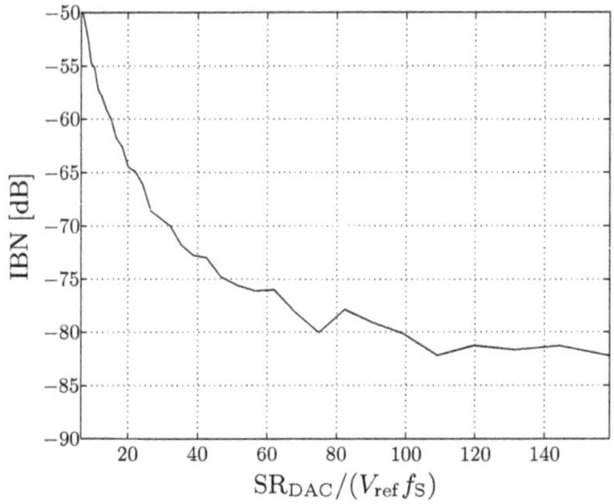

Fig. 7.13. Influence of the limited DAC slew rate on the modulator performance. DAC slew rate normalized to $V_{\mathrm{ref}}f_{\mathrm{S}}$ and swept from 10 to 1,000 with an input amplitude of $0.5V_{\mathrm{ref}}$

where C_i denotes the value of the integration capacitance of the ith integrator. Equation (7.40) additionally takes the actual values of the reference voltages $V_{\mathrm{ref}\pm}$ into account, which are assumed here to be 1.3 and 0 V. Any scaling of these reference voltages affects the feedback resistances in the same way.

As already mentioned in Sect. 4.4, NRZ-DACs are subject to intersymbol interference. To avoid performance degradation due to intersymbol interference, the DAC has to provide a certain slew rate as well as no difference between the positive and negative slope for a given modulator resolution. To determine the DAC specifications, behavioral simulations concerning both the slew rate and differences between the two slopes have been performed. Figure 7.13 shows the demanded DAC slew rate (normalized to $V_{\mathrm{ref}}f_{\mathrm{S}}$) of about 110 for the implemented 12-bit ΣΔ modulator sampled at 2.4 MHz. Thus, the modulator requires a DAC slew rate higher than 160 V/μs. Provided that this latter condition is met, any difference between the positive and negative slope has negligible influence on the overall modulator performance as proven by simulations.

7.3.3 Modulator Design

A block diagram of the implemented third-order modulator is shown in Fig. 7.14. The implemented ΣΔ modulator architecture is fully differential in order to minimize even-order harmonics as well as common-mode noise. The output signal of the CT loop filter is sampled by a single bit quantizer whose output state is stored in a D-flip-flop. The sampling frequency of the ΣΔ modulator is 2.4 MHz leading to an oversampling ratio of 48 for the 25 kHz passband.

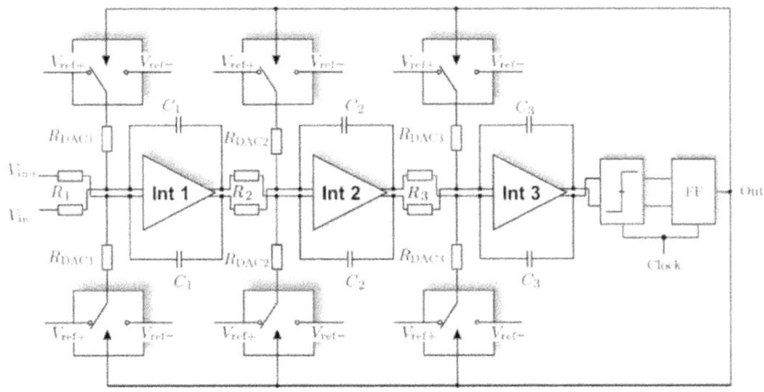

Fig. 7.14. Implementation of the third-order CT ΣΔ modulator

The CT loop filter was implemented using three active RC-integrators (Int 1–3). As presented in Sect. 5.3 the influence of time constant mismatch is of minor importance in single-loop ΣΔ modulators, if stability is guaranteed. Thus, tuning of the on-chip time constant has not been considered [105, 255]. The resistors R_i and $R_{\mathrm{DAC}i}$ implement the feedforward as well as the feedback coefficients according to (7.36).

The absolute values of the resistors R_1 and R_{DAC1} of the first integrator are determined by the low-power requirement, so that these are maximized in terms of the maximum tolerable in-band noise limit. Another benefit that comes along with this optimization approach is the improved integrator linearity [34, 128]. Concurrently, the integrator capacitor C_i can be minimized according to $C_i = k_i/(Rf_{\mathrm{S}})$. From Sect. 5.6.2, the input-referred noise of the first integrator is given by (5.46), if it is assumed that the amplifier noise contribution is negligible:

$$\bar{v}_{\mathrm{in}}^2 \approx 16KTf_{\mathrm{B}}\left(R_1 + R_{\mathrm{DAC}}\frac{R_1^2}{R_{\mathrm{DAC}}^2}\right). \tag{7.41}$$

Under the condition $R_1 = R_{\mathrm{DAC}}$, (7.41) results in:

$$\bar{v}_{\mathrm{in}}^2 \approx 32KTf_{\mathrm{B}}R_1\Big|_{R_1=R_{\mathrm{DAC}}}. \tag{7.42}$$

Using (7.42) and considering a certain resolution of B-bit, with (2.7) the maximum value of R_1 can be written as:

$$R_1 \approx \frac{V_{\mathrm{in}}^2/2}{32KTf_{\mathrm{B}}\,3\times 2^{2B-1}}, \tag{7.43}$$

which results in $R_1 < 2.7\,\mathrm{M\Omega}$ for this desired modulator ($V_{\mathrm{in}} = 0.65\,\mathrm{V},\ f_{\mathrm{B}} = 25\,\mathrm{kHz}, B = 12$-bit). The implemented resistor values are $R_1 = 2.08\,\mathrm{M\Omega},\ R_2 = 416\,\mathrm{k\Omega}$, and $R_3 = 416\,\mathrm{k\Omega}$. Thus, for the small scaling factor $k_1 = 0.05$ the first integration capacitor C_1 is only 4 pF. The subsequent integration capacitors are reduced to $C_2,\ C_3 = 1\,\mathrm{pF}$.

Fig. 7.15. Chip photo [37], courtesy of IEEE

The nominal supply voltage of the modulator is 1.5 V. Thus, analog gnd is chosen to be $V_{\mathrm{CM}} = 650\,\mathrm{mV}$ for maximal signal swing according to Sect. 7.3.2. Furthermore, the positive and negative reference voltages ($V_{\mathrm{ref}+}$, $V_{\mathrm{ref}-}$) are set to 975 mV and 325 mV, respectively, instead of 1.3 and 0 V to be able to halve the magnitude of the feedback resistors $R_{\mathrm{DAC}i}$ [34], which improves the overall chip area but worsens the noise contribution, see (7.41). The resulting feedback resistors are $R_{\mathrm{DAC1}} = 1.04\,\mathrm{M\Omega}$, $R_{\mathrm{DAC2}} = 694\,\mathrm{k\Omega}$ and $R_{\mathrm{DAC3}} = 325\,\mathrm{k\Omega}$. The ΣΔ modulator was implemented in a 3.3 V, 0.5 µm triple-metal standard analog CMOS technology ($V_{\mathrm{Tn}} = 0.58\,\mathrm{V}$, $V_{\mathrm{Tp}} = 0.62\,\mathrm{V}$). The required area is less than $1\,\mathrm{mm}^2$ with the first integrator consuming more than 50% of the total power consumption and area [37]. The chip photo is shown in Fig. 7.15.

Design for Testability

The designed 25 kHz ΣΔ modulator was implemented twice, Fig. 7.15. The lower one was especially designed for testability, which offers a special test mode, in which six additionally implemented analog buffers [256] are connected through switches to the differential outputs of each integrator. This access through the buffers allows on-chip monitoring of the output voltages, the integrator signal swings as well as verifying the operation of the comparator.

Figure 7.16 shows the measured output voltage levels at the sampling instants of the three integrators. The applied input signal amplitude was −12 dB below the overload level. As expected, the voltage swing of the first integrator is limited to only ±70 mV, while the second and third integrator feature an output voltage swing of ±180 mV and ±400 mV. The limited positive output swing of the third integrator (V_{out3}) is due to the limited swing of the additionally implemented output buffer.

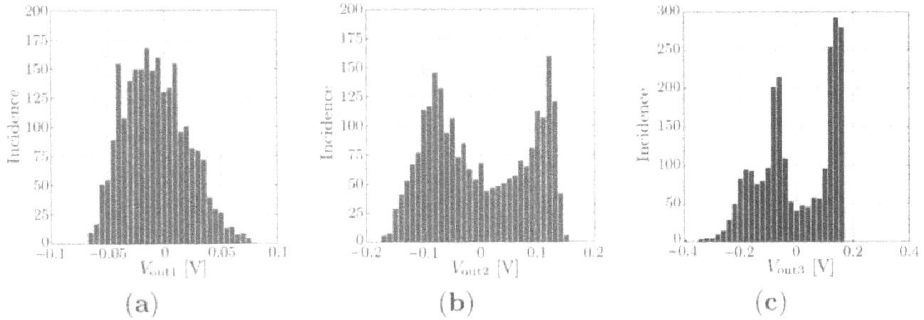

Fig. 7.16. Measured output voltage levels of the three integrators at the sampling instants. The input amplitude was 0.25 referred to analog ground. (**a**) First integrator; (**b**) second integrator; (**c**) third integrator

7.3.4 Measurements

Experimental Results

The nominal measurement conditions were at a supply voltage of 1.5 V and a positive and negative reference voltage of 975 mV and 325 mV, respectively. Since the analog and digital power supplies are separated, an individual power measurement was possible. The first integrator consumes more than 55% of the total power. The second and third integrator share a power drain of roughly 27%. The remaining parts are split accordingly to 10% for the bias and 8% for the digital part [34, 37].

Figure 7.17 shows the measured spectrum of the third-order modulator, using a 5 kHz, −15 dB input signal. Figure 7.17a shows the entire spectrum, the in-band noise and the upcoming distortion powers, whereas Fig. 7.17b presents the passband spectrum as well as again the accumulated in-band

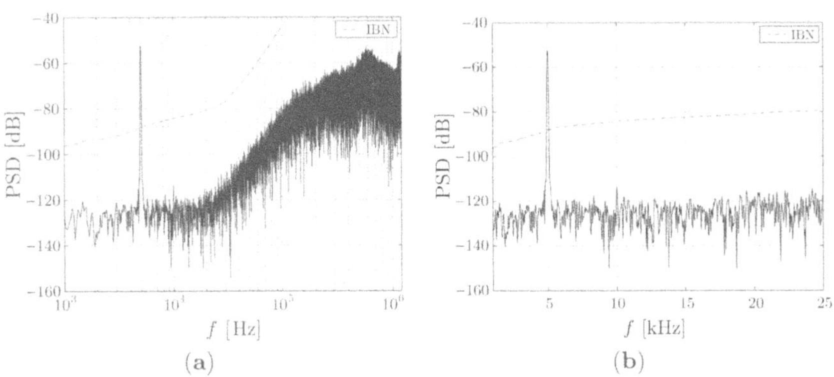

Fig. 7.17. Measured spectrum of the third-order modulator. $f_{sig} = 5\,\text{kHz}$, $P_{sig} = -15\,\text{dB}$. (**a**) Entire spectrum; (**b**) in-band spectrum

noise, here −79 dB. Furthermore, Fig. 7.18a shows the resulting signal-to-noise and signal-to-noise plus distortion ratio versus the normalized input signal V_{in}/V_{ref}. The applied input signal frequency is 5 kHz. The measured dynamic range and the peak SNR are 80 dB and 73 dB, respectively. This corresponds to a resolution of 12-bit. The measured peak SNDR is 70 dB.

As a summary, the obtained measurement results demonstrate the expected performance of the third-order ΣΔ modulator. Nevertheless, slight deviations from the ideal behavior still exist. A comparison of the theoretically achievable performance metrics (Fig. 7.8) with the obtained measurement results exhibits a 3 dB loss in dynamic range and SNR caused by the slightly increased noise floor. The measured peak SNDR of 70 dB suffers first of all from the increased noise floor and secondly from the reduced (by 2 dB) maximum stable input amplitude.

The measured IBN, SNDR, power consumption P and the resulting FOM as a function of the supply voltage are shown in Fig. 7.18b. In order to get comparable results, the used measurement conditions remain the same (as described above). However, for supply voltages of $V_{dd} \leq 1.3$ V the bias conditions are increased by approximately 50% in order to keep the accuracy constant; note that this increases the total power consumption. The modulator operates even with a supply voltage of only 1.3 V without performance degradation. This illustrates the low-voltage capability of the realized CT modulator. Furthermore, it is apparent that the minimal FOM is obtained at a supply voltage of 1.4 V and increases correspondingly to the overall power consumption P.

From Fig. 7.18b it is evident that this proposed design strategy results in a good power efficiency, compare [5]. A complete performance summary is given in Table 7.2. The power efficiency of various A/D converters with different resolutions and sampling rates can be compared by using again the

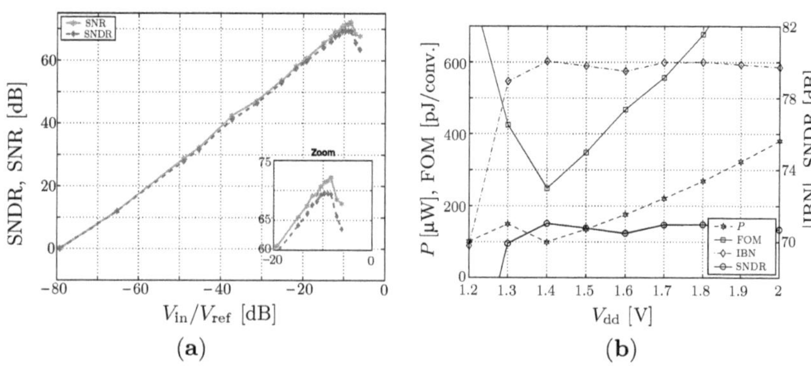

Fig. 7.18. (**a**) Measured signal-to-noise plus distortion ratio and (**b**) measured IBN, SNDR, power consumption P and resulting FOM as a function of V_{dd}, after [37], courtesy of IEEE

Table 7.2. Measured performance summary of the third-order ΣΔ modulator [37]

Supply voltage	1.3 V	1.5 V
Signal bandwidth	25 kHz	
Sampling frequency	2.4 MHz	
Dynamic range	79 dB	80 dB
IBN at $\sigma_t = 1\% T_S$	55 dB	
Peak SNR	72 dB	73 dB
Power consumption	150 μW	135 μW
FOM	0.42 pJ/conv.	0.35 pJ/conv.
Technology	3.3 V, 0.5 μm CMOS, $V_{Tn} = 0.58$ V, $V_{Tp} = 0.62$ V	

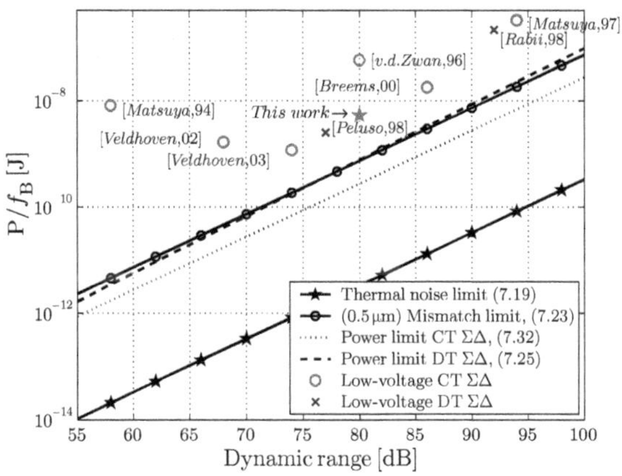

Fig. 7.19. Power consumption over the signal bandwidth vs. dynamic range of recent low-voltage ΣΔ A/D converters

FOM. Figure 7.19 plots the power consumption over the signal bandwidth versus the dynamic range for some recently published CMOS CT and DT ΣΔ modulators. The power dissipation used in the computation does not include the neither decimation filter, nor the antialiasing filter. Obviously, the realized third-order ΣΔ modulator exhibits an excellent performance in comparison with other recently published low-power CT modulators in spite of the used 0.5 μm CMOS technology.

Clock Jitter Test Setup

The third-order CT ΣΔ modulator has been measured also concerning its clock jitter performance with a measurement setup as shown in Fig. 7.20 [33].

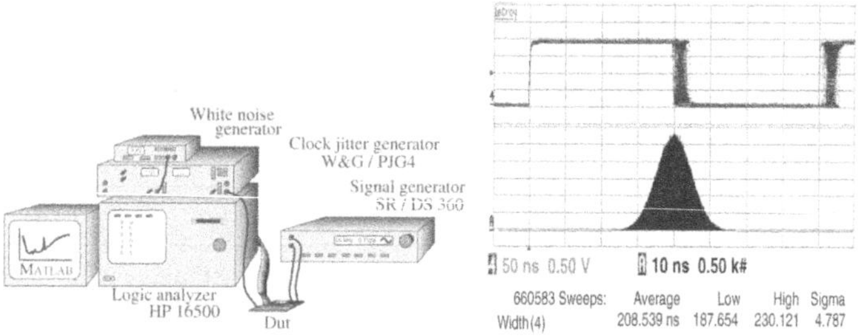

(a) (b)

Fig. 7.20. Measurement setup and jitter measurement [31, 33, 37]. (**a**) Measurement setup; (**b**) clock jitter distribution

A white noise controlled jittered clock has been supported using an externally modulated clock jitter generator (W&G, PJG4) while the adjusted clock jitter was observed using a digital oscilloscope (LeCroy, WavePro™ 950). The digital output bit stream has been recorded using a logic analyzer. Finally, a fast Fourier transformation using a Blackman–Harris window has been applied with Matlab.

Jitter Measurements

In Fig. 7.21a the measured modulator output spectrum is presented with and without clock jitter. The applied clock jitter was $\sigma_t = 1\% T_S$. As expected, the injected clock jitter appears as additional noise at the modulator output.

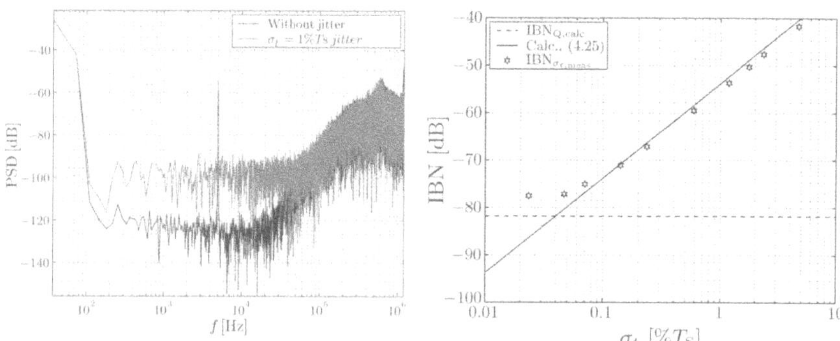

(a) (b)

Fig. 7.21. (**a**) Measured power spectral density (PSD) and (**b**) theoretical and measured in-band noise (IBN) for the implemented CT third-order ΣΔ modulator under clock jitter influence, after [37]. (**a**) PSD under clock jitter influence; (**b**) IBN under clock jitter influence

Figure. 7.21b shows the theoretical and measured in-band noise of the third-order modulator under clock jitter influence as well as the calculated quantization noise. It can be seen that the measured results match well with the calculated. The in-band noise of the ΣΔ modulator is dominated by circuit noise up to a clock jitter of $\sigma_t \approx 0.05\% T_S$, followed by an in-band noise increase induced by clock jitter of 20 dB/dec. [33, 34, 37].

7.4 Implementation Example II: A CT ΣΔ Modulator with SCR-Feedback

The clock jitter measurements of the former CT design example confirm the high sensitivity to any sampling uncertainty. As expected, the major source of error are the feedback DACs. A suitable technique to overcome this lack is the use of pulse-shape feedback DACs as explained in Sect. 4.3.4 and as proposed in, e.g., [136] and thereafter in [19, 31, 33, 130, 200]. Thus, this section presents two implemented ΣΔ modulator prototypes with reduced clock jitter sensitivty by using exponentially shaped DACs. In order to be able to compare the clock jitter measurement results, both modulator types, NRZ and the exponentially shaped (SCR and SCR-I) DAC use the same analog core as shown through Sect. 7.3, merely the feedback DACs are adapted.

7.4.1 SCR-Feedback Implementation

Principally the SCR feedback DAC, as schematically illustrated in Fig. 7.22b, has two modes of operation: first charging the capacitors C_{R_i} to either the positive or negative reference voltage, depending on the actual digital feedback signal V_{out}, and then discharging the capacitors over the resistors R_{R_i} to the input of the integrators. To simplify the system design, these phases were chosen equal to the system clock phases. Additionally, a finite time is required for the internal comparator to have a stable output signal V_{out} and another portion of the sampling time T_S is required to charge the feedback capacitors on either positive or negative reference voltage, depending on V_{out}. Therefore the first half of the clock cycle is used to charge the feedback capacitors, depending on the comparators output, while they are discharged in the second half, translating into a feedback pulse position $\{\alpha = 1/2,\ \beta = 1\}$.

The corresponding feedback circuit is shown in Fig. 7.22b. Here the capacitor C_R is charged on either positive or negative reference voltage when Charge $=$ Clock $V_{\mathrm{out}}(n)$ is high, where $V_{\mathrm{out}}(n)$ is the output bit of the quantizer at the nth sampling instant. It is discharged over R_R to the integrator input when Discharge $= \overline{\text{Clock}}$ is high. This structure has been inserted into a fully differential, third-order CT modulator architecture shown in Fig. 7.22c, where the feedback circuits are as given in Fig. 7.22b.

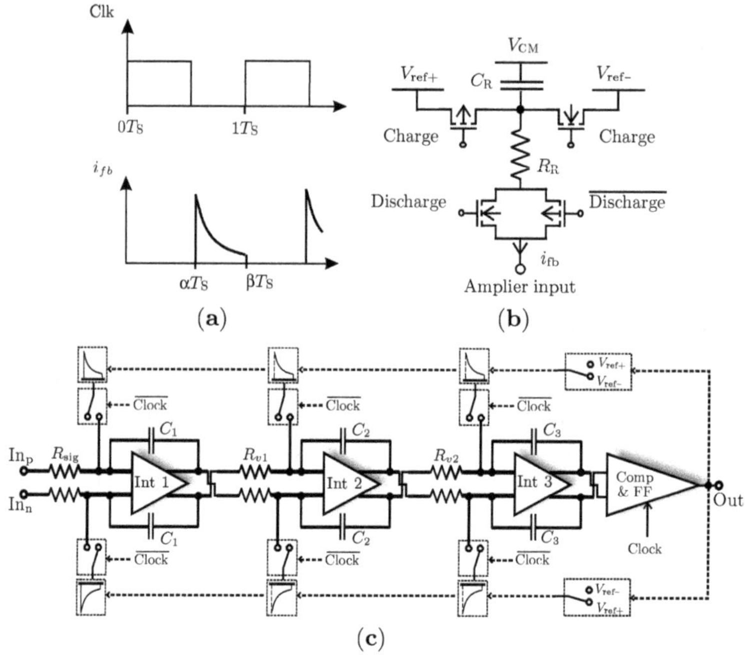

Fig. 7.22. Third-order modulator architecture with SCR feedback, after [31, 33], courtesy of IEEE. (**a**) SCR-pulseform; (**b**) SCR-feedback; (**c**) third-order modulator architecture

7.4.2 SCR Time Constant and Loop Filter Scaling

The basis for the CT SCR-modulator was the previously designed low voltage, low power ΣΔ modulator with NRZ feedback [36]. A calculation of the CT scaling coefficients considering the SCR feedback was exemplarily presented for a second-order modulator in Sect. 3.2.2 and can be similarly obtained for the third-order architecture. However, the CT scaling coefficients k_{SCR_i} are still a function of the SCR-DAC time constant τ_{DAC_i}.

The final step, the choice of the feedback time constants is essential since they mostly determine the clock jitter sensitivity. In Sect. 5.4.4 it was shown that finite gain-bandwidth of the integrators results in a nonideal discharging of the capacitance C_{R} in the feedback SCR-circuit. This leads to an increased feedback time constant and therewith increased jitter sensitivity. Since the intention was to build a very low power modulator, the reduction of the amplifier speed is one design target. Therefore it is not beneficial to choose arbitrarily low feedback time constants τ_{DAC_i}.

Thus, the third-order modulator with common feedback architecture as in Fig. 3.12a has been implemented with SCR DAC and feedback time constants: $\tau_{\mathrm{DAC}_1} = 0.05\,T_{\mathrm{S}}$, $\tau_{\mathrm{DAC}_2} = 0.05\,T_{\mathrm{S}}$, $\tau_{\mathrm{DAC}_3} = 0.22\,T_{\mathrm{S}}$ [33]. The employed SCR

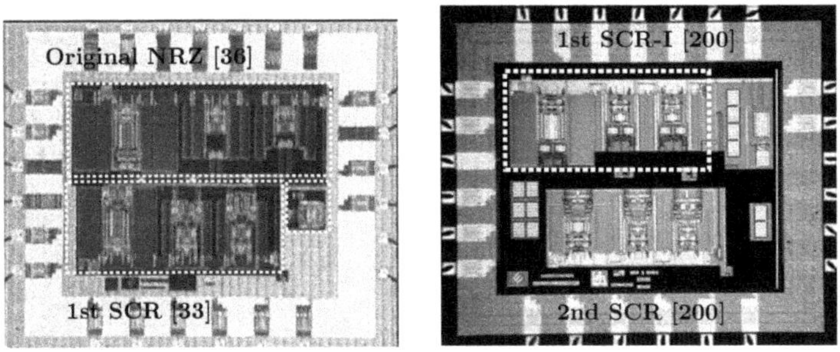

(a) (b)

Fig. 7.23. Chip photos of the SCR implementations, [31], courtesy of IEEE. (**a**) First implementation; (**b**) second implementation

feedback pulse has been positioned in the second half of the clock, i.e., $\alpha = 0.5$, $\beta = 1$, which simplifies the implementation of a unit SCR feedback DAC cell, which is shown in Fig. 7.22b. Simulations regarding the residual clock jitter sensitivity showed an improvement of more than one order of magnitude [33]. The calculation of the equivalent CT loop filter of the original DT third-order loop filter in (7.34) (with the same modification due to stability reasons as was outlined in Sect. 7.3.1), and this time employing the SCR feedback, yields the feedback scaling coefficients:

$$k_{1_{\mathrm{SCR}}} \approx 1, \quad k_{2_{\mathrm{SCR}}} \approx 6, \quad k_{3_{\mathrm{SCR}}} \approx 3.5\,, \tag{7.44}$$

while the forward coefficients were chosen to unity and the signal scaling is calculated to be $k_{\mathrm{sig}} = 0.05$. The values of the passive devices can be calculated using:

$$R_{R_i} = \frac{R_0}{k_{i_{\mathrm{SCR}}}} = \frac{1}{k_{i_{\mathrm{SCR}}} f_{\mathrm{S}} C_i}, \quad C_{R_i} = \frac{\tau_{\mathrm{DAC}_i}}{R_{R_i}}\,, \tag{7.45}$$

where C_i is the integrating capacitor, while C_{R_i} and R_{R_i} are the SCR feedback devices as in Fig. 7.22b.

This third-order modulator was implemented twice in a 3.3 V, 0.5 µm CMOS technology with only 1.5V supply voltage and a minimum power consumption of 150 µW without performance degradation [36]. In the first implementation [33], the same sampling and baseband frequency as in [36] was chosen, $f_{\mathrm{S}} = 2.4\,\mathrm{MHz}$, $f_{\mathrm{B}} = 25\,\mathrm{kHz}$. The second implementation changed the feedback and scaling devices such as to operate at $f_{\mathrm{S}} = 9.6\,\mathrm{MHz}$, $f_{\mathrm{B}} = 100\,\mathrm{kHz}$, but used the same active circuitry with only slightly increased current consumption (≈30%). It was implemented on the same chip as a modified version of the SCR technique presented in [200]. The chip photos of both SCR modulators are shown in Fig. 7.23.

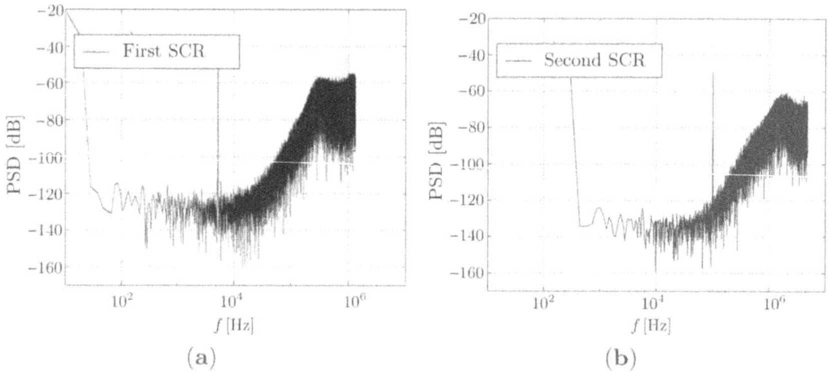

Fig. 7.24. Measured power spectrum of both SCR modulators, after [31], courtesy of IEEE. (**a**) First SCR, $f_S = 2.4\,\text{MHz}$ [33]; (**b**) second SCR, $f_S = 9.6\,\text{MHz}$ [200]

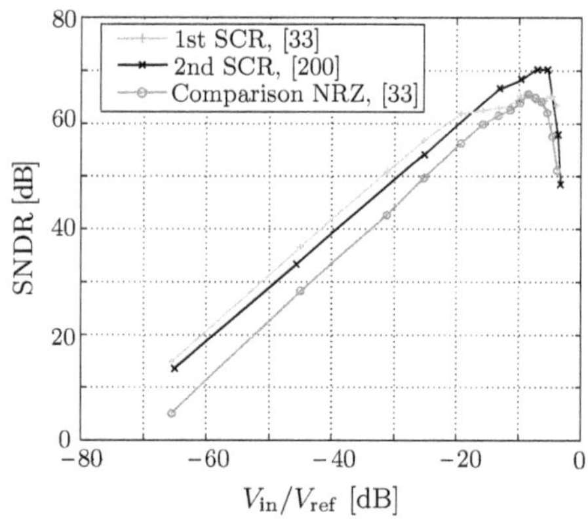

Fig. 7.25. SNDR of the NRZ, the first and the second SCR implementation, after [31], courtesy of IEEE

7.4.3 Experimental Results

In Fig. 7.24 the measured output spectrum is presented. The third-order noise shaping of both modulators and in addition the extended in-band of $f_B = 100\,\text{kHz}$ of the second implementation are clearly seen. The measured signal–noise-distortion ratio in Fig. 7.25 shows the results for both SCR modulators and also the measurements of the previously reported, original NRZ feedback implementation [36]. Since the first SCR modulator was designed with a severe coefficient mismatch in $k_{3_{SCR}}$, its maximum SNR could not be achieved [33].

In contrast, the second SCR implementation shows an almost ideal behavior and gives a large increase of both the maximum SNR and the dynamic range over the NRZ feedback implementation, since it lacks the large feedback resistors and their noise contribution in the NRZ modulator and additionally

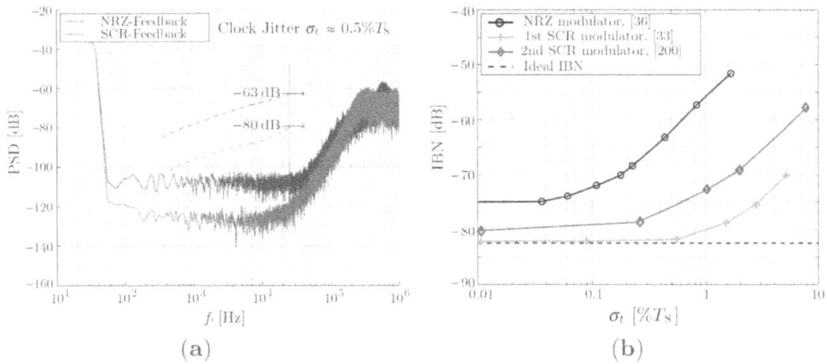

Fig. 7.26. Measured spectrum and IBN of the original NRZ and the SCR modulators under clock jitter influence, after [31] (**a**) measured spectrum; (**b**) measured IBN

it is not prone to intersymbol interference in the feedback. Figure 7.26 shows the measured spectrum and the corresponding integrated IBN for the original NRZ feedback modulator and the first implementation of the SCR modulator [33] under the influence of white clock jitter with $\sigma_t = 0.5\% T_S$. Obviously, the in-band noise floor of the NRZ modulator is severely increased. In contrast, the SCR feedback shows still the ideal in-band noise floor due to circuit noise from Fig. 7.24. In Fig. 7.26b the measured IBN over clock jitter is plotted for the NRZ feedback modulator [36] as well as for the first [33] and the second SCR modulator.

The NRZ modulator starts to decrease from its ideal resolution for $\sigma_t \geq 0.05\% T_S$. In contrast, the SCR feedback modulators show a factor of 10 or even a factor of 20 less clock jitter sensitivity. Compared to the ideal jitter behavior, the increased measured sensitivity of the SCR modulators arises from the finite GBW, low power integrator implementation [33], as discussed in Sect. 5.4.4. The integrators were similarly biased as in [36]. This is also the reason, why the second-SCR modulator with the higher sampling frequency, but with only significantly increased GBW of the integrators, shows an increased jitter sensitivity compared to the first implementation.

7.4.4 CT ΣΔ Modulator with SCR-I-Feedback

In addition to the modulator with SCR-feedback, the exponential feedback DAC was also implemented as current source DAC, i.e., using the SCR-I feedback. As discussed throughout Sect. 5.4.4, using a sloping feedback voltage from a capacitor, which is discharged over a resistor to the input of a CT integrator has some disadvantages concerning the discharge process to the virtual ground node of the OpAmps [124]. It has therefore been proposed to use a current mode feedback, whose control voltage is generated as a decaying

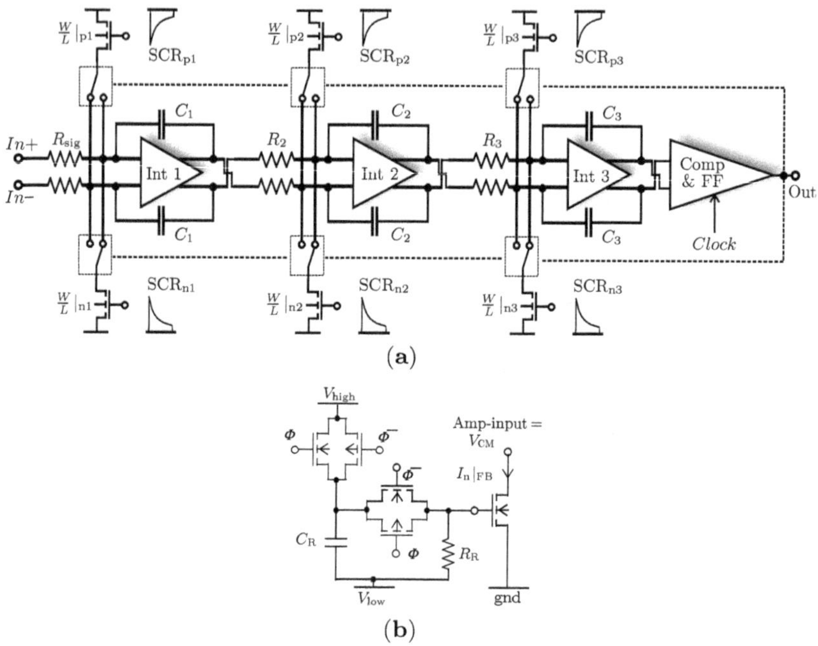

Fig. 7.27. Third-order modulator architecture with SCR-I feedback [200]. (**a**) Fully differential modulator architecture; (**b**) n-SCR-I-circuit

slope [200]. Thus, the feedback signal is independent from the virtual ground node and the jitter sensitivity is not affected by the OpAmp bandwidth. An illustrative schematic of this technique is shown in Fig. 7.27.

From there it is obvious that the feedback signals ideally are the same as for the pure SCR-feedback in Fig. 7.22b. Thus, also the DT-to-CT conversion can be performed the same way. The main difference is the generation of the feedback signal itself, i.e., the incorporation of a controlled current source on the transistor level as ideal as possible. To simplify matters, simple MOS-transistors have been used in the first implementation, whose output current ideally follows the input in a quadratic manner, until it is switched off at the transistor threshold. The implemented third-order CT modulator with SCR-I feedback is illustrated in Fig. 7.27a: The feedback current signals are generated by transistor current sources, while the control voltages, which steer the transistor gates, are generated as decaying slopes. To be able to compare the results to the SCR and NRZ feedback modulators [33, 34], exponential slopes were chosen. These are obtained through a capacitor–resistor arrangement, which performs similar as the pure SCR-feedback circuit above. In the first half of the clock period (Φ), the capacitor C_{R} in Fig. 7.27b is charged to a upper voltage V_{high}, which defines the feedback current peak. In the second half ($\overline{\Phi}$) of the clock, the capacitor is exponentially discharged to a lower reference voltage V_{low} with a time constant defined by $\tau = R_{\mathrm{R}}C_{\mathrm{R}}$ of the feedback current. Thus by choosing the correct values of V_{high} and V_{low} of the feedback current, together with the appropriate slope an equivalent feedback signal is obtained as for an ideal pure SCR-circuit.

The measurement results as well as implementation details were presented in [200]. The ideal performance could not be achieved, first of all due to the feedback current source noise, and additionally due to the nonideal voltage controlled current source, which was only a MOS transistor. Nonetheless, a more sophisticated implementation could obviously improve the jitter performance as well as the usage of other than the exponentially decaying slope.

7.5 Implementation Example III: A 12-Bit 2 MHz 1 V CT ΣΔ Modulator

The use of spread spectrum techniques in communication protocols such as UMTS or W-CDMA requires high-speed baseband circuits (few MHz) with moderate dynamic range (typically 12–13-bit). A basic building block of receivers is an analog-to-digital converter [257]. Therefore, in [36] an architecture of a wideband, medium resolution CT ΣΔ modulator was proposed.

7.5.1 Modulator Architecture

In order to combine 12-bit resolution and low-power consumption with high speed requirements, the used oversampling ratio should be as small as possible. Thereby, the integrator specifications and consequently the total power drain remain low [63]. The first objective is now to find a power-efficient modulator implementation for the desired dynamic range of 12-bit within the signal bandwidth of 2 MHz. According to Sect. 7.1.1, the best FOM is obtained by a second and third-order single-bit topology, but still at very high oversampling ratios of >100 and 50. Comparable power efficiencies are shown by the second and third-order structure with a multibit internal quantizer.

The use of a B_{int}-bit ADC as the internal quantizer offers beside a better loop stability (additional performance gain due to higher noise attenuation [36, 63]) also the advantage that the in-band quantization noise is a factor of $(2^{B_{int}} - 1)^2$ lower in comparison with a simple two level comparator. Thus, the better loop stability improves the quantization noise by 24 dB and additionally a 9-level ADC by another 17 dB. Therefore, an overall quantization noise improvement of 41 dB is achieved in comparison with the single-bit third-order modulator presented in Sect. 7.3. This lowers the demanded oversampling ratio for the same 12-bit dynamic range performance from 48 to 12, which is obviously much more feasible for the 2 MHz baseband. Besides, by using a small negative feedback γ around pairs of integrators (as shown in Fig. 7.28) it is possible to move the NTF zeros away from dc. As a result, the noise attenuation is spread over the baseband, and thus the in-band quantization noise and the oversampling ratio is further reduced, see Sect. 2.17. In this case,

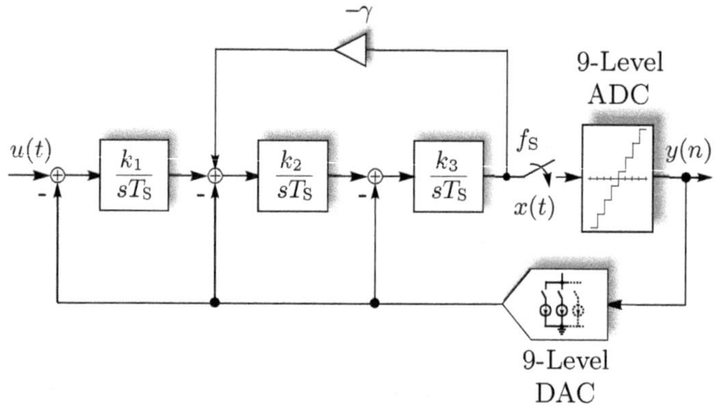

Fig. 7.28. Implemented 9-level third-order modulator architecture [36]

for the third-order modulator an additional noise attenuation of roughly 8 dB is obtained [4].

Indeed, these considerable performance enhancements still raise the circuit complexity of the ADC and DAC. In this context, circuit nonidealities due to the ADC are again substantially suppressed by the modulator loop gain. In contrast, the digital–analog conversion (DAC) errors add directly to the modulator input [4]. These DAC errors appear at the modulator output as additional noise and distortion. This requires the implementation of additional hardware, which attenuates or eliminates these errors. Accordingly, a careful circuit design and a DAC linearisation technique are required.

7.5.2 Loop Filter

The originally used third-order DT loop filter consisting of a chain of three DT integrators $I(z) = 1/(z-1)$ is:

$$\mathrm{LF}(z) = -\frac{a_3(z^2 + (a_2-2)z + a_2(a_1-1)+1)}{(z-1)((z-1)^2 + a_2 a_3 \gamma)}, \tag{7.46}$$

where a_i denote again the DT scaling factors and γ controls the placement of the complex loop filter poles. In order to achieve a dynamic range of more than 12-bit, the OSR must be at least 10, when the scaling coefficients are chosen as $\{a_1, a_2, a_3, \gamma\} = \{0.5, 0.75, 2.25, 0.035/(a_2 a_3)\}$ [4, 240]. Accordingly, this results in a sampling frequency of $f_S = 40\,\mathrm{MHz}$ for the desired 2 MHz passband.

Determined by the overall receiver system, multiples of 26 MHz were available as sampling frequency. Thus, the next sampling frequency of 52 MHz was chosen, leading to an oversampling ratio of 13. By employing again an NRZ feedback DAC, the corresponding DT-to-CT transformation of the DT loop filter in (7.46) yields the following CT scaling coefficients: $\{k_1, k_2, k_3, \gamma\} = \{1, 0.5,\ 1.64,\ 0.035/(k_2 k_3)\}$, if the CT architecture is build as in Fig. 7.28. The implemented integrator resistors and capacitors yield:

$$C_1 = 1\,\mathrm{pF}\,, \qquad C_2 = 1.6\,\mathrm{pF}\,, \qquad C_3 = 0.68\,\mathrm{pF}\,, \tag{7.47}$$

$$R_1 = 16\,\mathrm{k\Omega}\,, \qquad R_2 = 24\,\mathrm{k\Omega}\,, \qquad R_3 = 19\,\mathrm{k\Omega}\,. \tag{7.48}$$

The resulting dynamic performance of the DT and CT third-order modulator with spread zeros is presented in Figs. 7.29 and 7.30a. Evidently, the spectrum shows a noise increase of 20 dB per decade up to the notch located at $\sqrt{a_2 a_3 \gamma} f_S/(2\pi) = 1.54\,\mathrm{MHz}$. This is followed by a noise increase of 60 dB per decade. The accumulated error power results in IBN = −86 dB.

The simulated SNDR as a function of the input signal exhibits a peak SNDR of 81 dB. The dynamic range exhibits also 76 dB, which suffers from the increased in-band noise at very low input amplitudes, Fig. 7.30b. The increased in-band noise for small input signal amplitudes is caused by the zero transfer gain of the 9-level ADC at the midpoint. Nevertheless, both SNDR curves approach each other for medium input amplitudes. Thus, the slope of the SNDR curve for the 9-level mid-tread third-order ΣΔ modulator is larger than one. The major benefit of using the 9-level mid-tread ADC is the extended stable input range and higher peak SNDR (Fig. 7.30b), which are in this case −5 dB and 80 dB, respectively. This is due to the fact that in ΣΔ modulators employing a mid-tread ADC the maximum feedback signal is FS/2, while in mid-rise ADCs the maximum level is $\mathrm{FS}/2 - \Delta/2$. As a result, ΣΔ modulators using a mid-tread ADC operate stable for larger input level. If the resolution of the internal ADC is increased to 4- or 5-bit, this effect alleviates.

Influence of Circuit Nonidealities

The influence of a number of circuit nonidealities was analytically determined and proved by behavioral simulations in order to minimize their effect on the

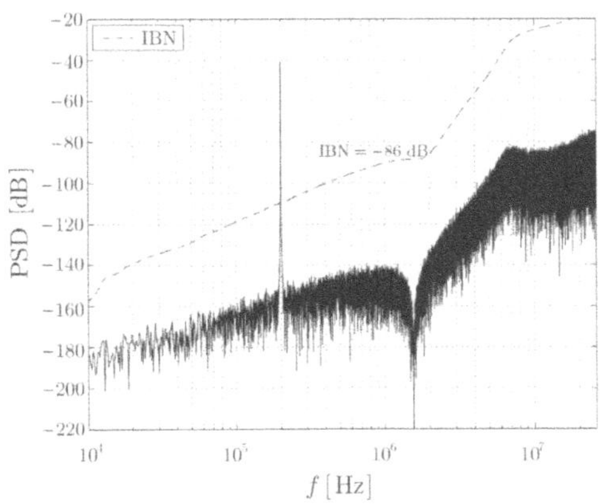

Fig. 7.29. Simulated spectrum and accumulated noise of the CT third-order modulator using a −15 dB, 200 kHz input signal and 2^{18}-points [36]

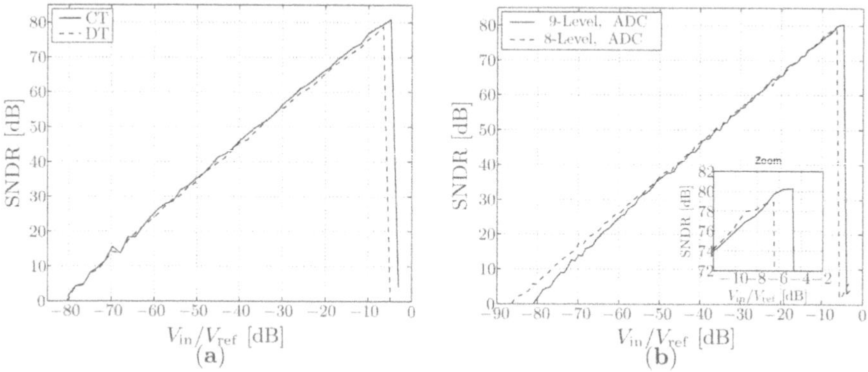

Fig. 7.30. (**a**) Simulated SNDR as a function of applied input level for a DT and CT loop filter; (**b**) simulated SNDR as a function of applied input level for a 8-level mid-rise and 9-level mid-tread internal ADC

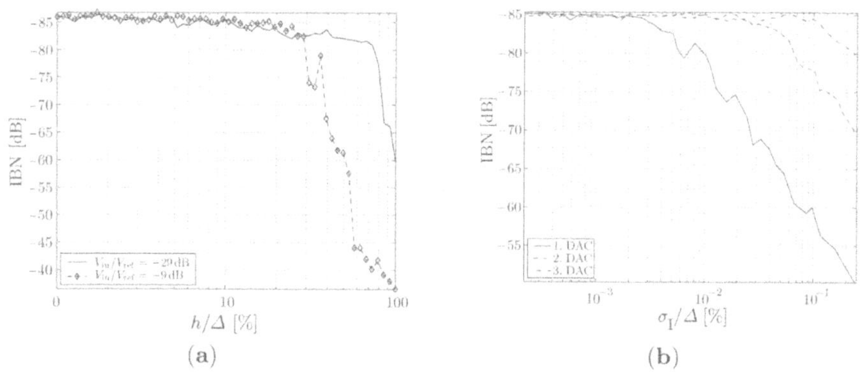

Fig. 7.31. The effect of (**a**) comparator hysteresis and (**b**) nonlinearity in the DACs on the in-band noise

total modulator performance, particularly the effect of finite gain-bandwidth of the used amplifiers, time constant variations, excess loop-delay τ_d, hysteresis h in the comparators and the nonlinearity of the used DACs σ_I. Behavioral simulations show that the amplifier dc gain and gain-bandwidth requirements are moderate, which is in agreement with Chap. 5. To preserve the overall resolution, a gain-bandwidth product equal to the sampling frequency f_S is sufficient. However, to avoid any problems due to parasitic capacitances, process variations or temperature effects a GBW of $2f_S$ was implemented.

Hysteresis h in the comparator causes additional noise, which is fortunately shaped in the same way as quantization noise. Figure 7.31a shows a circuit level simulation, where a hysteresis as high as $h \approx 0.2$ LSB can be tolerated without significant performance degradation, which confirms the discussion in

Chap. 6. Additionally, the effect of offset in the comparators is not a serious design issue. Nevertheless, in high resolution implementations (>16-bit), both effects become more and more critical [4, 63, 231].

In contrast, nonlinearity errors in the feedback DACs are the most demanding design issue. To evaluate the linearity requirements of the three DACs, a Monte Carlo simulation with 1000 bins was performed, whose results are shown in Fig. 7.31b. For the DAC in the outermost feedback branch, a $\sigma_{\mathrm{I}}/\mathrm{LSB} \leq 3 \times 10^{-5}$ is required for the simulated 14-bit dynamic range, while the second and third feedback DAC only demand a $\sigma_{\mathrm{I}}/\mathrm{LSB} \leq 3 \times 10^{-4}$ and 1×10^{-3}, respectively. Accordingly, this shows clearly that the non-idealities entering at the input of the second and third integrator are considerably suppressed by the gain of the preceding integrators, whereas the linearity demands on the first DAC are as high as on the total modulator. Since, the desired dynamic range is merely 12-bit (Sect. 7.5), the demanded linearity can be reduced to approximately $\sigma_{\mathrm{I}}/\mathrm{LSB} \leq 1.5 \times 10^{-4}$. To meet this design constraint, some resourceful design techniques in combination with sophisticated switching schemes and layout techniques have to be adapted. As shown in [67, 258, 259], it is possible to implement current source (CS) DACs with a linearity up to 14-bit [259] without using any trimming or calibration techniques. Additionally, linearization techniques are frequently used as was discussed in Sect. 4.5.

7.5.3 Circuit Implementation

Integrator Design

The power dissipation of the first integrator is the major contributor to the overall power dissipation in ΣΔ modulators, therefore a substantial quantity of power can be saved by a proper circuit design. Single-stage amplifiers, employing a cascode configuration to enhance the dc gain, are very power-efficient, since no frequency compensation as well as no additional current branches are required (Sect. 7.3). But the very low supply voltage of only 1 V raises the problem of the limited output swing (OS). As a consequence, the circuit noise has to be reduced to maintain a constant signal-to-noise ratio, since $\mathrm{SNR} \propto \mathrm{OS}^2$.

In order to fulfill the high output swing requirements, a fully differential two-stage amplifier has been chosen, see Fig. 7.32. The input common mode voltage is set to V_{CM}=300 mV leading to a peak-to-peak OS ≈ 400 mV. The currents in the input stage are chosen similar to the former cascode design, $2I_{\mathrm{M5}} = I_{\mathrm{M1}} = I_{\mathrm{M7}}$. A short performance summary of the amplifier as used in the first integrator, is represented in Table 7.3. The transistor device sizes are chosen in terms of a low power consumption, high output swing as well as a low noise contribution. The input referred noise spectral density of the amplifier yields:

$$\overline{v}^2_{\mathrm{n,in}} \approx 2\left[\overline{v}^2_{\mathrm{n},M3} + \left(\frac{gm5}{gm3}\right)^2 \overline{v}^2_{\mathrm{n},M5} + \left(\frac{gm7}{gm3}\right)^2 \overline{v}^2_{\mathrm{n},M7}\right]. \tag{7.49}$$

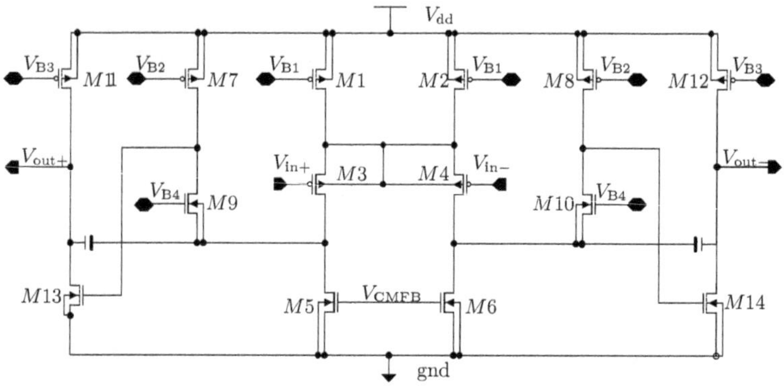

Fig. 7.32. Schematic of the employed amplifier in the first integrator [188]

Table 7.3. The amplifier performance summary

$C_L = 1\text{pF}\|\|24\,\text{k}\Omega$	
GBW	105 MHz
A_V	90 dB
φ_{M}	45°
$\bar{v}_{\text{in,rms}}$ (in-band)	29.5 µV
$\bar{v}_{\text{n},1/f}$ *at* 1 kHz	116 nV/$\sqrt{\text{Hz}}$
$\bar{v}_{n,\text{thermal}} at 1$ MHz	19 nV/$\sqrt{\text{Hz}}$
$\sum I = \sum I_{\text{OTA}} + \sum I_{\text{CMFB}}$	800 µA + 300 µA

When substituting the corresponding spectral density noise sources $\overline{v}^2_{\text{n},Mi}$ by means of both the thermal and $1/f$ noise contribution, the total input referred noise spectral density of the amplifier is obtained. According to the two-stage amplifier design, the common mode feedback (CMFB) circuit requires a sign inversion, making an additional error amplifier necessary. Hence, the overall power consumption increases by 37%, while the common mode stability and the linearity are improved [253].

The applied common mode feedback circuit consists of a resistive summing circuit and a PMOS differential pair. The overall power consumption of the first integrator is 1.2 mW. The three integrators were scaled corresponding to their capacitive load. A simplified schematic of the differential active RC-integrator with excess phase cancellation and common mode level shift is shown in Fig. 7.33. The small resistor R_Z moves the parasitic right half-plane zero resulting from the input transconductance gm and the integration capacitance C toward infinity.

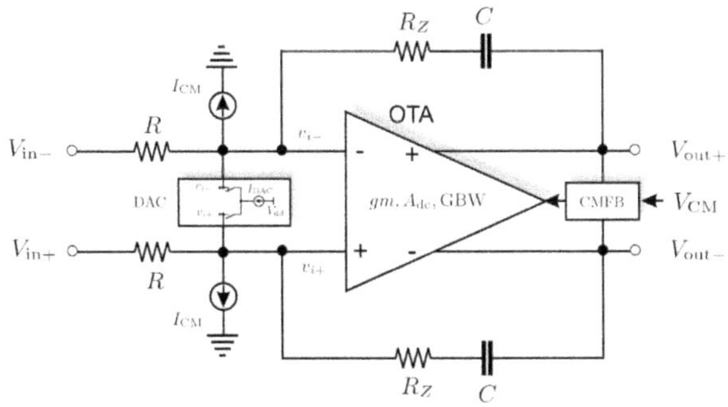

Fig. 7.33. Simplified schematic of the applied fully differential integrator where R_Z and I_{CM} perform excess phase cancellation and DC level shift, respectively

Because of the very low supply voltage of 1 V and the resulting low common mode voltage of 300 mV, only a unidirectional 3-bit DAC is practicable. The main reason for this is that a linear multibit current source DAC requires a very large output impedance [260], in order to limit harmonic distortion caused by data dependent output impedance variations. However, the very low common mode level does not permit to implement a cascode current source to increase the output impedance to the required value (see Sect. 7.5.3). By using a current shift of the half quantity of full scale, the implemented DAC is virtually bidirectional. Thus, the additional current sources I_{CM} in Fig. 7.33 conduct four times the DAC unit current I_{DAC} to ground, with the unidirectional DAC current delivered to the virtual ground nodes (v_{i+}, v_{i-}), shifted from $0.8 I_{DAC}$ to $\pm 4 I_{DAC}$. Fortunately, any mismatch in the shift current sources $I_{CM} \pm \Delta I$ affects the DAC transfer characteristic only as an offset, with almost negligible consequences on the overall ΣΔ modulator.

3-Bit 9-Level ADC

The overall architecture of the internal 9-level ADC is shown in Fig. 7.34. Since the converter must work at high sampling rates, a flash topology was chosen. Furthermore, this building block has also been implemented in a fully differential configuration. The used fully differential comparator with output buffer is illustrated in Fig. 7.35. The offset voltage of the comparator is given by:

$$\sigma^2_{\mathrm{offset}} = 4\sigma^2_{M1} + 2\left(\frac{gm3}{gm1}\right)^2 \sigma^2_{M3} + 2\left(\frac{gm7}{gm1}\right)^2 \sigma^2_{M7} \tag{7.50}$$

$$= 4\frac{A^2_{V_{T_p}}}{W_1 L_1}\left[1 + \frac{1}{2}\left(\frac{L_1}{L_3}\right)^2 + \frac{1}{2}\left(\frac{A_{V_{T_n}}}{A_{V_{T_p}}}\frac{L_1}{L_7}\right)^2\right] \tag{7.51}$$

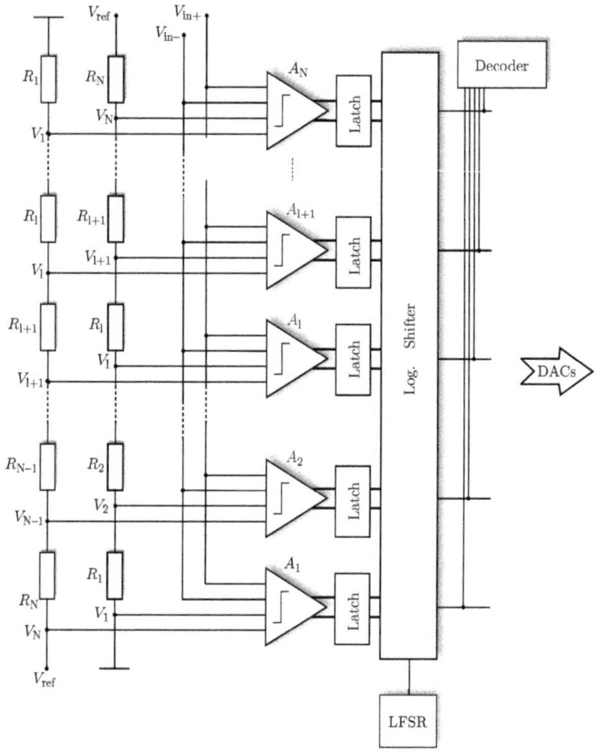

Fig. 7.34. Internal 9-level ADC

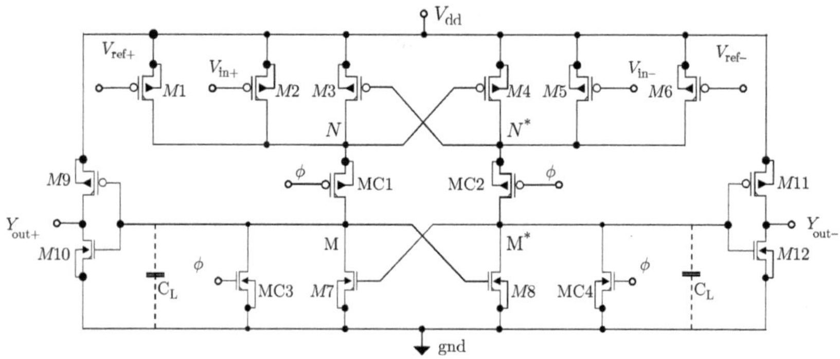

Fig. 7.35. Fully differential comparator

if only V_T mismatch is considered [50]. With the chosen comparator device size the offset voltage is <1 mV.

The nonidealities of the comparators in a flash ADC introduce substantial errors in the reference voltages resulting from the capacitive feed through at the input of each comparator (gate–source capacitance C_{gs}) and the resistor ladder ($R_1, \ldots, R_N$) [232]. Thus, capacitive feed trough over the gate–source capacitance will result in deterioration of the reference voltages. From an upper bound of the tolerable reference voltage shift, the maximum ladder impedance can be calculated. By using a simplified model [232], the maximum total reference ladder impedance R is given by:

$$R = \frac{4\varphi_{\mathrm{max}}}{\pi 2^{B_{\mathrm{int}}} f_{\mathrm{S}} C}, \tag{7.52}$$

where φ_{max} denotes the maximal allowed feed through in volt, B_{int} is the number of bits and C the total coupling capacitance between the input and the ladder.

It is evident form (7.52) that this parasitic effect can be minimized for a fixed sampling frequency by lowering the RC-time constant. On the other hand, the comparator input transistors ($M1$, $M2$ and $M5$, $M6$) have been sized for achieving the offset requirements, (7.50). Therefore, the gate–source capacitance is fixed. If the resistor value is now reduced, the signal feed through will be minimized at the expense of a higher power consumption, caused by the resistor ladder. The implemented unit resistance was chosen to be 1 kΩ. To avoid any coupling between the two differential inputs, two separate reference ladders were used [37, 261].

The requirements on the linearity of the resistor ladder regarding a linear gradient (i.e., a linear variation in doping or width from one end to the other [262]) and nonlinearity resulting from random resistor mismatch [263] are given in (7.53) and (7.54):

$$\mathrm{INL}_{\mathrm{max,grad}} \approx 2^{B_{\mathrm{int}}} V_{\mathrm{ref}} \frac{\Delta R}{8R} \approx V_{\mathrm{ref}} \frac{\Delta R}{R}, \tag{7.53}$$

$$\mathrm{INL}_{\mathrm{max,rand}} \approx \frac{\Delta R}{R\sqrt{2^{B_{\mathrm{int}}+2}}} V_{\mathrm{ref}}. \tag{7.54}$$

The ADC reference voltage was set to $V_{\mathrm{ref}} = 600\,\mathrm{mV}$, leading to a static power drain of the reference ladder of $2 \times 150\,\mu\mathrm{W}$, which is 10% of the total modulator power consumption!

3-Bit 9-Level DAC with 12-Bit Linearity

As shown in Sect. 7.5.2, the requirements on the D/A converter are high, especially in terms of linearity, because the DAC must have at least the overall resolution of the ΣΔ modulator. A current-steering thermometer decoded DAC architecture was employed, because the glitches are low (especially at midcode transition), they are intrinsically fast and can be designed in a standard digital CMOS technology. The goal is to scale the current sources (CS) in order to fulfill a resolution of 12-bit with a supply voltage of 1 V. Therefore, several mismatch effects have to be taken into account: random device

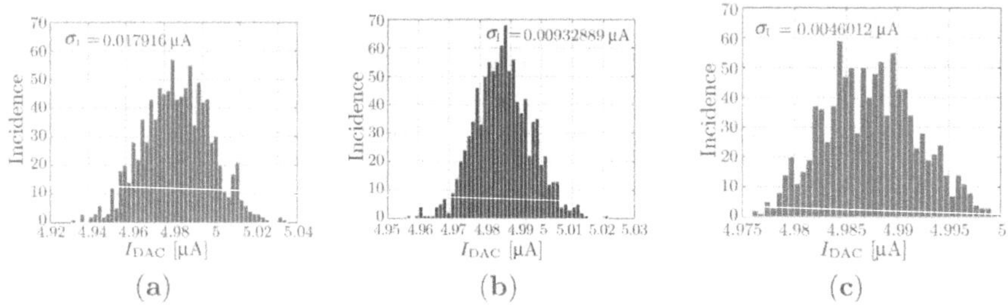

Fig. 7.36. Monte Carlo simulation results for a gate overdrive voltage ($V_{\mathrm{GS}} - V_{\mathrm{T}}$) of 500 mV and different WL transistor dimensions: (**a**) $10 \times 15\,\mu\mathrm{m}^2$, (**b**) $20 \times 30\,\mu\mathrm{m}^2$, (**c**) $40 \times 60\,\mu\mathrm{m}^2$

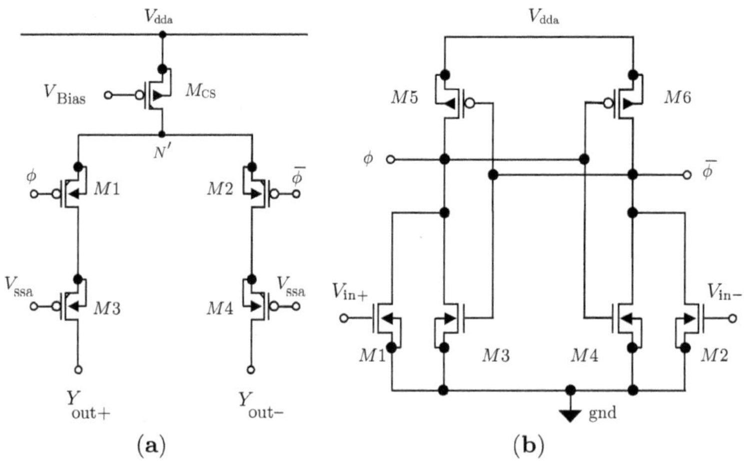

Fig. 7.37. Implemented unit current cell and corresponding deglitch driver. (**a**) DAC unit; (**b**) DAC unit driver

mismatch, systematic errors like data dependent output impedance, gradient errors like doping and oxide thickness variations, voltage drops in the supply lines etc. [67, 259, 264, 265]. To match the linearity requirements concerning the random error, the model proposed in [246], was applied:

$$\left(\frac{\sigma_I}{I}\right)^2 = \frac{A_\beta^2}{WL} + \frac{4A_{V_\mathrm{T}}^2}{WL(V_{\mathrm{GS}} - V_{\mathrm{T}})^2}, \tag{7.55}$$

$$W^2 = \frac{I}{2\,\beta_p\left(\frac{\sigma_I}{I}\right)^2}\left[\frac{A_\beta^2}{(V_{\mathrm{GS}} - V_{\mathrm{T}})^2} + \frac{4A_{V_\mathrm{T}}^2}{(V_{\mathrm{GS}} - V_{\mathrm{T}})^4}\right], \tag{7.56}$$

$$L^2 = \frac{\beta_p}{2\,I\left(\frac{\sigma_I}{I}\right)^2}\left[A_\beta^2(V_{\mathrm{GS}} - V_{\mathrm{T}})^2 + 4A_{V_\mathrm{T}}^2\right]. \tag{7.57}$$

Indeed, the threshold voltage mismatch dominates in (7.55). Thus, a high gate overdrive voltage $V_{GS} - V_T$ reduces the area, consumed by the CS array. However, the value is limited by the low supply voltage. A trade-off has to be made. A gate overdrive voltage of 500 mV was chosen for the Monte Carlo simulations shown in Fig. 7.36a–c with respect to different transistor gate areas. Obviously, for every increase of the current source transistor area by a factor of four, the matching improves by one bit. Finally, Fig. 7.36c shows a linearity of 0.092% which matches very well with the calculated one 0.079% for the used 0.18 µm CMOS technology ($A_{\beta_n} = 0.91\%\,\mu m$, $A_{V_{T_n}} = 6.8\,mV\,\mu m$ and $A_{\beta_p} = 1.2\%\,\mu m$, $A_{V_{T_p}} = 5.5\,mV\,\mu m$). A yield estimation of the 9-level DAC featuring a 12-bit linearity was performed, similar as proposed in [266]. It was found that for a yield of >99.7% the demanded current source accuracy is given by $\sigma_I/I = 0.02\%$.

A high ratio between the CS output impedance Z_{out} and the load impedance (here approximately $Z_L \approx 1/gm$, where gm is the transconductance of the respective integrator) reduces significantly the systematic errors and accordingly the INL and the SFDR. The relation between the INL and the output impedance is given by [262]:

$$\mathrm{INL} = \frac{I_{unit} Z_L^2 2^{2B_{int}}}{4 Z_{out}} \tag{7.58}$$

with I_{unit} the LSB current. Therefore, a high CS output impedance is required, which must be maintained up to f_S. This can only be accomplished by means of a cascode configuration [67, 259, 265, 267]. Due to the low supply voltage, no headroom for an additional transistor is available, thus the current source switches ($M1$, $M2$) are biased in saturation to form a high output impedance cascode configuration as shown in Fig. 7.37a:

$$Z_{out} = gm1\, r_{ds1}\, r_{ds_{CS}} \,. \tag{7.59}$$

Due to the large value of $L = 60\,\mu m$ of the unit CS transistor M_{CS}, the output impedance Z_{out} of the cascode configuration of the unit current source and the current switch transistor is sufficiently high to avoid any impact on the linearity. The used deglitch driver (Fig. 7.37b) in front of the DAC unit cell synchronizes the control signals, reduces the drain voltage fluctuations and defines the crossing point of the control signals, since voltage fluctuations at the node N' in Fig. 7.37a significantly degrade the dynamic performance (e.g., glitches, settling time) of the whole DAC [259, 267].

The cross-point voltage of the DAC control signals (ϕ, $\overline{\phi}$), is of real importance because its influence on the glitch energy of the DAC is very high [267]. Glitches represent an important dynamic error source: they generate harmonic distortion and thus significantly reduce the DAC spurious free dynamic range. Therefore, the key issue of the driver is to guarantee that when one of the switching transistors begins to turn off (e.g., $M1$), the other transistor (e.g., $M2$) is on the edge of turning on.

Charge feed through to the output lines due to the parasitic gate–drain capacitance of the switching transistors is also a source of glitches generating distortion. A minimization can be performed by means of a minimum control voltage swing [67] (here unsuitable because of the low supply voltages) or instead of isolating the current source outputs ($Y_{\text{out}\pm}$) from the switching transistors ($M1,2$) by means of two cascaded transistors ($M3,4$), as it is shown in Fig. 7.37a.

Random noise of the DAC unit cell is not significant, as the following calculation exhibits [267]. The major source of random noise in CS DACs is the current source transistor (M_{CS}). The mean-square current noise $\overline{i_{\text{n}}^2}$ of a transistor in saturation is given by:

$$\overline{i_{\text{n}}^2} = \left(\frac{8KT\,gm}{3} + \frac{K_{\text{f}} I_{\text{DS}}}{C_{\text{ox}} L^2 f} \right) \Delta f\,, \qquad (7.60)$$

where K is the Boltzmann constant, T is the absolute temperature, $gm = -2I_{\text{DS}}/(V_{\text{GS}} - V_{\text{T}})$ is the transistor small signal transconductance, K_{f} is the flicker-noise coefficient ($K_{\text{f}} = 2 \times 10^{-28}$ FA in the CMOS 0.18 μm technology), and Δf is the output signal bandwidth. For $T = 300$ K, $I_{\text{DS}} = 5\,\mu\text{A}$, $(V_{\text{GS}} - V_{\text{T}}) = 0.5$ V, and $\Delta \text{f} = 2$ MHz the RMS value $\overline{i_{\text{n}}}$ is

$$\overline{i_n} \approx 0.9\,\text{nA}\,. \qquad (7.61)$$

Note, that the $1/f$ noise term contributes only with <0.1 nA, due to the large channel length $L = 60\,\mu\text{m}$ and the large signal bandwidth. The $\overline{i_{\text{n}}}$ value in (7.61) is roughly 5,300 times smaller than the nominal value of the unit current $I_{\text{unit}} = 5\,\mu\text{A}$, resulting in an SNR for a maximum amplitude sinusoidal signal of approximately 92 dB. Therefore, random circuit noise can be totally ignored.

In order to compensate gradient and symmetrical systematic errors, each CS is divided into four sources, each placed in one quadrant and switched for spatial averaging (centroid sequence) as proposed in [67, 264, 265]. The principle is switching the current sources symmetrically placed away from the geometrical point(s) where the error source has its average value. It is important to note that by using the symmetrical switching sequence, one obtains a maximum INL error which is independent of the total number of elements, and depend only on the graded error peak-to-peak amplitude [267]. An even better linearity up to 14-bit is archived by the "Q^2 random walk" scheme [259, 264], however, the required decoder complexity is then higher.

A graphical representation of the used CS array, the double centroid switching scheme as well as dummy CS is given in Fig. 7.38. Two dummy rows and columns have been added to avoid edge effects, increasing the overall DAC area by roughly 111%. The double centroid switching scheme improves the matching but at the expense of increased parasitic capacitances and decoder complexity. To ensure the desired modulator [55, 57]. All of these methods consume additional power and silicon area. Since, the DAC should provide

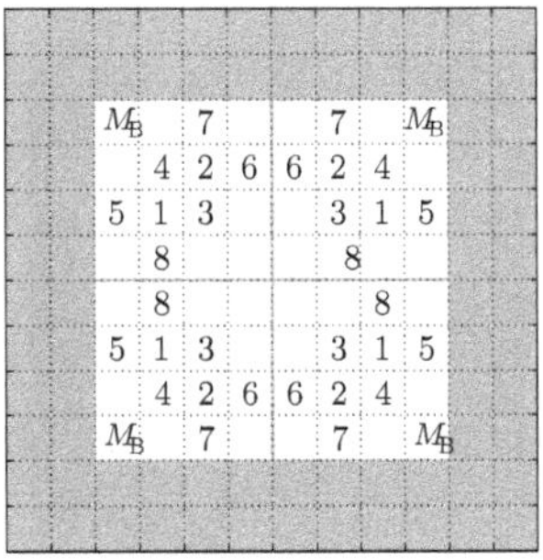

Fig. 7.38. Current source floor plan and the used double centroid switching scheme

the required linearity of 12-bit, an almost simple randomization (area efficient) technique consisting of a logarithmic shifter and a linear shift feedback register (LFSR) has been applied, as shown in Fig. 7.34.

7.5.4 Layout Consideration

The ΣΔ modulator was implemented in a 0.18 μm analog CMOS technology. The required area is less than 1mm^2 with the first integrator consuming more than 45% of the total power consumption and area. The whole chip layout is shown in Fig. 7.39. The employed C_{MIM} interconnection capacitors feature different stray capacitors at each input. Thus, all integration capacitors are laid out in such a way that the stray capacitances of either plate are equal. Thus a symmetrical structure is figured out.

Each integration capacitor was split into four unit capacitors, as shown in Fig. 7.40a. For minimizing even-order distortion, two of them are placed below the amplifier and two above it, as in a common centroid way, surrounded by ten dummy capacitors.

Both input resistors R are also divided into unit elements, placed alternating with dummy devices at the borders. In Fig. 7.40b, the two comparator resistor strings with dummy devices and the eight fully differential comparators including the latches, are shown. The succeeding switching matrix in combination with the LFSR performs the random switching sequence. The implemented thermometer-to-binary coder minimizes the number of required output drivers to four. Additionally, as visible from the chip photo in Fig. 7.39, digital and analog parts are separated to reduce the substrate noise impact.

7.5.5 Experimental Results

Measurements were performed at supply voltages of 1.2 and 1 V. The input common mode voltage was set to $V_{\mathrm{CM}} = 300$ mV, which leads to a maximum

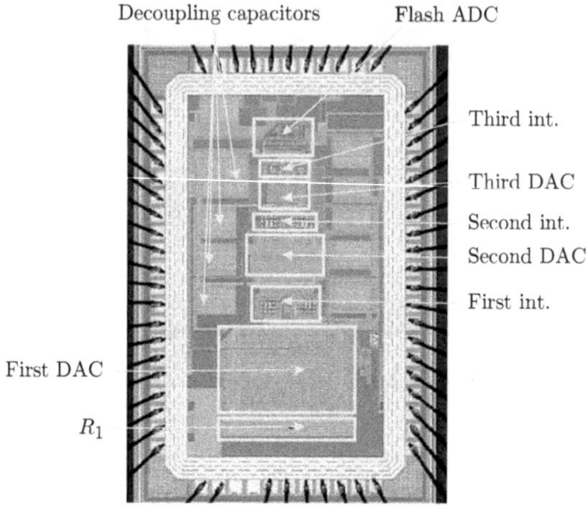

Fig. 7.39. Chip photo of the third-order, 2 MHz ΣΔ modulator

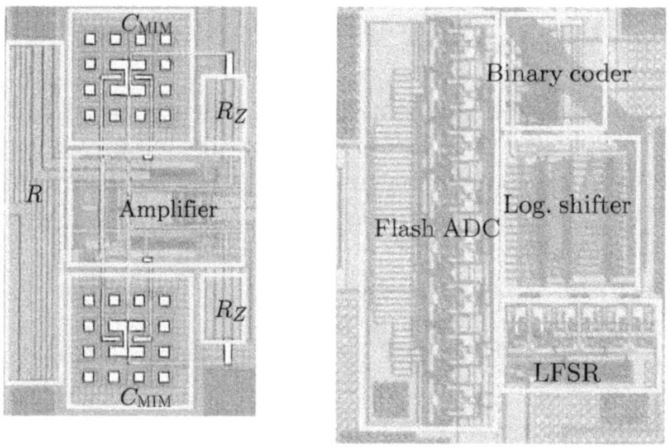

Fig. 7.40. (**a**) Chip photo of the first integrator including the C_{MIM} capacitors and excess phase cancellation resistors R_{Z}. (**b**) Chip photo of the implemented flash ADC including the succeeding LFSR and switching matrix

peak-to-peak input voltage of 400 mV. Furthermore, the used ADC reference voltage is $V_{\mathrm{ref}} = 600$ mV.

The measured power spectrum and accumulated in-band noise of the implemented modulator is presented in Fig. 7.41. It is clearly visible that the obtained PSD features a third-order noise shaping characteristic. Furthermore, the 9-level ΣΔ modulator shows a noise floor of $\approx -135\,\mathrm{dB}$ as well as an accumulated IBN $\approx -75\,\mathrm{dB}$. Figure 7.42 shows the resulting signal-to-noise and signal-to-noise plus distortion ratio vs. the normalized input signal $V_{\mathrm{in}}/V_{\mathrm{ref}}$ for a supply voltage of 1.2 V as well as 1 V. The applied input signal

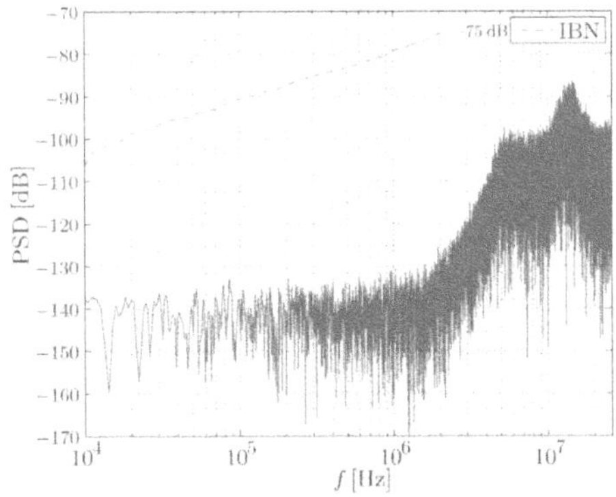

Fig. 7.41. Measured power spectrum of the 9-level third-order ΣΔ modulator at a supply voltage of 1.2 V

frequency was 200 kHz. The measured dynamic range is 75 dB and 72 dB; the peak SNR is 68 dB and 62 dB, respectively. Finally, the measured peak SNDR is 67 dB and 61 dB Table 7.4.

In the following, a short discussion concerning the theoretically achievable and the experimentally obtained results is presented. The dynamic range and SNR measurements performed at a supply voltage of 1.2 V (Figs. 7.41 and 7.42a) feature a 5 dB increased noise floor in comparison with the ideal values illustrated in Fig. 7.29. The maximum stable input amplitude of −5 dB fits very well with the expected value. However, if the supply voltage is further reduced to 1 V, additional performance degradations occur. Apparent in

Table 7.4. Performance summary of the 9-level third-order ΣΔ modulator

Signal bandwidth	2 MHz	
Sampling frequency	52 MHz	
Technology	0.18 μm, 1.8 V CMOS, $V_{T_n} = 0.48$ V, $V_{T_p} = 0.51$ V	
Core area	1 mm^2 CMOS	
$V_{in_{pp}}$	0.4 V	
Supply voltage	1.2 V	1 V
Dynamic range	75 dB	72 dB
Peak SNDR	67 dB	61 dB
Power consumption	<3.1 mW	<2.8 mW
FOM	0.17 pJ/conv.	0.21 pJ/conv.

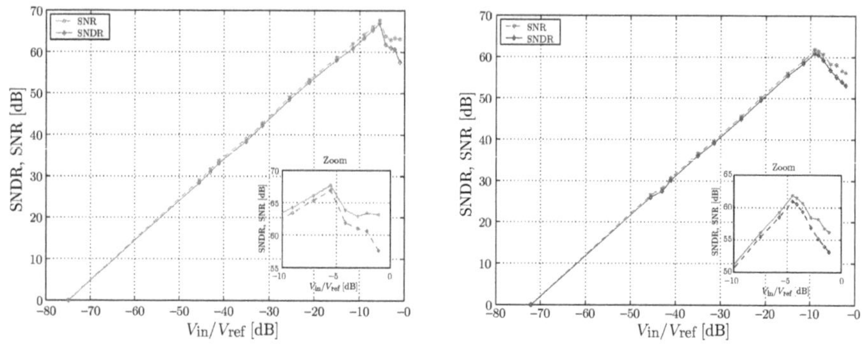

(**a**) SNR/SNDR at 1.2 V (**b**) SNR/SNDR at 1 V

Fig. 7.42. Measured dynamic modulator performance of the 9-level third-order ΣΔ modulator using a supply voltage of 1.2 and 1 V

Fig. 7.42b is the increased noise floor as well as the reduced maximum stable input amplitude. These performance losses are mainly caused by the reduced (−17%) supply voltage, which drastically increases the excess loop delay in the modulator, since the operation speed (e.g., propagation delay) of the digital blocks is inversely proportional to V_{dd}^2. As described in Sect. 4.2, excess loop delay causes performance degradations on both, the dynamic range and the maximal stable input amplitude, which is especially seen in NRZ implementations, if no compensation approach is used.

7.6 Implementation Example IV: A 2–1–1 Cascaded CT ΣΔ Modulator

In Sect. 3.2.4 different cascaded CT ΣΔ modulator architectures have been derived, e.g., the fourth-order, 2–1–1 structure. The corresponding system is given in Fig. 3.8b. Its optimal CT scaling coefficients were presented in Table 3.4, which resulted from the DT-to-CT transformation. The 2–1–1 architecture is popularly implemented in the DT domain due to its high order, almost ideal noise-shaping. Nonetheless, it is well known to be very sensitive to circuit nonidealities [190]. This could be seen especially in Sect. 5.3 as well as Fig. 5.3b and (5.15), where integrator gain errors of less than 1% remarkably decreased the ideal modulator performance. In the CT domain, integrator gain variations due to RC mismatch of more than 20% are typically seen. Thus, the 2–1–1 modulator is a predestinated example to be used for the demonstration of the possible gain-error correction presented above. There it has been shown that by digitally multiplying correction factors Corr_i to the output bitstreams of the different stages, an almost ideal behavior can be obtained with regard to integrator gain variations. The condition to be fulfilled in the case of the 2–1–1 modulator can be found in (5.19).

7.6.1 Circuit Realization of the Cascaded Modulator

To be able to measure the influence of integrator gain errors and thus to see the influence of a variation of several passive circuit components over a wide range, the modulator in Fig. 3.8b has been implemented on a circuit board [97]. This enables the exchange of the integrator capacitors and resistors and thus a controllable gain error. To account for the limited speed of discrete electronic components as the comparators and switches, as well as the large parasitic capacitors and the lack of low integrator capacitors, the modulator has been realized with a sampling frequency of $f_S = 1\,\mathrm{MHz}$. The oversampling ratio is chosen to OSR = 24 and the feedback reference voltage $V_{\mathrm{ref}} = \Delta/2 = \pm 1\,\mathrm{V}$. The corresponding baseband frequency yields $f_B = 20.8\,\mathrm{kHz}$. Following (2.35), this yields an ideal IBN of:

$$\mathrm{IBN} = \frac{\pi^8}{9}\frac{\Delta^2}{12}\frac{1}{c_1^2 c_2^2 \mathrm{OSR}^9} \approx -92\,\mathrm{dB}\,, \qquad (7.62)$$

where $c_1 = 1/2$, $c_2 = 1$ are the interstage coupling coefficients [6]. Considering the feedback waveform of the CT implementation, a RZ DAC is very advantageous, because it gives the internal ADC time to settle, before the value is fed back to the integrators [132, 192]. Additionally, nonequal rise and fall times of the feedback DAC are of much less importance, if an RZ pulse form is chosen [3]. Consequently, the relative feedback pulse position has been chosen to be $\{\alpha = 0,\ \beta = 1/2\}$. The realized electronic circuit is shown in Fig. 7.43. Here, every CT integrator is implemented as an inverting RC-integrator and an inverting voltage follower. The comparator is latched by the system clock and its output fed to the feedback DAC. The feedback DAC consists of three multiplexers, where the first selects the actual feedback state, the second generates the RZ pulse while the third avoids glitches in the feedback signal. The delayed clock of the second multiplexer allows the comparator to settle before the new feedback signal is generated. This delayed clock has been obtained from the delay time of two comparators.

The passive capacitors could be chosen with an absolute accuracy of 2%. This intrinsic relative component mismatch suggests that the ideal performance will rather not be achieved. The resulting circuit board was shown in [32, 97].

7.6.2 Measured Ideal Modulator Performance

The circuit shown in Fig. 7.43 has been measured concerning its general performance. Therefore the outputs $y_i(n)$ of the modulator have been recorded

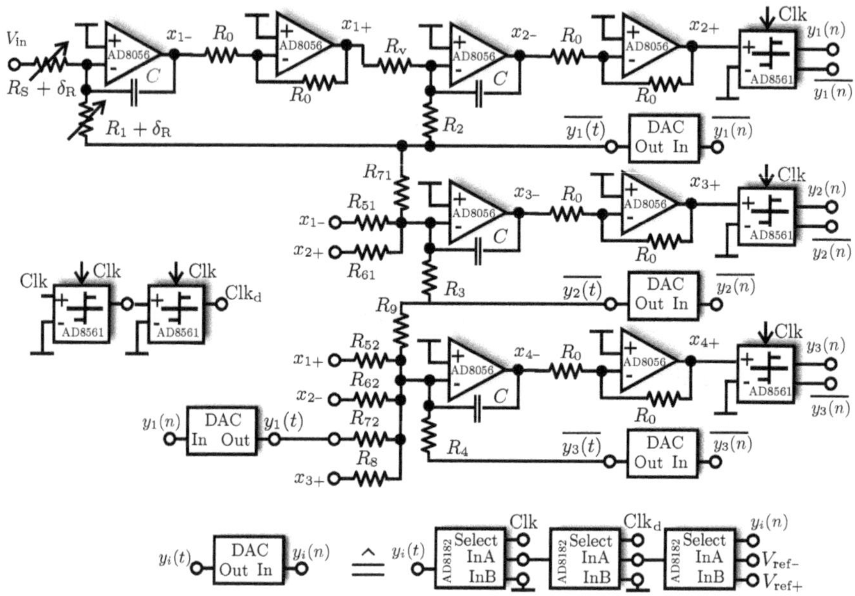

Fig. 7.43. Printed circuit board implementation of the 2–1–1 modulator, after [32, 97], courtesy of IEEE

using a logic analyzer, while the digital cancellation logic DF_i has been implemented in MATLAB according to [6]. Finally, a FFT has been used to obtain the power spectral density of the combined output. To be able to show the enhancement through the cascade, the results are shown separately for the second-order single-stage, for the third-order two-stage (2–1) and finally for the complete fourth-order, three-stage (2–1–1) modulator.

In Fig. 7.44a the spectra are shown for all three modulators for a $-23\,\text{dB}$ input signal at $f_{\text{sig}} = 10\,\text{kHz}$. The perfect enhancement of the modulator order and thus of the noise-shaping behavior can be seen clearly. The first-stage, with only second-order noise shaping, achieves an almost ideal measured $\text{IBN}_{\text{Mod2}} \approx -60\,\text{dB}$; also the two-stage cascaded structure behaves almost ideally with $\text{IBN}_{21} \approx -74\,\text{dB}$. The complete 2–1–1 modulator shows fourth-order noise shaping, and in spite of the inaccuracy of the passive components achieves an $\text{IBN}_{211} \approx -86\,\text{dB}$, which is only 6 dB below the calculated result.

In Fig. 7.44b the measured SNDR of all three modulators is presented: the first stage and also the cascaded 2–1 modulator show almost ideal behavior with $\text{SNDR}_{\text{p|Mod2}} \approx 50\,\text{dB}$, while this increases to $\text{SNDR}_{\text{p|21}} \approx 67\,\text{dB}$. In spite of white noise limitation as well as intrinsic coefficient mismatch the additional third stage increases the resolution to more than $\text{SNDR}_{\text{p|211}} \approx 75\,\text{dB}$, while a dynamic range of approximately $\text{DR}_{211} \approx 80\,\text{dB}$ is obtained.

As discussed in Sect. 3.5, a favorable feature of CT ΣΔ modulators is the implicit antialiasing behavior; this becomes especially important for low over-

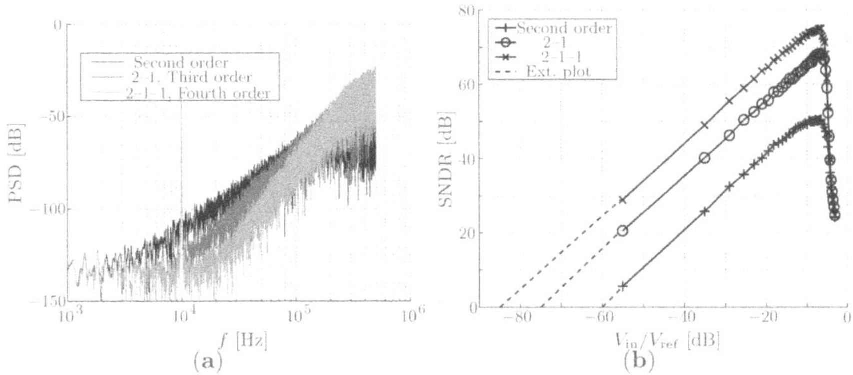

Fig. 7.44. Measured spectrum and SNDR of the CT second-order, 2–1 and 2–1–1 modulator, [32] courtesy of IEEE

sampling ratios, where the AAF is a nontrivial circuit. Usually, in single-loop CT modulators the AAF is of the same order than the modulator [105]. This enhances in cascaded structures, because the aliased component enters the system at the quantizer [20, 97, 147]. Thus, ideally together with the quantization noise, also the aliased components should be further suppressed by the higher cascades, which was presented in Sect. 3.5.2. This behavior is confirmed in Fig. 7.45, where together with a 5 kHz input signal an additional high frequency signal with +5 dB at 990 kHz has been applied. Obviously, the high frequency signal is aliased into the signal band at 10 kHz, but heavily suppressed by approximately 80 dB by the second-order modulator. When regarding the output of the complete 2–1–1 cascade, the suppression is improved by approximately −30 dB. The complete behavior of the 2–1–1 antialiasing filter is shown in Fig. 7.45c, where also the simulated, as well as the calculated results are presented [147]. The calculated expression was similarly derived as presented in Sect. 3.5.2.

7.6.3 Verification of the Digital Gain-Error Cancellation

As outlined above, the circuit board implementation was motivated by the intention to be able to measure the sensitivity of the cascaded 2–1–1 modulator to integrator gain errors. Consequently, in the circuit realization in Fig. 7.43 the resistors of the first integrator, R_1 and R_S, have been implemented exchangeable on the board. Thus, it is possible to introduce a variable gain error GE_1 in the first integrator by a shift of its resistors by δ_{R1}. A corresponding measurement is shown in Fig. 7.46a, where the feedback resistor has been changed within a [−60%, +100%] range.

Obviously, the single-loop modulator does not suffer too much over a wide range, but already the performance of the 2–1 modulator and even more

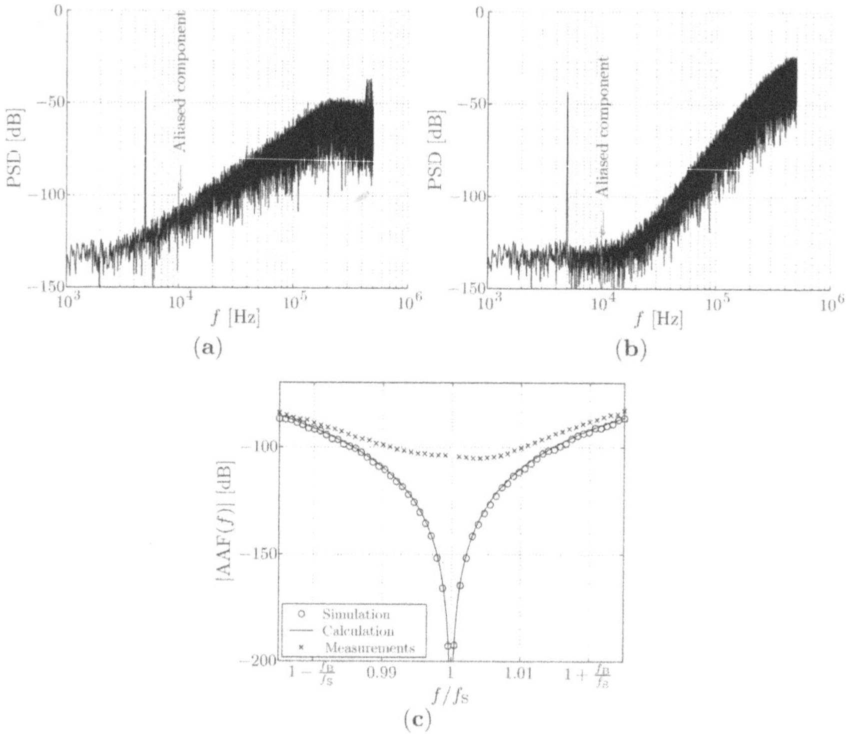

Fig. 7.45. Measured AAF behavior of the second-order and 2–1–1 modulator, after [32, 97], courtesy of IEEE (**a**) second-order AAF; (**b**) 2–1–1 AAF; (**c**) 2–1–1 AAF [147]

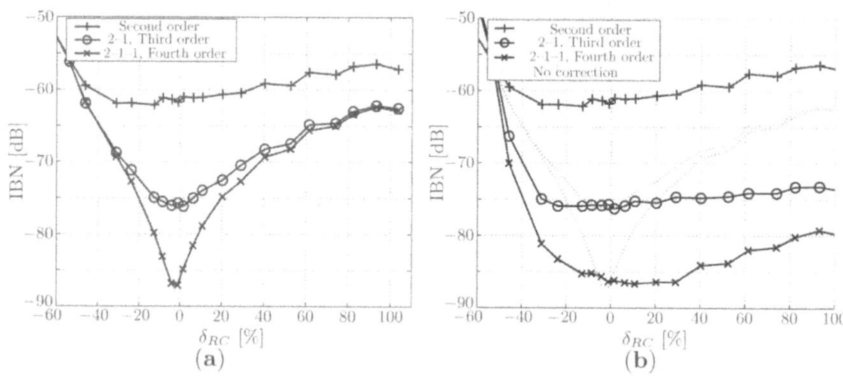

Fig. 7.46. IBN of the second-order, 2–1 and 2–1–1 modulator under RC-variation (GE_1) (**a**) without and (**b**) with digital correction in DF_1, after [97]

the 2–1–1 modulator drastically drops at very low gain errors. These measurements match almost perfectly with the simulations and calculations in Fig. 5.3b and (5.15).

As discussed in Sect. 5.3.6, a complete correction of a gain error is mathematically possible, if the condition in (5.19) is fulfilled. In the realization presented in Fig. 7.43, only the first integrator is affected, i.e., $\mathrm{GE}_{12} = 1$ and $\mathrm{GE}_{21} = 1$, while $\mathrm{GE}_{11} = 1/(1 + \delta_R)$. Figure 7.46b shows the resulting modulator behavior with correction in the first stage, i.e., $\mathrm{Corr}_1 = \mathrm{GE}_{11}$, $\mathrm{Corr}_2 = \mathrm{Corr}_3 = 1$.

As desired, the modulator performance is almost insensitive from the integrator gain error over a very large range. As expected and already seen in Fig. 7.44b, due to the intrinsic relative mismatch of the passive components of up to 2%, the optimal performance of the 2–1–1 modulator is about 5–6 dB below the ideal value. With *RC*-variation, the digital error correction keeps the performance almost constant: for a −3 dB loss of performance the 2–1–1 modulator with purely digital correction can tolerate a variation from −20% up to +50%, which is clearly enough for an integrated circuit.

Different possibilities exist for an implementation of the gain-error correction, either trying to determine the gain error in the analog domain [125], or by doing some kind of blind error correction through analyzing the output bit stream of the ΣΔ ADC, e.g., [268]. Since the latter requires significant DSP effort [20], recently also a blind error correction technique has been proposed, which minimizes the noise in the output bitstream of the ΣΔ ADC by minimizing the output bit activity [269]. This technique is currently only applicable without applied input signal, i.e., during startup or in regular calibration intervals, but has the potential to be extended to a more general automatic correction procedure.

A

Program Code

The calculation of the spectrum *psdy* of a DT modulator output bit-stream y_a of length n_p sampled at f_S has been done as proposed in [270], whose MATLAB source code is shortly outlined. In addition, the proposed code was slightly extended for IBN and SNR/SNDR calculation:

```
W=blackman(np); %Define a window of length np
[ys,f] = psd(ya,np,fs,W,[ ]);
%Use build-in Matlab 'psd' function
psdy = ys*norm(W)^2/sum(W)^2/(fs/np);
%Reverse norm and windowing
```

Subtraction of the input signal at f_sig from the PSD:

```
fsignr = round(fsig/(fs/(np));
%  Which bin contains the signal peak
%Find approx. number of bins (n) with leakage
comp = psdy(fsignr); n = 1; %Start at signal frequency
while (min(psdy((fsignr+n):(fsignr1+n+10)))<vergl),
comp=psdy(fsignr+n); n=n+1; % Enlarge n while there is
end;                        % still signal over the noise
psdy2 = psdy; %Copy the PSD-vector
psdy2(fsignr-n:fsignr+n) =
~~~~~~~~~~~~~~(psdy(fsignr-n)+psdy(fsignr-n))/2;
%Replace signal peak by medium
%  PSD-noise-magnitude around it
```

Determination of the integrated in-band noise $[0, f_\mathrm{B}]$ from the PSD:

```
psdy2_bin = psdy2*(fs/np);
%Calculate the power of each bin
fbnr = round(fb/(fs/(np));
%Which bin is at the bandwidth's edge
IBN = sum(psdy2_bin(0:fbnr))
```

Calculation of the output signal-to-noise ratio from the PSD:

```
psdy_bin = psdy*(fs/np); %Calculate power of each bin,
                         %here including the signal peak
SNR = sum(psdy_bin) - sum(psdy2_bin);
%Total integrated power minus
% total integrated power without signal
```

B

General Loop Filter Pole Transformation for the Exponential Feedback

For the conversion of a first-order pole, one starts with a simple DT first-order transfer function $H(z)$ with a pole at z_k, whose step-response is given by:

$$H(z) = \frac{1}{z - z_k} \quad \bullet\!\!-\!\!\circ \quad h(n) = z_k^{n-1} \,. \tag{B.1}$$

An equivalent CT transfer function should also be a first-order system with a pole at s_k and a gain g:

$$H(s) = \frac{g}{s - s_k} \quad \bullet\!\!-\!\!\circ \quad h(t) = g\,\mathrm{e}^{s_k t} \,. \tag{B.2}$$

Together with the impulse response of the SCR-feedback DAC $r_{\mathrm{SCR}}(t)$ in (3.5) this is introduced into (3.12) to solve the condition for the time-invariant transformation:

$$z_k^{n-1} = \int\limits_{\alpha T_{\mathrm{S}}}^{\beta T_{\mathrm{S}}} \mathrm{e}^{\alpha T_{\mathrm{S}} - \tau/\tau_{\mathrm{DAC}}}\, g\,\mathrm{e}^{s_k (n T_{\mathrm{S}} - \tau)} \mathrm{d}\tau \Big|_{t = n T_{\mathrm{S}}} \,. \tag{B.3}$$

Note the difference that τ_{DAC} is the time constant of the exponential feedback pulse, while τ is the operator for the convolution.

This equation can be solved with regard to powers of n. Therefore, both sides are subject to an ln operation to bring the n from an exponential operation to a product. The placement of the continuous-time filter pole yields:

$$\ln(z_k) = s_k T_{\mathrm{S}} \,. \tag{B.4}$$

For the CT loop filter gain g in (B.2) follows:

$$g = \frac{(1 + s_k \tau_{\mathrm{DAC}})\mathrm{e}^{s_k T_{\mathrm{S}}(\beta + \alpha - 1)}}{\tau_{\mathrm{DAC}}\left(\mathrm{e}^{(T_{\mathrm{S}}/\tau_{\mathrm{DAC}})(\alpha - \beta)}\mathrm{e}^{s_k T_{\mathrm{S}} \alpha} - \mathrm{e}^{s_k T_{\mathrm{S}} \beta}\right)} \,. \tag{B.5}$$

Inserting (B.5) into (B.2) gives the equivalent first-order loop filter pole with exponential feedback DAC as illustrated in Fig. 3.3a. A table with the resulting higher-order poles is given in [29, 31].

C

On the CT Integrator, Sampling Frequency f_S and the Amplifier GBW

As an outcome of many discussion, there exist regular confusion about the corner frequency of the used continuous-time integrators, the sampling frequency and the relation between both and the required gain-bandwidth product of the amplifiers in the integrators. Therefore, in the following, the nomenclature and dependencies among these variables is outlined. In almost all publications, the CT integrators within a $\Sigma\Delta$ modulator are defined with a transfer function according to:

$$I(s) = k_i \frac{f_S}{s} , \tag{C.1}$$

where k_i is the integrator scaling coefficient, s the Laplace operator and f_S the modulator sampling frequency in Hz. This is confusing, since a transfer function usually contains corner frequencies, which are denoted in rad/s

$$I(s) = \frac{\omega_I}{s} . \tag{C.2}$$

Therefore, we illustrate in the following, why (C.1) is still correct:

1. Starting with a simple switched capacitor DT integrator, charge conservation between both sampling phases reveals the integrator transfer function [5, 71, 72] as:

$$I(z) = \frac{C_1}{C_2} \frac{z^{-1}}{(1 - z^{-1})} = \frac{a_1 z^{-1}}{(1 - z^{-1})} = \frac{a_1}{(z - 1)} , \tag{C.3}$$

   where z is the $\mathcal{Z}$-domain variable, s the Laplace domain variable and $T_S = 1/f_S$ the sampling period in s. Replacing now z with $z = e^{sT_S}$ and subsequently adopting the Taylor series expansion (for low frequencies $f \ll f_S$) the equivalent CT transfer function yields:

$$I(s)|_{eq} = \frac{a_1}{(\underbrace{e^{(sT_S)}}_{\substack{\text{Taylor} \\ \approx 1 + sT_S}} - 1)} \approx \frac{a_1}{(1 + sT_S - 1)} = \frac{a_1}{sT_S} = a_1 \frac{f_S}{s} . \tag{C.4}$$

Obviously, this approximation reveals that the chosen CT integrator transfer function is directly equivalent to the DT integrator!

2. Another way to explain equation (C.1) is as follows: Using the impulse-invariant transform for the DT–CT conversion of a $\Sigma\Delta$ loop filter, the objective is to figure out the CT equivalents of a given DT loop filter, most easiest, a first-order filter given by:

$$H(z) = \frac{1}{z - z_k}, \tag{C.5}$$

where z_k is the pole of this first-order $\mathcal{Z}$-domain transfer function. For the DT-to-CT conversion a certain DAC transfer function is used. As an example a rectangular pulse form is chosen (Fig. 3.2a). Thus the following DAC impulse response $r_{\mathrm{DAC}}(t)$ is obtained

$$r_{\mathrm{DAC}}(t) = h(t - \alpha T_S) - h(t - \beta T_S), \tag{C.6}$$

where α and β determine the starting and end point of the rectangular pulse referenced to the sampling period T_S. Please note, that $T_S = 1/f_S$ is the sampling period in s. In general, the impulse invariant transform shown in (C.7) has to be fulfilled

$$\mathcal{Z}^{-1}\{H(z)\} = \mathcal{L}^{-1}\{H(s)\, R_{\mathrm{DAC}}(s)\}. \tag{C.7}$$

Here, $H(z)$ denotes the original $\mathcal{Z}$-domain transfer function, $H(s)$ is the required equivalent CT transfer function and R_{DAC} is the transfer function corresponding to (C.6). Equivalent to the first-order DT transfer function $H(z)$ given in (C.5), also for the equivalent CT transfer function a first-order filter is assumed

$$H(s)|_{\mathrm{eq}} = k_1 \frac{1}{(s - s_k)}, \tag{C.8}$$

using k_1 as a scaling coefficient and s_k for the $\mathcal{S}$-domain pole. Solving (C.7) yields:

$$s_k = \frac{\ln(z_k)}{T_S}, \tag{C.9}$$

$$k_1 = \frac{1}{(\beta - \alpha)\, T_S}, \tag{C.10}$$

where $T_S = 1/f_S$ is again the sampling period, s_k is the CT pole and z_k the DT pole. Assuming next a DT integrator with z_k=1. Accordingly, the equivalent CT transfer function becomes:

$$I(s)|_{\mathrm{eq}} = \frac{f_S}{(\beta - \alpha)\, s} \propto \frac{f_S}{s}, \tag{C.11}$$

which is again the same result as above in (C.4) (the same was found in [132] and illustrated in Table 3.6. Again the integrator transfer function in a CT $\Sigma\Delta$ modulator is $I(s) = k_i f_S/s$, where f_S is the sampling frequency in Hz, and s is the Laplace domain variable.

When realizing such an integrator, the confusion starts again: the transfer function of a simple RC-integrator is (compare Chap. 3)

$$I(s)|_{RC} = \frac{1}{sRC}, \tag{C.12}$$

where RC is the time constant of this CT integrator or $1/RC$ the integrator corner frequency ω_I. Nonetheless, by realizing the wanted CT integrator transfer function as shown in (C.1) the RC-integrator gain given in (C.12) yields:

$$\begin{aligned} \frac{1}{sRC} &= k_1 \frac{f_S}{s}, \\ k_1 f_S &= \frac{1}{RC} = \omega_I . \end{aligned} \tag{C.13}$$

In (C.13), f_S denotes the sampling frequency in Hz, k_1 is the integrator scaling coefficient, and ω_I the RC-integrator corner frequency in rad/s. As a result, the numerical value of the corner frequency in rad/s has to be set equal to the sampling frequency $f_S = 1/T_S$ of the $\Sigma\Delta$ modulator.

To complete the confusion, there is another issue regarding the frequency behavior in the $\Sigma\Delta$ modulator, which is the gain-bandwidth product GBW of the amplifiers adopted in the integrators. Usually, looking into a data sheet reveals the gain-bandwidth product in Hz. (Note, the gain-bandwidth product is sometimes also misleadingly called the "bandwidth", which is not correct, since the bandwidth is the 3 dB corner frequency of the amplifier, i.e., it is the position of the dominant pole $\omega_A = 2\pi f_A$, while GBW [rad/s] $= A_{dc}\omega_A$). On the other hand, the calculation of a gain-bandwidth product, e.g., for the folded cascade amplifier in Fig. 7.10, yields the GBW

$$\mathrm{GBW} = \frac{gm_1}{C_L}, \tag{C.14}$$

where GBW is in rad/s, gm_1 is the transconductance of the input transistors and C_L is the load capacitor. For illustration, the gain-bandwidth requirements of the amplifiers in the integrators of $\Sigma\Delta$ modulators is often related to the sampling frequency f_S (in Hz) of the modulator. In order to compare measures with the same unity, the GBW in Hz should be compared to the sampling frequency in Hz, thus resulting in an exemplary requirement of: "the bandwidth needs to be two times as high as the sampling frequency" or similar. Therefore, the relative bandwidth measure in (C.15) is normalized to rad/s, through:

$$\mathrm{GBW}\ [\mathrm{rad/s}] = 2\pi c f_S , \tag{C.15}$$

where $f_S = 1/T_S$ denotes again the sampling frequency in Hz. An equivalent measure to (C.15) would be

$$\mathrm{GBW}\ [\mathrm{Hz}] = c f_S , \tag{C.16}$$

with again $f_S = 1/T_S$ being the sampling frequency in Hz.

References

1. Inose, H., Yasuda, Y., Murikami, J.: A telemetering system by code modulation–$\Sigma\Delta$ modulation. IRE Trans. Space Electron. Telemetry **SET-8**, 204–209 September (1962)
2. Candy, J.C.: A use of double integration in Sigma-Delta modulation. IEEE Trans. Commun. **COM-33**, 249–258 March (1985)
3. Cherry, J.A., Snelgrove, W.M.: Continuous-time Delta-Sigma Modulators for high-speed A/D Conversion. Chapter 3 Kluwer Academic Publisher, Dordrecht (1999)
4. Northworthy, S., Schreier, R., Temes, G.: Delta-Sigma Data Converters, Chapters 2, 4, 6, 7, 11. IEEE Press, Piscataway, NJ (1997)
5. Medeiro, F., Perez-Verdu, B., Rodriguez-Vazquez, A.: Top-Down Design of high-performance Sigma-Delta Modulators. Kluwer Academic Publisher, Dordrecht (1999).
6. Marques, A., Peluso, V., Steyaert, M.S., Sansen, W.M.: Optimal parameters for $\Sigma\Delta$ modulator topologies. IEEE Trans. Circuits Syst. II **45**, 1232–1241 September (1998)
7. Peluso, V., Steyaert, M., Sansen, W.: Design of low-voltage low-power CMOS Delta-Sigma A/D converters. Kluwer Academic Publisher, Dordrecht (1999)
8. Geerts, Y., Steyaert, M., Sansen, W.: Design of Multi-Bit Delta-Sigma A/D Converters. Kluwer Academic Publisher, Dordrecht (2002)
9. van der Zwan, E.J., Dijkmans, E.C.: A 0.2-mW CMOS $\Sigma\Delta$ modulator for speech coding with 80 dB dynamic range. IEEE J. Solid-State Circuits **31**(12) December (1996)
10. Schreier, R., Zhang, B.: Delta-sigma modulators employing continuous-time circuitry. IEEE Trans. Circuits Syst. I **43**(4), 324–332 April (1996)
11. Tao, H., Toth, L., Khoury, J.M.: Analysis of timing jitter in bandpass sigma-delta modulators. IEEE Trans. Circuits Syst. II **46**(8), 991–1001 August (1999)

12. Cherry, J.A., Snelgrove, W.M.: Clock jitter and quantizer metastability in continuous-time delta-sigma modulators. IEEE Trans. Circuits Syst. II **46**(7), 661–676 June (1999)
13. Berkovitz, A., Rusnak, I.: FFT processing of randomly sampled harmonic signals. IEEE Trans. Signal Process. **40**, 2816–2819 November (1992)
14. Benabes, P., Keramat, M., Kielbasa, R.: A methodology for designing continuous-time sigma-delta modulators. Proc. IEEE Eur. Des. Test Conf. 46–50 (1997)
15. Benabes, P., Aldebert, P., Kielbasa, R.: A matlab based tool for bandpass countinuous-time sigma-delta modulators design. Proc. IEEE Int. Symp. Circuits Sys. **6**, 274–277 (1998)
16. Yahia, A., Benabes, P., Kielbasa, R.: Influence of the feedback dac delay on a continuous-time bandpass $\Delta\Sigma$ converter. Proc. IEEE Int. Symp. Circuits Sys. **2**, 648–651 (2002)
17. Breems, L.J., van der Zwan, E.J., Huijsing, J.H.: A 1.8-mW CMOS $\Sigma\Delta$ modulator with integrated mixer for A/D conversion of IF signals. IEEE J. Solid-State Circuits **43**(4), 468–475 April (2000)
18. Breems, L., Huijsing, J.H.: Continuous Time Sigma Delta Modulation for A/d Conversion in Radio Receivers. Springer-Verlag, Heidelberg, Germany (2001)
19. van Veldhoven, R.: A tri-mode continuous-time $\Sigma\Delta$ modulator with switched-capacitor feedback DAC for a GSM-EDGE, CDMA2000, UMTS receiver. IEEE Int. Solid-State Circuits Conf. **46**, 60–61 (2003)
20. Breems, L.J.: A cascaded continuous-time $\Sigma\Delta$ modulator with 67 dB dynamic range in 10 MHz bandwidth. Proc. Int. Solid-State Circuits Conf. 72–73 February (2004)
21. Doerrer, L., et al. A 10-Bit, 3 mW continuous-time sigma-delta ADC for UMTS in a 0.12 µm CMOS process. Proc. Eur. Solid-State Circuits Conf. 245–248 (2003)
22. Doerrer, L.: A 3 mW 74 dB SNR 2 MHz CT $\Sigma\Delta$ ADC with a tracking-ADC-quantizer in 0.13 m CMOS. Proc. Int. Solid-State Circuits Conf. 72–73 February (2005)
23. Paton, S., DiGianmenico, A., Hernandez, L., Wiesbauer, A., Poetscher, T.: A 70 mW 300 MHz CMOS continuous-time sigma-delta ADC with 15 MHz bandwidth and 11-bit of resolution. IEEE J. Solid-State Circuits **39**(7), 1056–1063 July (2004)
24. Oliaei, O.: Jitter effects in continuous time $\Sigma\Delta$ modulators with delayed return-to-zero feedback. Proc. IEEE Int. Conf. Electron. Circuits Syst. 351354 (1998)
25. Oliaei, O.: Design of continuous-time sigma-delta modulators with arbitrary feedback waveform. IEEE Trans. Circuits Syst.-II **50**(8), 437–444 August (2003)
26. Oliaei, O.: Sigma-delta modulator with spectrally shaped feedback. IEEE Trans. Circuits Syst.-II **50**(9), 518–530 September (2003)

27. Hernandez, L., Wiesbauer, A., Paton, S., DiGiandomenico, A.: Modelling and optimization of low pass continuous-time ΣΔ modulators for clock jitter noise reduction. Proc. IEEE Int. Symp. Circuits and Syst. 1072–1075 (2004)
28. Hernández, L., Rombouts, P., Prefasi, E., Paton, S., Garcia, M., Lopez, C.: A jitter insensitive continuous-time ΣΔ modulator using transmission lines. Proc. IEEE Int. Conf. Electron. Circuits Syst. 109–112 (2004)
29. Ortmanns, M.: Error compensation in continuous-time ΣΔ A/D converters. PhD Thesis, Albert-Ludwigs-University, Freiburg, Germany (2004)
30. Ortmanns, M., Gerfers, F., Manoli, Y.: Compensation of finite gain-bandwidth induced errors in continuous-time sigma-delta modulators. IEEE Trans. Circuits Syst. II **51**(6), 1088–1100 June (2004)
31. Ortmanns, M., Gerfers, F., Manoli, Y.: A continuous-time sigma-delta modulator with reduced sensitivity to clock jitter through SCR-feedback, to be published. IEEE Trans. Circuits Syst. I **52**(5), 875–884 May (2005)
32. Ortmanns, M., Gerfers, F., Manoli, Y.: A case study on a 2-1-1 cascaded continuous-time sigma-delta modulator, to be published. IEEE Trans. Circuits Syst. I (2005)
33. Ortmanns, M., Gerfers, F., Manoli, Y.: A continuous-time sigma-delta modulator with reduced jitter senitivity. Proc. Eur. Solid-State Circuits Conf. 287–290 (2002)
34. Gerfers, F., Ortmanns, M., Manoli, Y.: A 12-bit power efficient continuous-time ΣΔ modulator with 220 μW power consumption. Proc. Eur. Solid-State Circuits Conf. 536–539 (2001)
35. Gerfers, F., Ortmanns, M., Manoli, Y.: A design strategy for low-voltage low-power continuous-time ΣΔ A/D converters. Des. Autom. Test Conf. 361–368 (2001)
36. Gerfers, F., Ortmanns, M., Manoli, Y.: A 1 V, 12-bit wideband continuous-time ΣΔ modulator for UMTS applications. Proc. IEEE Int. Sym. Circuits Syst. **1**, 921–924 May (2003)
37. Gerfers, F., Ortmanns, M., Manoli, Y.: A 1.5-V, 12-bit power efficient continuous-time third-order ΣΔ modulator. IEEE J. Solid-State Circuits **38**(8), 1343–1352 August (2003)
38. Gerfers, F.: Design strategy, limits and implementation of low-voltage low-power continuous-time ΣΔ modulators. PhD Thesis, Albert-Ludwigs-University, Freiburg, Germany (2005)
39. Oppenheim, A., Schafer, R.: Discrete-Time Signal Processing. Prentice-Hall, US (1989)
40. Bennett, W.: Spectra of quantized signals. Bell Syst. Tech. J. **27**, 446–472 (1948)
41. Widrow, B.: A study of rough amplitude quantization by means of Nyquist sampling theory. IRE Trans. Circuit Theory **CT-3**, 266–276 (1956)

42. Cline, D.W.: Noise, speed and power trade-offs in pipelined analog to digital converters. PhD Thesis, University of California at Berkeley (1995)
43. Welch, P.D.: The use of fast Fourier transfrom for the estimation of power spectra: A method based on time averagin over shot, modified periodograms. IEEE Trans. Audio Electroacoust. **AU-15**, 70–73 (1967)
44. Inc. The mathworks. Matlab R13 and Simulink. The MathWorks, Inc. Natick, MA (2003)
45. Harris, F.J.: On the use of windows for harmonix analysis with the discrete Fourier transfrom. Proc. IEEE, 51–83 (1978)
46. Europractice and esprit mixed-signal design cluster. Embeddable data converters for mixed-signal ASICs. EUROPRACTICE Analog Thematic Training and ESPRIT Mixed-Signal Design Cluster, July (1999)
47. Rabii, S., Wooley, B.A.: The Design of Low-Voltage Low-Power Sigma-Delta Modulators. Kluwer Academic Publisher, Dordrecht (1999)
48. Gerfers, F., Soh, K.M., Ortmanns, M., Manoli, Y.: Figure of merit based design strategy for low-power continuous-time ΣΔ modulators. Proc. IEEE Int. Symp. Circuits Syst. 233–236 (2002)
49. Gielen, G., Francken, K., Martens, E., Vogels, M.: An analytical integration method for the simulation of continuous-time $\Delta\Sigma$ modulators. IEEE Trans. Comput. Aided Des. **23**(3), 389–399 March (2004)
50. Pelgrom, M.J., Duinmaijer, A.C.J., Welbers, A.P.G.: Matching properties of MOS transistors. IEEE J. Solid-State Circuits **24**(5), 1433–1439 October (1989)
51. Brooks, T.L., Robertson, D.H., Kelly, D.F., Del Muro, A., Harston, S.W.: A cascaded sigma-delta pipeline A/D converter with 1.25 MHz signal bandwidth and 89 dB SNR. IEEE J. Solid-State Circuits **32**(12), 1896–1906 December (1997)
52. Aziz, P.M., Sorensen, H.V., van der Spiegel, J.: An overview of sigma-delta converters. IEEE Signal Process. Mag. **13**(1), 61–84 January (1996)
53. Candy, J.C., Benjamin, O.J.: The structure of quantization noise form sigma-delta modulation. IEEE Trans. Commun. **COM-29**, 1316–1323 September (1981)
54. van de Plassche, R.J.: A sigma-delta modulator as an A/D converter. IEEE Trans. Circuits Syst. **25**(7), 510–514 July (1978)
55. Sarhang-Nejad, M., Temes, G.C.: A high-resolution multibit ΣΔ ADC with digital correction and relaxed amplifier requirements. IEEE J. Solid-State Circuits **28**(6), 648–660 June (1993)
56. Fujimori, I., Sugimoto, T.: A 1.5 V, 4.1 mW dual-channel audio delta-sigma D/A coverter. IEEE J. Solid-State Circuits **33**(12), 1863–1870 December (1998)
57. Yasuda, A., Tanimoto, H., Iida, T.: A third-order ΣΔ modulator using second-order noise-shaping dynamic element matching. IEEE J. Solid-State Circuits **33**(12), 1879–1886 December (1998)
58. Jiang, R., Fiez, T.: A 1.8 V 14 b ΣΔ A/D converter with 4 M Samples/s conversion. IEEE Int. Solid-State Circuits Conf. **42**, 220–221 (2002)

59. Vittoz, E.: Limits to low-power/low-volatge analog cicuit design. Advanced Engineering Course on Low-Power, Low-Volatge Analog CMOS IC Design (1998)
60. Marques, A., Peluso, V., Steyaert, M., Sansen, W.: A 15-bit 2 MHz Nyquist rate ΣΔ ADC in a 1 µm CMOS technology. IEEE J. Solid-State Circuits **33**(7), 1065–1075 July (1998)
61. Geerts, Y., Steyaert, M.S.J., Sansen, W.M.: A 3.3-V, 15-bit delta-sigma ADC with a signal bandwidth of 1.1 MHz for ADSL applications. IEEE J. Solid-State Circuits **34**(7), 927–936 July (1999)
62. Medeiro, F., Perez-Verdu, B., Rodriguez-Vazquez, A.: A 13-bit, 2.2-MS/s, 55-mW multibit cascade ΣΔ modulator in CMOS 0.7-µm single-poly technology. IEEE J. Solid-State Circuits **34**(6), 748–760 June (1999)
63. Geerts, Y., Steyaert, M.S.J., Sansen, W.M.: A high-performance multibit ΣΔ CMOS ADC. IEEE J. Solid-State Circuits **35**, 1829–1840 December (2000)
64. Fujimori, I., Longo, L., Hairapetian, A., Seiyama, K., Kosic, S., Cao, J., Chan, S.: A 90 dB SNR , 2.5 MHz output rate ADC using cascaded multibit ΣΔ modulation at 8x oversampling ratio. IEEE Int. Solid-State Circuits Conf. 238–239 February (2000)
65. Adams, R.: Design and implementation of an audio 18-bit analog-to-digital converter using oversampling techniques. J. Audio Eng. Soc. 153–166 March (1986)
66. Hauser, M.W., Brodersen, R.W.: Circuit and technology considerations for MOS delta-sigma A/D converters. Proc. IEEE Int. Symp. Circuits Syst. 1310–1315 (1986)
67. Bastos, J., Marques, A.M., Steyaert, M.J., Sansen, W.: A 12-Bit intrinsic accuracy high-speed CMOS DAC. IEEE J. Solid-State Circuits **33**(12), 1959–1969 December (1998)
68. Carley, L.R.: A noise-shaping coder topology for 15+ bit converters. IEEE J. Solid-State Circuits **SC-28**, 267–273 April (1989)
69. Baird, R.T., Fiez, T.S.: Linearity enhancement of multibit ΣΔ A/D and D/A converters using data weighted averaging. IEEE Trans. Circuits Syst. II **42**, 753–762 December (1995)
70. Kiss, P., Un-Ku Moon, Steensgaard, J., Stonick, J.T., Temes, G.: Multibit ΣΔ ADC with mixed-mode DAC error correction. Proc. IEEE Int. Symp. Circuits Syst. 280–283 (2001)
71. Peluso, V., Steyaert, M.S.J., Sansen, W.: A 1.5 V-100 µW ΣΔ modulator with 12-b dynamic range using the switched-opamp technique. IEEE J. Solid-State Circuits **32**(7), 943–952 July (1997)
72. Peluso, V., Vancorenland, P., Marques, A., Steyaert, M.S., Sansen, W.M.: A 900-mV low-power ΣΔ A/D converter with 77-dB dynamic range. IEEE J. of Solid-State Circuits **33**(12) December (1998)
73. Ferguson, P.F., Ganesan, A., Adams, R.W.: One bit higher order sigma-delta A/D converters. Proc. IEEE Int. Symp. Circuits Syst. **2**, 890–893 (1990)

74. Maulik, P.C., Chadha, M.S., Lee, W.L., Crawley, P.J.: A 16-Bit 250-kHz delta-sigma modulator and decimation filter. IEEE J. Solid-State Circuits **35**(4) April (2000)
75. Lee, W.L., Sodini, C.G.: A topology for higher order interpolative coders. Proc. IEEE Int. Symp. Circuits Syst. 459–462 (1987)
76. Ritchie, G.R.: Higher order interpolation analog-to-digital converters. PhD Thesis, Universiy of Pennsylvania (1977)
77. Stephen Au, Bosco, H., Leung, A.: 1.95-V, 0.34-mV, 12-b sigma-delta modulator stabilized by local feedback loops. IEEE J. Solid-State Circuits **32**(3), 321–328 March (1997)
78. Welland, D.R., Del Signore, B.P., Swanson, E.J.: A Stereo 16-Bit delta-sigma A/D converter for digital audio. J. Audio Eng. Soc. **37**, 365–374 June (1989)
79. van der Zwan, E.J., Philips, K., Bastiaansen, C.A.A.: A 10.7-MHz IF-to-baseband ΣΔ A/D conversion system for AM/FM radio receivers. IEEE J. Solid-State Circuits **43**(12), 1810–1819 December (2000)
80. Op't Eynde, F., Sansen, W.: Analog Interfacess for Digital Signal Processing Systems. Kluwer Academic Publisher, Dordrecht (1993)
81. Risbo, L.: ΣΔ modulators-stability and design optimization. PhD Thesis, Technical University of Denmark (1994)
82. Baird, R.T., Fiez, T.S.: Stability analysis of high-order delta-sigma modulation for ADC's. IEEE Trans. Circuits Syst. II **41**(1), 59–62 January (1994)
83. Benabes, P., Aldebert, P. Kielbasa, R.: Analog-to-digital sigma-delta converters modeling for simulation and synthesis. The 4th workshop on ADC Modelling and Testing, 3–14 (1999)
84. Schreier, R.: ΣΔ Toolbox. http:// www.mathworks.com/matlabcentral/fileexchange/
85. Agrawal, B.P., Shenoi, K.: Design methodology for sigma-delta modulation. IEEE Trans. Commun. **COM-31**, 360–370 March (1983)
86. Ardalan, S.H., Paulus, J.J.: An analysis of nonlinear behavior in delta-sigma modulators. IEEE Trans. Circuits Syst. **34**(6), 593–603 June (1987)
87. Williams(III), L.A., Wooley, B.A.: Third-order cascaded sigma-delta modulators. IEEE Trans. Circuits Syst. II **38**(5), 489–498 May (1991)
88. Maguire, P.T., Huang, Q.: Quantiser gain in Nth-order ΣΔ modulator linear models: its determination based on constant output power criterion. Proc. IEEE Int. Symp. Circuits Syst. **5**, 333–336 (1994)
89. Hayashi, T., Inabe, Y., Uchimura, K., Kimura, T.: A multi stage delta-sigma modulator without double integration loop. Proc. IEEE Int. Solid-State Circuits Conf. 182–183 (1986)
90. Ribner, D.B.: A comparison of modulator networks for high-order oversampled ΣΔ analog-to-digital converters. IEEE Trans. Circuits Syst. **38**(2), 145–159 February (1991)

91. Fischer, G., Davis, A.J.: Alternative topologies for sigma-delta modulators — A comparative study. IEEE Trans. Circuits Syst. II **44**(10), 789–797 October (1997)
92. Matsuya, Y. et al. A 16-bit oversampling A-to-D conversion technology using triple-integration noise shaping. IEEE J. Solid-State Circuits **SC-22**, 237–244 December (1987)
93. Longo, L., Copeland, M.: A 13 bit ISDN-band oversampled ADC using two-stage third order noise shaping. IEEE Proc. Custom IC Conf. 21.2.1–21.2.4 (1988)
94. Yin, G., Sansen, W.: A high-frequency and high-resolution fourth-order ΣΔ A/D converter in BiCMOS technology. IEEE J. Solid-State Circuits **29**(8), 857–865 August (1994)
95. Brandt, B.P., Wooley, B.A.: A 50-MHz multibit sigma-delta modulator for 12-b 2-MHz A/D conversion. IEEE J. Solid-State Circuits **26**(12), 1746–1756 December (1991)
96. Ortmanns, M., Gerfers, F., Manoli, Y.: On the synthesis of cascaded continuous-time ΣΔ modulators. Proc. IEEE Int. Symp. Circuits Syst. 419–422 (2001)
97. Ortmanns, M., Gerfers, F., Manoli, Y.: A cascaded continuous-time $\Sigma\Delta$ modulator with 80 dB dynamic range. Proc. IEEE Int. Symp. Circuits Syst. **1**, 405–408 (2004)
98. Matsuya, Y., Yukhi, N., Akazawa, Y.: Digital correction technique for multi-stage noise-shaping with an RC-analog integrator. IEICE Trans. Elektron **E77-C**(12), 1912–1919 (1994)
99. Oliaei, O.: Analysis of multirate sigmadelta modulators. Proc. IEEE Int. Symp. Circuits Syst. **1**, 448–451 (2001)
100. Colodro, F., Torralba, A.: Multirate ΣΔ modulators. IEEE Trans. Circuits Syst.-II **49**(3), 170–176 March (2002)
101. Ortmanns, M., Gerfers, F., Manoli, Y.: Multirate cascaded continuous-time ΣΔ modulators. Proc. IEEE Int. Conf. Electron. Circuits Syst. **4**, 1049–1052 (2002)
102. Malcovati, P., Maloberti, F.: Optimization of the integrator output swing in low-voltage sigma-delta modulators. Proc. IEEE Int. Symp. Circuits Syst., 607–610 May (2001)
103. Coban, L., Allen, P.A.: A 1.5 V, 1.0 mW audio ΣΔ modulator with 98 dB dynamic range. IEEE Int. Solid-State Circuits Conf. **39**, 50–51 (1999)
104. Matsuya, Y., Yamada, J.: 1 V power supply, low-power consumption A/D conversion techique with swing-suppression noise shaping. IEEE J. Solid-State Circuits **29**(12), 1524–1530 December (1994)
105. Shoaei, O.: Continuous-time delta-sigma A/D converters for high speed applications. PhD Thesis, Carleton University (1995)
106. van Engelen, J., van de Plassche, R.: Bandpass sigma delta modulators. Kluwer Academic Publisher, Dordrecht (1999)

107. Baird, R.T., Fiez, T.S.: A low oversampling ratio 14-b 500-kHz ΣΔ ADC with a self-calibrated multibit DAC. IEEE J. Solid-State Circuits **31**(3), 312–320 März (1996)
108. Nedved, J., Vanneuville, J., Gevaert, D., Sevenhans, J.: A transistor-only switched current sigma-delta A/D converter for a CMOS speech codec. IEEE J. Solid-State Circuits **30**(7), 819–822 July (1995)
109. Luh, L., Choma, J., Draper, J.: A 50-MHz continuous-time switched-current ΣΔ modulator. Proc. IEEE Int. Symp. Circuits Syst. 579–582 (1998)
110. Boser, B.E., Wooley, B.A.: The design of sigma-delta modulation analog-to-digital converters. IEEE J. Solid-State Circuits **23**(6), 1298–1308 December (1988)
111. Candy, J.C., Ninke, W.H., Wooley, B.A.: A per-channel A/D converter having 15-segment u-255 companding. IEEE Trans. Commun. **COM-24**, 33–42 January (1976)
112. Koch, R., Heise, B., Eckbauer, F., Engelhardt, E., Fischer, J.A., Parzefall, F.: A 12-bit sigma-delta analog-to-digital converter with a 15-MHz clock rate. IEEE J. Solid-State Circuits **21**(6), 1003–1009 December (1986)
113. Jensen, J.F., Raghavan, G., Cosand, A.E., Walden, R.H.: A 3.2-GHz second-order delta-sigma modulator implemented in InP HBT technology. IEEE J. Solid-State Circuits **30**(10), 1119–1127 October (1995)
114. Bazarjani, S., Snelgrove, W.M.: A 160 MHz fourth-order double sampled SC bandpass sigmadelta modulator. IEEE Trans. Circuits Syst. II **45**, 547555 May (1998)
115. Burger, T., Huang, Q.: A 13.5-mW 185-M sample/s ΣΔ modulator for UMTS/GSM dual-standard IF reception. IEEE J. Solid-State Circuits **36**(12), 1868–1878 December (2001)
116. Balmelli, P., Huang, Q.: A 25 MS/s 14 b 200 mW ΣΔ modulator in 0.18 μm CMOS. ISSCC Digest of Technical Papers, 74–75 (2004)
117. Naiknaware, R., Fiez, T.: Power optimization of ΣΔ analog-to-digital converters based on slewing and partial settling consideration. Proc. IEEE Int. Symp. Circuits Syst. 360–364 (1998)
118. Balmelli, P.: Broadband sigma-delta A/D converters. PhD Thesis, Swiss Federal Institute of Technology Zurich (2003)
119. Ortmanns, M., Gerfers, F., Manoli, Y.: Increased jitter sensitivity in continuous- and discrete-time $\Sigma\Delta$ modulators due to finite OpAmp settling speed. Proc. IEEE Int. Symp. Circuits Syst. (2005)
120. Marques, A., Peluso, V., Steyaert, M.S., Sansen, W.M.: Analysis of the trade-off between bandwidth, resolution, and power in ΣΔ analog to digital converters. Proc. IEEE Int. Conf. Electron. Circuits Syst. **2**, 153–156 September (1998)
121. Chan, K.T., Martin, K.W.: Components for a GaAs delta-sigma modulator oversampled analog-to-digital converter. Proc. IEEE Int. Symp. Circuits Syst. 1300–1303 (1992)

122. Wongkomet, N.: A comparison of continuous-time and discrete-time sigma-delta modulators. Master's Thesis, University of California at Berkeley (1995)
123. del Signore, B.P., Kerth, D.A., Sooch, N.S., Swanson, E.J.: A monolithic 20-b delta-sigma A/D converter. IEEE J. Solid-State Circuits **25**(6), 1311–1317 December (1990)
124. Ortmanns, M., Gerfers, F., Manoli, Y.: Influence of finite integrator gain bandwidth on continuous-time $\Sigma\Delta$ modulators. Proc. IEEE Int. Symp. Circuits Syst. **1**, 925–928 May (2003)
125. Ortmanns, M., Gerfers, F., Manoli, Y.: Successful design of cascaded continuous-time $\Sigma\Delta$ modulators. Proc. IEEE Int. Conf. Electron. Circuits Syst. 321–324 (2001)
126. Yan, S., Sanchez-Sinencio, E.: A continuous-time $\Sigma\Delta$ modulator with 88-dB dynamic range and 1.1-MHz signal bandwidth. IEEE J. Solid-State Circuits **39**(1), 75–86 January (2004)
127. Kaiser, A.: A micropower CMOS continuous-time low-pass filter. IEEE J. Solid-State Circuits **24**(3), 736–743 March (1989)
128. Breems, L.J., van der Zwan, E.J., Huijsing, J.H.: Design for optimum performance-to-power ratio of a continuous-time $\Sigma\Delta$ modulator. Proc. Eur. Solid-State Circuits Conf. 318–321 September (1999)
129. Samid, L., Gerfers, F., Ortmanns, M., Manoli, Y.: A new kind of low power multibit third order ct-lowpass modulator. Proc. IEEE Int. Symp. Circuits Syst. **3**, 293–296 (2002)
130. Luschas, S., Lee, H.S.: High-speed $\Sigma\Delta$ modulators with reduced timing jitter sensitivity. IEEE Trans. Circuits Syst.-II **49**(11), 712–720 November (2002)
131. Cherry, J.A., Snelgrove, W.M.: Approaches to simulating continuous-time delta sigma modulators. Proc. IEEE Int. Symp. Circuits Syst. **1**, 587–590 June (1998)
132. Cherry, J.A., Snelgrove, W.M.: Excess loop delay in continuous-time delta-sigma modulators. IEEE Trans. Circuits Syst. II **46**(4), 376–389 April (1999)
133. Oliaei, O., Aboushady, H.: Jitter effects in continuous time $\Sigma\Delta$ modulators with delayed return-to-zero feedback. Proc. Int. Conf. Electron. Circuits Syst. 351–354 September (1998)
134. Gerfers, F., Manoli, Y.: A 1.5 V low-power third-order continuous-time lowpass $\Sigma\Delta$ A/D converters. Proc. Int. Symp. Low-Power Electron. Des. 219–221 July (2000)
135. Aboushady, H.: Design for reuse of current-mode continuous-time $\Sigma\Delta$ analog-to-digital converter. PhD Thesis, University of Paris VI (2002)
136. Ortmanns, M., Gerfers, F., Manoli, Y.: Jitter insensitive feedback DAC for continuous-time $\Sigma\Delta$ modulators. Proc. IEEE Int. Conf. Electron. Circuits Syst. 1049–1052 (2001)

137. Gardner, F.M.: A transformation for digital simulation of analog filters. IEEE Trans. Commun. **COM-34**, 676–680 July (1986)
138. Waterloo Maple6. www.maplesoft.com.
139. Jury, E.I.: Theory and applications of the Z-transform method. John Wiley & Sons New York (1964)
140. Gao, W., Shoaei, O., Snelgrove, W.M.: Excess loop delay effects in continuous-time delta-sigma modulators and the compensation solution. Proc. IEEE Int. Sym. Circuits Syst. **1**, 65–68 June (1997)
141. Bronstein, I.N., Semendjajew, K.A.: Taschenbuch der Mathematik. B. G. Teubner Verlag, Verlag Nauka Moskau (1991)
142. Torralba, A.: Multirate-cascade sigma-delta (MC-SD) modulators. Proc. IEEE Int. Symp. Circuits Syst. **1**, 384–387 (2001)
143. Cherry, J.A., Snelgrove, W.M.: Loop delay and jitter in continuous-time delta sigma modulators. Proc. IEEE Int. Symp. Circuits Syst. **1**, 596–599 June (1998)
144. Matsuya, Y., Yukhi, N., Akazawa, Y.: Digital correction technique for multi-stage noise-shaping with an RC-analog integrator. IEICE Trans. Electron. **E77-C**(12), 1912–1919 December (1994)
145. Lin, C.H., Ismail, M.: Synthesis and analysis of high-order cascaded continuous time ΣΔ modulators. Proc. Int. Conf. Electron. Circuits Syst. 1693–1696 September (1999)
146. Ramón Tortosa, José M., de la Rosa, Angel Rodríguez-Vázquez, Francisco, V. Fernández. A direct synthesis method of cascaded continuous-time sigma-delta modulators. Proc. IEEE Int. Symp. Circuits Syst. 5585–5589 May (2005)
147. Keller, M., Buhmann, A., Ortmanns, M., Gerfers, F., Manoli, Y.: On the implicit anti-aliasing filter in continuous-time single stage and multi-stage noise shaping sigma delta modulators. IEEE Trans. Circuits Syst. I, submitted to (2005)
148. Lindfors, S., Halonen, K., Ismail, M.: A 2.7-V elliptical MOSFET only gmC-OTA filter. IEEE J. Solid-State Circuits **47**(2), 89–95 February (2000)
149. Silva-Martinez, J., Steyaert, M., Sansen, W.: Design techniques for high-performance full-CMOS OTA-RC continuous-time filters. IEEE J. Solid-State Circuits **31**(7), 993–1001 July (1992)
150. Willingham, S.D., Martin, K.W., Ganesan, A.: A BiCMOS low-distortion 8-MHz low-pass filter. IEEE J. Solid-State Circuits **28**(12), 1234–1245 December (1993)
151. Das, A., Hezar, R., Byrd, R., Gomez, G., Haroun, B.: 4th-order 86 dB CT ΣΔ ADC with two amplifiers in 90 nm CMOS. Proc. IEEE Int. Solid-State Circuits Conf. 496–497 (2005)
152. Yoo, C., Lee, S.W., Kim, W.: A ± 1.5-V, 4 MHz CMOS continuous-time filter with a single-integrator based tuning. IEEE J. Solid-State Circuits **33**(1) January (1998)

153. Jaganathan, S., Krishnan, S., Mensa, D., Mathew, T., Betser, Y., Wei, Y., Scott, D., Urtega, M., Rodwell, M.: An 18-GHz continuous-time $\Sigma\Delta$ analog-digital converter implemented in InP-transferred substrate HBT. IEEE J. Solid-State Circuits **36**(9), 166–172 September (2001)
154. Raghavan, G., Jensen, J., Walden, R., Posey, W.: A bandpass $\Sigma\Delta$ modulator with 92 dB SNR and center frequency continuously programmable from 0 to 70 MHz. IEEE Int. Solid-State Circuits Conf. **37**, 214–215 (1997)
155. Cherry, J.A., Snelgrove, W.M., Gao, W.: On the design of a fourth-order continuous-time LC delta-sigma modulator for UHF A/D conversion. IEEE J. Solid-State Circuits **33**(5), 723–732 May (1998)
156. Schreier, R., Lloyd, J., Singer, L., Paterson, D., Timko, M., Hensley, M., Patterson, G., Behel, K., Zhou, J., Martin, W.J.: A 50 mW bandpass $\Sigma\Delta$ ADC with 333 kHz BW and 90 dB DR. Proc. IEEE Int. Solid-State Circuits Conf. 216–217 (2002)
157. Gao, W., Snelgrove, W.M.: A 950-MHz IF second-order integrated LC bandpass delta-sigma modulator. IEEE J. Solid-State Circuits **33**(5), 723–732 May (1998)
158. Benabid, S., Aghdam, E.N., Benabes, P., Guessab, S., Kielbasa, R.: CMOS design of a multibit bandpass continuous-time sigma delta modulator running at 1.2 GHz. Proc. IEEE Int. Caracas Conf. Devices, Circuits Syst., 51–55 (2004)
159. Dagher, E., Stubberud, P., Masenten, W., Conta, M., Dinh, T.: A 2 GHz analog-to-digital delta-sigma modulator for CDMA receivers with 79 dB signal-to-noise ratio in 1.23 MHz bandwidth. IEEE J. Solid-State Circuits **39**(11), 1819–1828 November (2004)
160. Tsividis, Y.P.: Integrated continuous-time filter design - An overview. IEEE J. solid-state circuits **29**(3), 166–176 March (1994)
161. Zele, R.H., Allstot, D.J.: Low-power CMOS continuous-time filter. IEEE J. Solid-State Circuits **31**(2), 157–168 February (1996)
162. Aboushady, H., Louerat, M.-M.: Systematic design of high-linearity current-mode Integrators for low-power cotinuous-time $\Sigma\Delta$ modulators. Proc. IEEE Int. Conf. Electron. Circuits Syst. **1**, 963–966 September (2001)
163. Cong, Y., Geiger, R.L.: Performance association and discrimination between current-mode and voltage-mode operation of monolithic linear circuits. Proc. Southwest Symp. Mixed-Signal Des. 131–134 (2001)
164. Punzenberger, M., Enz, C.C.: A 1.2-V low-power BiCMOS class AB log-domain filter. IEEE J. Solid-State Circuits **32**(12), 1968–1978 December (1997)
165. Punzenberger, M., Enz, C.C.: A compact low-power BiCMOS log-domain filter IEEE J. Solid-State Circuits **33**(7), 1123–1129 July (1998)
166. Python, D., Punzenberger, M., Enz, C.C.: A 1-V CMOS log-domain integrator. Proc. IEEE Int. Symp. Circuits Syst. 685–688 May (1999)

167. Python, D., Punzenberger, M., Enz, C.C.: A micropower class AB CMOS log-domain filter for DECT applications. Proc. Eur. Solid-State Circuits Conf. September (2000)
168. Huising, J., van de Plassche, R., Sansen, W.: Embeddable data converters for mixed-signal ASICs, Chapter 2. Kluwer Academic Publishers, Dordrecht (1999)
169. van der Zwan, E.J.: A 2.3 mW CMOS ΣΔ modulator for audio applications. IEEE J. Solid-State Circuits, 220–221 (1997)
170. Wittman, R., Schardein, W., Hosticka, B.J., Burbach, G., Arndt, J.: Trimless high precision ratioed resistors in D/A and A/D converters. IEEE J. Solid-State Circuits **30**(8), 935–939 August (1995)
171. Razavi, B.: CMOS technology characterization for analog and RF design. IEEE J. Solid-State Circuits **34**(3), 268–276 March (1999)
172. Sansen, W.: Distortion in elementary transistor circuits. IEEE Trans. Circuits Syst.-II **46**(3), 315–325 March (1999)
173. Song, B.S.: CMOS RF circuits for data commnications applications. IEEE J. Solid-State Circuits **21**(4), 310–317 April (1986)
174. Moon, U.K., Song, B.S.: Design of a low-distortion 22 kHz fifth-order bessel filter. IEEE J. Solid-State Circuits **28**(12), 1254–1264 December (1993)
175. Cheung, V.S.L., Luong, H., Chan, M.: A 0.9-V 0.2-μW CMOS single-OpAmp-based switched-Opamp ΣΔ modulator for pacemaker applications. Proc. IEEE Int. Symp. Circuits Syst. 185–188 May (2002)
176. van Veldhoven, R., Philips, K., Minnis, B.: A 3.3 mW ΣΔ modulator for UMTS in 0.18 μm CMOS with 70 dB dynamic range in 2 MHz bandwidth. IEEE Int. Solid-State Circuits Conf. 222–223 February (2002)
177. Gaggl, R., Wiesbauer, A., Schranz, C., Pessl, P.: A 14-Bit ΣΔ modulator for ADSL-CO applications in 0.18 μm CMOS. Proc. Eur. Solid-State Circuits Conf. 583–587 (2002)
178. Sauerbrey, J., Tille, T., Schmitt-Landsiedel, D., Thewes, R.: A 0.7 V MOS-FET-only switched Opamp ΣΔ modulator. IEEE Int. Solid-State Circuits Conf. **42**, 310–311 (2002)
179. Senderowicz, D., Nicollini, G., Pernici, S., Nagari, A., Confalonieri, P., Dallavalle, C.: Low-voltage duoble-sampled ΣΔ converters. IEEE J. Solid-State Circuits **32**(12), 1907–1919 December (1997)
180. Tille, T., Sauerbrey, J., Thewes, R., Schmitt-Landsiedel, D.: Ultra low-voltage MOS-FET-only switched Opamp ΣΔ modulators using depletion-mode MOS-capacitors. Proc. Eur. Solid-State Circuits Conf. 587–590 (2002)
181. Aboushady, H., Montaudon, F., Paillardet, F., Louerat, M.-M.: A 5 mW, 100 kHz bandwidth, current-mode continuous-time ΣΔ modulators with 84 dB dynamic range. Proc. Eur. Solid-State Circuits Conf. 283–286 September (2002)

182. Gomez, G., Haroun, B.: A 1.5 V 2.4/2.9 mW 79/50 dB DR ΣΔ modulator for GSM WCDMA in 0.13 µm digital process. IEEE Int. Solid-State Circuits Conf. **42**, 306–307 (2002)
183. Matsuya, Y., Terada, J.: 1.2-V, 16-bit audio A/D converter with suppressed latch error noise. Symp. VLSI Circuits, 19–20 (1997)
184. Rabii, S., Wooley, B.A.: A 1.8-V digital-audio sigma-delta modulator in 0.8-µm CMOS. IEEE J. Solid-State Circuits **32**(6), 783–796 June (1997)
185. Nys, O., Henderson, R.K.: A 19-bit low-power multibit sigma-delta ADC based on data weighted averaging. IEEE J. Solid-State Circuits **32**(7), 933–941 July (1997)
186. Li, B., Tenhunen, H.: A second order sigma delta modulator using semi-uniform quantizer with 81 dB dynamic range at 32x OSR. Proc. Eur. Solid-State Circuits Conf. 579–582 September (2002)
187. Chen, F., Leung, B.: A high resolution multibit sigma-delta modulator with individual level averaging. IEEE J. Solid-State Circuits **30**(4), 453–460 April (1995)
188. Baschirotto, A., Castello, R.: A 1 V 1.8 MHz CMOS switched Opamp SC filter with rail-to-rail output swing. IEEE J. Solid-State Circuits **32**(12), 22–33 December (1997)
189. Del Signore, B.P., Kerth, D.A., Sooch, N.S., Swanson, E.J.: A monolithic 20-b delta-sigma A/D converter. IEEE J. Solid-State Circuits **25**(6), 1311–1316 December (1990)
190. Medeiro, F., Perez-Verdu, B., de la Rosa, J.M., Rodriguez-Vazquez, A.: Fourth-order cascade SC ΣΔ modulators: A comparative study. IEEE Trans. Circuits Syst. I **45**(10), 1041–1051 October (1998)
191. Adams, R., Nguyen, K., Sweetland, K.: A 113-dB SNR oversampling DAC with segmented noise-shaped scrambling. IEEE J. Solid-State Circuits **33**(12), 1871–1878 December (1998)
192. van Engelen, J., van der Plassche, R.J., Stikvoort, E., Venes, A.G.: A sixth-order continuous-time bandpass sigma-delta modulator for digital radio IF. IEEE J. Solid-State Circuits **34**(12), 1753–1764 December (1999)
193. Benabes, P., Keramat, M., Kielbasa, R.: A methodolgy for designing countinuous-time sigma-delta modulators. Analog. Int. Circuits Signal Process. **123**(3), 189–200 June (2000)
194. Weiler, D., van den Boom, T., Hostica, B.J.: Resolution prediction for bandpass-ΣΔ-modulator using SIMULINK behavior simulation. Proc. IEEE Int. Symp. Circuits Syst. May (2003)
195. Ortmanns, M., Gerfers, F., Manoli, Y.: Fundamental limits of jitter insensitivity in discrete and continuous-time sigma delta modulators. Proc. IEEE Int. Symp. Circuits Syst. **1**, 1037–1040 May (2003)
196. Da Dalt, N., Harteneck, M., Sandner, C., Wiesbauer, A.: On the jitter requirements of the sampling clock for analog-to-digital converters. IEEE Trans. Circuits Syst. -I **49**(9), 1354–1360 (2002)

197. Geerts, Y.: Design of high-performance CMOS delta-sigma A/D converters. PhD Thesis, University Leuven (2002)
198. Samid, L., Manoli, Y.: The nonidealities of multibit continuous time $\Sigma\Delta$ modulators. Proc. IEEE Int. Conf. Electron. Circuits Syst. **2**, 790–793 (2003)
199. Gerfers, F., Ortmanns, M., Samid, L., Manoli, Y.: Implementation of a 1.5 V low-power clock-jitter insensitive continuous-time $\Sigma\Delta$ modulator. Proc. IEEE Int. Symp. Circuits Syst. 652–655 (2002)
200. Ortmanns, M., Gerfers, F., Manoli, Y.: A continuous-time sigma-delta modulator with switched capacitor controlled current mode feedback. Proc. Eur. Solid-State Circuits Conf. (2003)
201. Gerfers, F., Ortmanns, M., Schmitz, P., Manoli, Y., Soh, K.M.: A clock jitter insensitive multibit DAC architecture for high-performance low-power continuous-time $\Sigma\Delta$ modulators. Proc. IEEE Int. Conf. Electron. Circuits Syst. **1** December (2003)
202. Younis, S.: Method and apparatus for eliminating clock jitter in continuous-time delta-sigma analog-to-digital converters. Int. Patent Appl. Qualcomm Inc. July (2000)
203. Luschas, S., Schreier, R., Lee, H.S.: Radio frequency digital-to-analog converter. IEEE J. Solid-State Circuits **39**(9), 1462–1467 September (2004)
204. Oliaei, O.: State-space analysis of clock jitter in continuous-time oversampling data converters. IEEE Trans. Circuits Syst.-II **50**(1), 31–37 January (2003)
205. Herzel, F., Razavi, B.: A study of oscillator jitter due to supply and substrate noise. IEEE Trans. Circuits Syst. II **46**(1), 56–62 January (1999)
206. Da Dalt, N.: Effect of jitter on asynchronous sampling with finite number of samples. IEEE Trans. Circuits Syst. -II **51**(12), 660–664 (2004)
207. Dong, Y., Opal, A.: Fast time-domain noise simulations of sigma-delta converters and periodically switched linear networks. Proc. IEEE Int. Symp. Circuits Syst. 114–117 (1998)
208. Fakhfakah, A., Milet-Lewis, N., Deval, Y., Levi, H.: Study and behavioural simulation of phase noise and jitter in oscillators. Proc. IEEE Int. Symp. Circuits Syst. May (2002)
209. Benabes, P., Kielbasa, R.: Fast clock-jitter simulation in continuous-time delta-sigma modulators. IEEE Inst. Meas. Tech. Conf. 1587–1590 (2001)
210. Clara, M., Wiesbauer, A., Klatzer, W.: Nonlinear distortion in current-steering D/A-converters due to asymmetrical switching errors. Proc. IEEE Int. Symp. Circuits Syst. **1**, 285–288 (2004)
211. van de Plassche, R.J.: Dynamic element matching for high-accuracy monolithic DA converters. IEEE J. Solid-State Circuits **11**(6) December (1976)
212. Thanh, C.K., Lewis, S.H., Hurst, P.J.: A second-order double-sampled delta-sigma modulator using individual-level averaging. IEEE J. Solid-State Circuits **32**(8), 1269–1273 August (1997)

213. Kuo, T.H., Chen, K.D., Yeng, H.R.: A wideband CMOS sigma-delta modulator with incremental data weighted averaging. IEEE J. Solid-State Circuits **37**(1), 11–17 January (2002)
214. Chao, K., Nadeem, S., Lee, W.L., Sodini, C.G.: A higher order topology for interpolative modulators for oversampleed A/D converters. IEEE Trans. Circuits Syst. 309–318 March (1990)
215. Candy, J.C., Temes, G.C.: Oversampling delta-sigma data converters; therory, design and simulation. IEEE Press, Piscataway, NJ (1992)
216. Rebeschini, M., van Bavel, N., Rakers, P., Greene, R., Cladwell, J., Haug, J.R.: A 16-b 160-kHz CMOS A/D converter using sigma-delta modulation. IEEE J. Solid-State Circuits **25**(2), 431–440 April (1990)
217. Silva-Martinez, J., Steyaert, M., Sansen, W.M.: High-performance CMOS continuous-time filters. Kluwer Academic Publisher, Dordrecht (1993)
218. Shoaei, O., Snelgrove, W.M.: Design and implementation of a tunable 40 MHz, Gm-C bandpass $\Sigma\Delta$ modulator. IEEE Trans. Circuits Syst. II **44**(7), 521–530 July (1997)
219. Tao, J.H., Khoury, J.M.: A 400-Ms/s frequency translating bandpass sigma-delta modulator. IEEE J. Solid-State Circuits **34**(12), 1741–1752 December (1999)
220. Xia, B., Yan, S., Sanchez-Sinencio, E.: An RC time constant auto-tuning structure for high linearity continuous-time $\Sigma\Delta$ modulators and active filters. IEEE Trans. Circuits Syst.-I **51**(11), 2179–2188 November (2004)
221. Xia, B., Yan, S., Sanchez-Sinencio, E.: An auto-tuning structure for continuous-time sigma-delta AD converters and high precision filters. Proc. IEEE Int. Symp. Circuits Syst. V593–596 (2002)
222. Gregorian, R., Temes, G.C.: Analog MOS integrated circuits for signal processing. Wiley, New York (1995)
223. Medeiro, F., Perez-Verdu, B., Rodriguez-Vazquez, A., Huertas, J.L.: Modeling opamp-induced harmonic distortion for switched-capacitor $\Sigma\Delta$ modulator design. Proc. IEEE Int. Symp. Circuits Syst. **5**, 445–448 (1994)
224. Elgerd, O.I.: Control System Theory. McGraw, New York (1967)
225. Foellinger, O.: Regelungstechnik. Einfuehrung in die Methoden und ihre Anwendung. Huethig Verlag, Germany (1994)
226. Chao, K.C.H., Nadeem, S., Lee, W.L., Sodini, C.G.: A higher order topology for interpolative modulators for oversampling A/D converters. IEEE Trans. Circuits Syst. **37**, 309–318 March (1990)
227. Enz, C.C., Temes, G.C.: Circuit techniques for reducing the effects of Op-Amp imperfections: autozeroing, correlated double sampling and chopper stabilization. Proc. IEEE **84**(11), 1584–1612 November (1996)
228. Wambacq, P., Sansen, W.: Distortion analysis of analog integrated circuits. Kluwer Academic Publisher, Dordrecht (1998)
229. Razavi, B.: Design of Analog CMOS Integrated Circuits. McGraw-Hill, New York (1999)

230. Leuciuc, A.: On the nonlinearity of integrators in continuous-time delta-sigma modulators. Proc. IEEE Int. Conf. Electron. Circuits Syst. 862–865 (2001)
231. Geerts, Y., Steyaert, M.S.J.: Flash A/D specifiactions of multibit $\Sigma\Delta$ A/D converters. IEE ADDA, 50–53 July (1999)
232. Huising, J., van de Plassche, R., Sansen, W.: Analog Circuit Design; Low-Noise, Low-Power, Low-Voltage, Mixed-Mode Design with CAD Tools; Voltage , Current and Time References. Kluwer Academic Publishers, Dordrecht (1996)
233. Wang, F., Harjani, R.: Design of Modulators for Oversampled Converters. Kluwer Academic Publisher, Dordrecht (1998)
234. Lampinen, H., Vainio, O.: An optimization approach to designing OTAs for low-voltage sigma-delta modulators. Proc. IEEE Instrum. Meas. Technol. Conf. **2**, 1066–1070 June (2000)
235. Gerfers, F., Manoli, Y.: Ein verlustleistungsarmer zeitkontinuierlicher $\Sigma\Delta$ A/D Wandler fuer eine Versorgungsspannung von 1.5 V. Workshop Mikroelektronik fuer die Informationstechnik, VDE-ITG, 67–72 November (2000)
236. Strle, D.: Capacitor-area and power-consumption optimization of high-order $\Sigma\Delta$ modulators. Proc. IEEE Int. Symp. Circuits Syst. **1**, 331–334 June (2001)
237. Breems, L.J., van der Zwan, E.J., Huijsing, J.H.: A 1.8 mW CMOS $\Sigma\Delta$ modulator with integrated mixer for A/D conversion of IF signals. IEEE Int. Solid-State Circuits Conf. 52–53 February (1999)
238. van Engelen, J., van de Plassche, R.J.: New stability criteria for the design of low-pass sigma-delta modulators. Proc. Int. Symp. Low Power Electron. Des. 114–118 (1997)
239. Marques, A., Peluso, V., Steyaert, M.S., Sansen, W.M.: Optimal parameters for cascaded $\Sigma\Delta$ modulators. Proc. IEEE Int. Symp. Circuits Syst. **1**, 61–64 (1997)
240. Peluso, V., Marques, A., Steyaert, M.S., Sansen, W.M.: Optimal parameters for cascaded $\Sigma\Delta$ modulators. Proc. IEEE Int. Symp. Circuits Syst. **1**, 57–61 (1997)
241. Rabii, S.: Design of low-voltage low-power sigma-delta modulators. PhD Thesis, Stanford University (1998)
242. Becker, M., Heiber, K., Ortmanns, M., Manoli, Y.: A power optimized decimator architecture for cascaded sigma-delta analog-to-digital converters. Proc. IEEE Int. Conf. Electron. Circuits Syst. **1** December (2003)
243. Kiss, P.: Adaptive digital compensation of analog circuit imperfections for cascaded delta-sigma analog-to-digital converters. PhD Thesis, University of Timisoara, (1999)
244. Vittoz, E.: Low power design: ways to approach the limits. IEEE Int. Solid-State Circuits Conf. **34**, 14–18 (1994)

245. Annema, A.-J.: Analog circuit performance and process scaling. IEEE Trans. Circuits Syst. II **46**(6), 711–725 June (1999)
246. Kinget, P., Steyaert, M.S.: Impact of transistor mismatch on the speed–accuracy–power trade off of analog CMOS circuits. Proc. IEEE Custom Integr. Circuits Conf. 285–292 (1996)
247. Steyaert, M., Peluso, V., Bastos, J., Kinget, P., Sansen, W.: Custom analog low power design and the problem of low voltage and mismatch. Proc. IEEE Custom Integr. Circuits Conf. 285–292 (1997)
248. Gerfers, F., Manoli, Y., Ortmanns, M.: CMOS Telecom Data Converters: Chapter 10: Continuous-Time Sigma-Delta for IF. Kluwer Academic Publisher, Dordrecht first edition, January (2004)
249. Laker, K.R., Sansen, W.M.: Design of analog integrated circuits and systems. McGraw-Hill, Inc., New York (1994)
250. Gerfers, F., Hack, C., Ortmanns, M., Manoli, Y.: A 1.2 V rail-to-rail low-power OpAmp with replica amplifier gain enhancement. Proc. IEEE Int. Symp. Circuits Syst. 420–423 (2002)
251. Gerfers, F., Hack, C., Ortmanns, M., Manoli, Y.: A 1.2 V, 200 μW rail-to-rail OpAmp with 90 dB THD using replica gain enhancement. Proc. Eur. Solid-State Circuits Conf. 175–178 (2002)
252. Choi, T., Kaneshiro, R.T., Brodersen, R.W., Gray, P.R., Jett, W.B., Wilcox, M.: High-frequency CMOS swiched-capacitor filters for communication applications. IEEE J. Solid-State Circuits **18**, 652–664 (1983)
253. Duque-Carrillo, J.F.: Control of the common-mode component in CMOS continuous-time fully differential signal processing. Analog Integr. Circuits Signal Process. **4**, 131–140 (1993)
254. Cho, T.B., Gray, P.R.: A 10 b, 20 M sample/s, 35 mW pipeline A/D converter. IEEE J. Solid-State Circuits **30**(3), 166–172 March (1995)
255. Henkel, F., Langmann, U., Hanke, A., Heinen, S., Wagner, E.: A 1 MHz-bandwidth second-order continuous-time quadrature bandpass sigma-delta modulator for low-IF radio receivers. IEEE Int. Solid-State Circuits Conf. **45**, 214–215 (2002)
256. Duque-Carrillo, J.F., Perez-Aloe, R.: High-bandwidth CMOS test buffer with very small input capacitance. Electron. Lett. **26**, 540-543 (1989)
257. Feldman, A.R.: High-speed, low-power sigma-delta modulators for RF baseband channel applications. PhD Thesis, University of California at Berkeley (1997)
258. Bugeja, A.R., Song, B., Rakers, P.L., Gillig, S.F.: A 14-B, 100-MS/s CMOS DAC designed for spectral performance. IEEE J. Solid-State Circuits **34**(12) December (1999)
259. van der Plas, G., Vandenbussche, J., Sansen, W., Steyaert, M., Gielen, G.: A 14-Bit intrinsic accuracy Q^2 random walk CMOS DAC. IEEE J. Solid-State Circuits **34**(12) December (1999)
260. Wilkner, J.: Studies on CMOS digital-to-analog converters. PhD Thesis, Linkoepings University (2001)

261. Uyttenhove, K., Steyaert, M.: Speed-power-accuracy tradeoff in high-speed CMOS ADCs. IEEE Trans. Circuits Syst.-II **49**(4) April (2002)
262. Razavi, B.: Principles of Data Conversion System Design. IEEE Press Piscataway, NJ (1995)
263. Kuboki, S., Kato, K., Miyakawa, N., Matsubara, K.: Nonlinearity analysis of resistor string A/D converters. IEEE Trans. Circuits Syst. **29**(6) June (1982)
264. Cong, Y., Geiger, R.L.: Switching sequence optimization for gradient error compensation in thermometer-decoded DAC arrays. IEEE Trans. Circuits Syst.-II **47**(7), 585–595 (2000)
265. van den Bosch, A., Borremanns, M., Steyaert, M., Sansen, W.: A 10-Bit 1-G Sample/s Nyquist current-steering CMOS D/A converter. IEEE J. Solid-State Circuits **36**(3) March (2001)
266. van den Bosch, A., Steyaert, M., Sansen, W.: An accurate statistical yield model for CMOS current-steering D/A converters. Proc. IEEE Int. Symp. Circuits Syst. May (2000)
267. Jose Bastos. Characterization of MOS transistor mismatch for analog design. PhD Thesis, University of Leuven (1999)
268. Kiss, P., Silva, J., Wiesbauer, A., Sun, T., Moon, U.-K., Stonick, J.T., Temes, G.: New multirate bandpass sigma-delta modulators. IEEE Trans. Circuits Syst.-I **47**(7), 629–638 July (2000)
269. Ortmanns, M., Gerfers, F., Manoli, Y.: A new technique for automatic error correction in $\Sigma\Delta$ modulators. Proc. IEEE Int. Symp. Circuits Syst. (2005)
270. Europractice, Esprit Mixed-Signal Design Cluster Embeddable data converters for mixed-signal ASIC's. EUROPRACTICE Analog Thematic Training and ESPRIT Mixed-Signal Design Cluster, July (1999)

Index

www.ingramcontent.com/pod-product-compliance
Lightning Source LLC
Chambersburg PA
CBHW070748200825
31298CB00008B/7
9783540284062